Introduction to the Theory of Statistical Inference

CHAPMAN & HALL/CRC
Texts in Statistical Science Series

Series Editors

Francesca Dominici, *Harvard School of Public Health, USA*
Julian J. Faraway, *University of Bath, UK*
Martin Tanner, *Northwestern University, USA*
Jim Zidek, *University of British Columbia, Canada*

Introduction to Statistical Limit Theory
A.M. Polansky

Introduction to Statistical Methods for Clinical Trials
T.D. Cook and D.L. DeMets

Introduction to the Theory of Statistical Inference
H. Liero and S. Zwanzig

Large Sample Methods in Statistics
P.K. Sen and J. da Motta Singer

Linear Models with R
J.J. Faraway

Logistic Regression Models
J.M. Hilbe

Markov Chain Monte Carlo — Stochastic Simulation for Bayesian Inference, Second Edition
D. Gamerman and H.F. Lopes

Mathematical Statistics
K. Knight

Modeling and Analysis of Stochastic Systems, Second Edition
V.G. Kulkarni

Modelling Binary Data, Second Edition
D. Collett

Modelling Survival Data in Medical Research, Second Edition
D. Collett

Multivariate Analysis of Variance and Repeated Measures — A Practical Approach for Behavioural Scientists
D.J. Hand and C.C. Taylor

Multivariate Statistics — A Practical Approach
B. Flury and H. Riedwyl

Pólya Urn Models
H. Mahmoud

Practical Data Analysis for Designed Experiments
B.S. Yandell

Practical Longitudinal Data Analysis
D.J. Hand and M. Crowder

Practical Statistics for Medical Research
D.G. Altman

A Primer on Linear Models
J.F. Monahan

Principles of Uncertainty
J.B. Kadane

Probability — Methods and Measurement
A. O'Hagan

Problem Solving — A Statistician's Guide, Second Edition
C. Chatfield

Randomization, Bootstrap and Monte Carlo Methods in Biology, Third Edition
B.F.J. Manly

Readings in Decision Analysis
S. French

Sampling Methodologies with Applications
P.S.R.S. Rao

Statistical Analysis of Reliability Data
M.J. Crowder, A.C. Kimber, T.J. Sweeting, and R.L. Smith

Statistical Methods for Spatial Data Analysis
O. Schabenberger and C.A. Gotway

Statistical Methods for SPC and TQM
D. Bissell

Statistical Methods in Agriculture and Experimental Biology, Second Edition
R. Mead, R.N. Curnow, and A.M. Hasted

Statistical Process Control — Theory and Practice, Third Edition
G.B. Wetherill and D.W. Brown

Statistical Theory, Fourth Edition
B.W. Lindgren

Statistics for Accountants
S. Letchford

Statistics for Epidemiology
N.P. Jewell

Statistics for Technology — A Course in Applied Statistics, Third Edition
C. Chatfield

Statistics in Engineering — A Practical Approach
A.V. Metcalfe

Statistics in Research and Development, Second Edition
R. Caulcutt

Stochastic Processes: An Introduction, Second Edition
P.W. Jones and P. Smith

Survival Analysis Using S — Analysis of Time-to-Event Data
M. Tableman and J.S. Kim

The Theory of Linear Models
B. Jørgensen

Time Series Analysis
H. Madsen

Time Series: Modeling, Computation, and Inference
R. Prado and M. West

Texts in Statistical Science

Introduction to the Theory of Statistical Inference

Hannelore Liero

Silvelyn Zwanzig

CRC Press
Taylor & Francis Group
Boca Raton London New York

CRC Press is an imprint of the
Taylor & Francis Group an **informa** business

A CHAPMAN & HALL BOOK

CRC Press
Taylor & Francis Group
6000 Broken Sound Parkway NW, Suite 300
Boca Raton, FL 33487-2742

International Standard Book Number: 978-1-4398-5292-7 (Paperback)

Visit the Taylor & Francis Web site at
http://www.taylorandfrancis.com

and the CRC Press Web site at
http://www.crcpress.com

Contents

List of Figures

List of Tables

Preface

This textbook is written for students of mathematical statistics who already know most of the main statistical methods and now want to understand the fundamental classical principles of statistical inference. Since 2004 preliminary versions of this book have been used as a textbook for the course "Inference II" at Uppsala University, Sweden.

There exist a lot of textbooks. Why did we write our own textbook? We both studied statistics together in the seventies at the Humboldt University in Berlin. After years of teaching at different universities and in different countries we now know that we had had one of the best lecture courses. Of course this textbook is not just a copy of our own old handwritten lecture notes, but parts come from the lectures "Statistische Methoden" by our teacher Klaus Fischer. We include also examples, explanations and theorems we used in our own lecture courses.

Some years ago Gerhard Hübner, professor for stochastics at the University of Hamburg, started to draw illustrations for the examination papers of his students. We are very pleased that he agreed to draw illustrations to this textbook. All drawn cartoons are from him.

Acknowledgments

We wish to thank our colleagues at the University of Potsdam and the University of Uppsala who supported our project, especially Henning Läuter and Allan Gut. We both would like to also thank our students for contributing with their feedback—the students in Uppsala who took an active part in the courses underlying this textbook and the students at the University of Potsdam who followed up on the ideas of the material. In particular we would like to acknowledge Katja Fröhlich for her continued support of the manuscript. We thank Jesús E. Sánchez García and Dmitry Chibisov for their valuable comments.

We are very grateful to David Grubbs of CRC Press for giving us support to write this book, for his helpful and patient guidance.
Our families deserve the most thanks—for their encouragement and their understanding.

Potsdam and Uppsala
May 2011

Hannelore Liero
Silvelyn Zwanzig

Chapter 1

Introduction

The aim of the book is to present the core concepts of statistical inference. The development of modern computers and software allows us to realize computationally complex procedures. Sometimes these methods are only intuitively motivated, their properties are unknown and their application cannot be justified. Therefore it is necessary to understand the theoretic background of statistical methods. This is the starting point of our textbook. We want to convey basic ideas—it is not our goal to provide a catalog of procedures. Of course, aimed at students the book presents only the first steps into the statistical theory.

The basic approach in inferential statistics is to handle data as values of random variables. This approach leads to the notion of the statistical model as a set of suitable probability distributions. In Chapter 2 we give examples for finding an appropriate statistical model for a problem at hand. We show that the choice of the model depends on the knowledge of the underlying problem and on the question to be answered.

Furthermore, in Chapter 2 we introduce the exponential families. These families of probability distributions contain many well-known parametric distributions. Thus, results proven for an exponential family are true also for these distributions. Moreover, exponential families have many useful properties.

In Chapter 3 two basic principles of statistical inference for parametric models are explained: The likelihood and the sufficiency. The likelihood function is defined and its properties are discussed. We demonstrate that the likelihood function provides the information about the underlying parameter contained in the data. Based on this we define the Fisher information.

Sufficiency is a central notion in statistics. A sufficient statistic is a function of the data which contains all available sample information concerning the underlying parameter. We describe sufficiency in detail and present a criterion for verifying sufficiency. Moreover, we consider minimal sufficient statistics, that is, statistics which summarize the information as effectively as possible. Many statements in the following chapters about both the main topics of statistics—estimation of parameters and testing hypotheses—are based on the likelihood approach and the sufficiency principle.

Chapter 4 begins with the presentation of methods of estimation. Without assuming the type of the underlying distribution of the data one can apply the method of moments to estimate unknown parameters. The maximum likelihood method requires the specification of the underlying distribution. We discuss existence, uniqueness and problems of the computation of maximum likelihood estimators. Furthermore, we mention the M-estimators.

As a criterion how good an estimator is we consider the mean squared error. After presenting some examples for the bias-variance trade-off we restrict ourselves to unbiased estimators. Based on the results in Chapter 3 the Cramér–Rao bound for the variance of these estimators is established. The main result in Chapter 4 is the Lehmann–Scheffé theorem, which provides sufficient conditions for an estimator to be the best one of all unbiased estimators.

This chapter is closed by defining some convergence notions and presenting some asymptotic results to describe the properties of estimators for large sample sizes.

Chapter 5 treats the testing of hypotheses. We present the two approaches: Providing evidence and making decision. The first one, which is an intuitive approach, is shortly explained. Here the notion of a p-value is introduced and its properties are discussed. The second approach is the Neyman–Pearson theory for testing hypotheses. The Neyman–Pearson lemma is proved. The uniform most powerful tests are introduced. At the end of the chapter we consider the conditional tests. The optimality of the well-known t-test is shown.

In Chapter 6 we apply the core concepts of estimation and testing presented in the previous chapters to the linear model. We give a number of examples to demonstrate the size of the class of linear models. It includes the linear regression models, which describe the relationship between a random response variable and exploratory variables, but also models of the analysis of variances can be written as a linear model. For the estimation of the parameter the least squares method is introduced. This type of estimator is characterized as a projection estimator. The Gauss–Markov theorem states the optimality of this estimator under all linear unbiased estimators. Under the assumption that the data are from a normal distribution we consider maximum likelihood estimators. The F-test is derived and its geometrical properties are explained. Using results from Chapter 5 the optimality of the F-test is verified.

The book is aimed for advanced undergraduate students and graduate students in mathematics and statistics, but also for theoretically interested students from other disciplines.

Prerequisites for the book are calculus, algebra and basic probability theory. We do not assume that the reader is familiar with measure theory; however, the notion of conditional probability should be known. In Chapter 6 we apply generalized inverse matrices to invert a matrix which is not of full rank. It is advantageous to know this matrix algebra, but for understanding this chapter it is not required.

Some words about the **Structure of the book**: We present the results as theorems and corollaries. All theorems are proven. Additionally important statements are verbally formulated as guidelines in boxes.

Many figures complement the text. The graphics are produced by using the freely available R-package, see http://cran.r-project.org. The theorems are followed by **Special cases** and **Examples**. In Special cases we apply the theoretic results stated in the theorem to a special statistical model. In this sense Special cases are of their own interest. The aim of the examples is to illustrate the theory. Some of the examples go throughout the book and are marked by a cartoon.

Furthermore, exercises are given in the running text. The solution to these exercises are simple variations of the proof or the calculation just before. Each chapter ends with a list of problems. Detailed solutions to these problems are provided in Chapter 7.

The * indicates more theoretical parts or difficult problems. They can be skipped without missing the main ideas.

We use the abbreviation of observe (**OBS!**) as a warning symbol for unusual facts or frequent mistakes.

The notation used in the book is defined in the corresponding chapters; an overview of some of them is given in the following list.

Notation

X, Y	random variables (r.v.'s)
x, y	realization of random variables
$\mathbf{X}, \mathbf{x}$	sample of n r.v.'s (and its realization)
$\mathbf{X} = (X_1, \ldots, X_n)$	
$\mathcal{X}$	sample space
$\mathcal{P}$	statistical model
P_θ	probability distribution parameterized by $\theta \in \Theta$
$\mathbf{X} \sim \mathsf{P}_\theta$	$\mathbf{X}$ is distributed according to P_θ
$p(\cdot; \theta)$	probability function w.r.t. P_θ
$L(\cdot; \mathbf{x}), l(\cdot; \mathbf{x})$	likelihood and log likelihood function given $\mathbf{x}$
$f(\cdot; \theta)$	density of a continuous r.v.
$\mathsf{E}_\theta, \mathsf{Var}_\theta, \mathsf{Cov}_\theta$	expectation, variance and covariance w.r.t. P_θ
$I_t, \mathbb{1}_A$	indicator functions, i.e., $I_t(x) = 1$ if $t = x$, else $I_t(x) = 0$, and $\mathbb{1}_A(x) = 1$ if $x \in A$, else $\mathbb{1}_A(x) = 0$, respectively.
$I_{\mathbf{X}}(\theta)$	Fisher information contained in $\mathbf{X} \sim \mathsf{P}_\theta$
x^{T}	transposed vector $x \in \mathbb{R}^k$
I_n	$n \times n$ identity matrix
$l'(\theta_0; \mathbf{x}), (l''(\theta_0; \mathbf{x}))$	first (second) derivative (w.r.t. θ) of the log likelihood function at θ_0 if $\theta \in \mathbb{R}$
$\frac{\partial l(\theta; \mathbf{x})}{\partial \theta_j}\big|_{\theta=\theta_0}$	first partial derivative of the log likelihood function w.r.t. θ_j, $(j = 1, \ldots, k)$ at θ_0
$\frac{\partial^2 l(\theta; \mathbf{x})}{\partial \theta_j \partial \theta_r}\big|_{\theta=\theta_0}$	second partial derivative of the log likelihood function w.r.t. θ_j and θ_r $(j, r = 1, \ldots, k)$ at θ_0
$\mathsf{Ber}(\theta)$	Bernoulli distribution with $\theta \in (0, 1)$
$\mathsf{Bin}(n, \theta)$	Binomial distribution with $\theta \in (0, 1)$
$\mathsf{Mult}(n, p_1, \ldots, p_k)$	Multinomial distribution with k outcomes with parameters $p_1, \ldots, p_k$ satisfying $\sum_{j=1}^k p_j = 1$, $p_j \in (0, 1)$
$\mathsf{Poi}(\mu)$	Poisson distribution with $\mu \in \mathbb{R}_+$
$\mathsf{N}(\mu, \sigma^2)$	normal distribution with $\mu \in \mathbb{R}$ and $\sigma^2 \in \mathbb{R}_+$
$\mathsf{Exp}(\lambda)$	exponential distribution with $\lambda \in \mathbb{R}_+$
$\mathsf{U}[a, b]$	uniform distribution over the interval $[a, b]$, $a, b \in \mathbb{R}$

Chapter 2

Statistical Model

2.1 Data

Observations or data are the basis of a statistical investigation. These observations **x** are numbers or vectors, categorical quantities or functions. In general they are taken from a large set of individuals or objects under consideration. The possible observation results constitute a certain set, the so-called **sample space $\mathcal{X}$.**

In the following we give several examples for data and the corresponding sample spaces.

These examples go throughout the book. We will use them to illustrate the theory of the following chapters.

Example 2.1 (Ballpoint pens) A manufacturer of ballpoint pens randomly samples $n = 400$ units per day from the daily production of 40000 pens. The observations are the numbers of defective pens per day. For example,

Mon	Tue	Wed	Thu	Fri	Sat	Sun
8	5	9	4	6	8	10

and we write $\mathbf{x} = (8, 5, 9, 4, 6, 8, 10)$. Thus our sample space $\mathcal{X}$ consists of seven-dimensional vectors, where each component can take a value between 0 and 400, i.e., $\mathcal{X} = \{0, \ldots, 400\}^7$. □

Example 2.2 (Flowers) Snakes's head (Kungsängsliljan, *Fritillaria meleagris*) is one of the sights in Uppsala. The flowers were discovered in Uppsala surrounding Kungsängen in 1743. Probably the flowers had been imported from Holland to the gardens in Uppsala. To protect the flower the nature reserve Kungsängen was established in 1951. In May the whole field is covered by the flowers. There are three different colors: violet, white and pink. The observations are the colors of 20 chosen flowers; thus $\mathbf{x}$ is categorical quantity. The sample space is $\mathcal{X} = \{violet, white, pink\}^{20}$. $\square$

Example 2.3 (Pendulum) Suppose we carry out a physical investigation and are interested in the period of oscillation. So we take repeated measurements of the time required by the pendulum to reach a given position. The observation results are the following (in sec):

$$1.99, \ 2.00, \ 1.98, \ 2.00, \ 2.01, \ 2.02, \ 2.01, \ 1.98, \ 1.99, \ 2.00. \qquad (2.1)$$

The data consist of $n = 10$ real numbers x_i, $i = 1, \ldots, 10$ and $\mathcal{X} = \mathbb{R}^{10}$. $\square$

Example 2.4 (Dolphins) This example is taken from Augier et al. (1993). Environmental scientists studied the accumulation of toxic elements in marine mammals. The mercury concentrations (microgram/gram) in the livers of 28 male striped dolphins (*Stenella coeruleoalba*) are given in Table 2.1. Here our observed sample $\mathbf{x}$ consists of 28 real numbers x_i, $i = 1, \ldots, 28$ and $\mathcal{X} = \mathbb{R}^{28}$.

1.70	1.72	8.80	5.90	183.00	221.00
286.00	168.00	406.00	286.00	218.00	252.00
241.00	180.00	329.00	397.00	101.00	264.00
316.00	209.00	85.40	481.00	445.00	314.00
118.00	485.00	278.00	318.00		

Table 2.1: Mercury concentrations in dolphins (in μg/g)

$\square$

Example 2.5 (Soybeans) Consider a large field of soybean plants. During seven weeks each Monday five plants are randomly chosen. The average of the heights (in cm) is registered. The data are

$$\mathbf{x} = (5, 13, 16, 13, 23, 33, 40)$$

and the sample space is $\mathcal{X} = \mathbb{R}^7$. (This example is similar to an example in Weber (1972).)   □

Example 2.6 (Patients data) In a medical study different characteristics of patients are registered, for example: sex, age, weight (in kg), systolic blood pressure (in mm Hg), cholesterol level (in mg/dL) and pulse (beats per min). Table 2.2 is a part of a list taken for 200 patients in some hospital. Here the observation x_i for patient i is a six-dimensional vector, where the first component is a nominal quantity, the second is an integer and the others are real numbers. The observation for one patient can take values in $\{m,f\} \times \mathbb{N} \times \mathbb{R}^4$. The sample space $\mathcal{X}$ for the whole data of all patients is the set of (200×6)-matrices with rows in $\{m,f\} \times \mathbb{N} \times \mathbb{R}^4$.   □

Pat.-No.	Sex	Age	Weight	Blood pressure	Cholesterol	Pulse
111	m	54	73	130	200	75
112	f	37	65	125	225	65
113	m	50	78	145	210	60
114	m	62	84	160	240	80

Table 2.2: Part of the list of medical data

Example 2.7 (Friesian cows) In a study on protein biosynthesis of milk and the accompanying decomposition of nucleic acids into various constituents data on milk production (kg/day) and milk protein (kg/day) for Holstein-Friesian cows were reported. The following data are taken from Devore and Peck (1993). Table 2.3 gives pairs (x_i, y_i) where x_i is the milk production and y_i is the milk protein of cow i, $i = 1, \ldots, 14$. The sample space is $\mathcal{X} = (\mathbb{R} \times \mathbb{R})^{14}$.   □

No.	Milk production	Milk protein
1	42.7	1.20
2	40.2	1.16
3	38.2	1.07
4	37.6	1.13
5	32.2	0.96
6	32.2	1.07
7	28.0	0.85
8	27.2	0.87
9	26.6	0.77
10	23.0	0.74
11	22.7	0.76
12	21.8	0.69
13	21.3	0.72
14	20.2	0.64

Table 2.3: Milk production and milk protein of Friesian cows

Example 2.8 (Genotypes) In usual genetic models genes appear in pairs and each gene of a particular pair can be either of two types called A and a. The sample space in this case is formed by: $\mathcal{X} = \{\mathtt{AA}, \mathtt{Aa}, \mathtt{aa}\}$.   □

2.2   Statistical Model

Since an observation $\mathbf{x}$ is an outcome of a random experiment we consider $\mathbf{x}$ as a value (realization) of a random variable $\mathbf{X}$. Its distribution is at least partly unknown. Statistical inference is concerned with methods of using these observation $\mathbf{x}$ to obtain information about the distribution of $\mathbf{X}$. That means starting point of the statistical analysis is not one probability measure on $\mathcal{X}$, but a class of possible probability measures $\mathcal{P}$, parameterized by a parameter θ:

$$\mathcal{P} = \{\mathsf{P}_\theta : \theta \in \Theta\}.$$

The set Θ is called the **parameter space**. We call this class of probability measures on the sample space $\mathcal{X}$ a **statistical model.**

We very often consider repeated experiments, where the experiments do not influence each other. That is, our observations are realizations of independent random variables $X_1, \ldots, X_n$. In this case $\mathcal{P}$ is a family of product measures:

$$\mathsf{P}_\theta = \mathsf{P}_{1,\theta} \otimes \cdots \otimes \mathsf{P}_{n,\theta}.$$

Definition 2.1 A sample $\mathbf{X} = (X_1, \ldots, X_n)$ is a collection of independent random variables (r.v.'s), where X_i is distributed according to a distribution $\mathsf{P}_{i,\theta}$. The **sample size** n is the number of the r.v.'s.

An **i.i.d. sample** $\mathbf{X} = (X_1, \ldots, X_n)$ is a collection of independent random variables, where all X_i have the same distribution, i.e., $\mathsf{P}_{i,\theta} = \mathsf{P}_{1,\theta}$ for all i. Sometimes one can justify that the underlying distribution is known with exception of a finite-dimensional parameter. Then the statistical model is called a **parametric model** and we write

$$\mathcal{P} = \{\mathsf{P}_\theta : \theta \in \Theta \subseteq \mathbb{R}^k\}$$

for some integer k. Otherwise it is called a **nonparametric model.**

Since measure theory does not belong to the prerequisites of our book we introduce the **probability function** to handle discrete and continuous distributions together. Let $A \subseteq \mathcal{X}$. Then for a continuous r.v. $\mathbf{X}$ with density $f(\cdot; \theta)$

$$\mathsf{P}_\theta(A) = \int_A f(\mathbf{x}; \theta) d\mathbf{x}$$

and for a discrete r.v.

$$\mathsf{P}_\theta(A) = \sum_{\mathbf{x} \in A} \mathsf{P}_\theta(\{\mathbf{x}\}).$$

The probability function is defined by

$$p(\mathbf{x}; \theta) = \begin{cases} f(\mathbf{x}; \theta) & \text{if } \mathsf{P}_\theta \text{ is continuous} \\ \mathsf{P}_\theta(\{\mathbf{x}\}) & \text{if } \mathsf{P}_\theta \text{ is discrete} \end{cases}.$$

For simplicity of notation instead of $\mathsf{P}_\theta(\{\mathbf{x}\})$ we write $\mathsf{P}_\theta(\mathbf{x})$.

Example 2.9 (Ballpoint pens) The random variable X_i is the number of defective pens among the $n = 400$ pens sampled at each day i. In a quality control we interpret the number of defective parts as "the number of successes in a series of n trials." We consider as unknown parameter the number of defective pens $\theta = K$ out of the $N = 40000$ of the daily production. Since we draw without replacement the probability of the event $\{X_i = k\}$ is

$$\frac{\binom{K}{k}\binom{N-K}{n-k}}{\binom{N}{n}}, \qquad k = 0, \ldots, \min(n, K),$$

calculated from the hypergeometric distribution $\mathsf{H}(N, n, K)$. Hence we have a sample $\mathbf{X} = (X_1, ..., X_7)$ of independent identically distributed random variables $X_i \sim H(N, n, K)$. The parameter space is $\{0, ..., N\}$ and the statistical model is

$$\mathcal{P} = \{H(N, n, K)^{\otimes 7} : \quad K \in \{0, ..., N\}\}.$$

Since the daily production N is very large in comparison to the taken subset of size n it is justified to approximate the hypergeometric distribution $H(N, n, K)$ by the binomial distribution $\mathsf{Bin}(n, \theta)$ with $p = \frac{K}{N}$, where p is the unknown probability of successes. The parameter space is $\Theta = (0, 1)$ and the statistical model is

$$\mathcal{P} = \{\mathsf{Bin}(n, p)^{\otimes 7} : p \in (0, 1)\}. \tag{2.2}$$

OBS! The sample size is the number of days: 7. It is neither 400 nor 40000 ! □

Example 2.10 (Flowers) The observed color is given by the realization x defined by

$$x = \begin{cases} 1 & \text{if the flower is white} \\ 2 & \text{if the flower is violet} \\ 3 & \text{if the flower is pink} \end{cases}.$$

The statistical distribution is a three-point distribution. The probability of the event $\{X = x\}$ is

$$\mathsf{P}_\theta(x) = \begin{cases} p_1 & \text{if} \quad x = 1 \\ p_2 & \text{if} \quad x = 2 \\ p_3 & \text{if} \quad x = 3 \end{cases}, \tag{2.3}$$

where θ stands for the unknown probabilities. For $k = 1, 2, 3$ introduce the indicator function

$$I_k(x) = \begin{cases} 1 & \text{if} \quad x = k \\ 0 & \text{else} \end{cases},$$

then we can rewrite (2.3) as

$$\mathsf{P}_\theta(x) = p_1^{I_1(x)} p_2^{I_2(x)} p_3^{I_3(x)}.$$

We are interested in the unknown probabilities p_1, p_2, p_3. Note there exist no yellow or blue flowers. Only three colors are possible, thus we have

$$p_1 + p_2 + p_3 = 1 \quad \text{and} \quad p_3 = 1 - (p_1 + p_2).$$

Therefore the unknown parameter is $\theta = (p_1, p_2)$ and the statistical model for one observation is

$$\mathcal{P} = \left\{ p_1^{I_1} p_2^{I_2} p_3^{I_3} : \theta = (p_1, p_2) \in (0, 1)^2, p_3 = 1 - p_1 - p_2 \in (0, 1) \right\}. \qquad \square$$

Example 2.11 (Pendulum) This is a classical example of a physical experiment with repeated observations under equivalent conditions. The sources of randomness are the measurement errors, which can be assumed as normally distributed. Thus we have a sample of i.i.d. r.v.'s and the model is given by

$$\mathcal{P} = \{N(\mu, \sigma^2)^{\otimes 10} : (\mu, \sigma^2) \in \mathbb{R} \times \mathbb{R}_+\}.$$

□

Example 2.12 (Dolphins) The first four dolphins are younger than the others. The random variables are not identically distributed. If we suppose normal distributions for the mercury concentration, a variant of the statistical model can be

$$\mathcal{P} = \{N(\mu_1, \sigma_1^2)^{\otimes 4} \otimes N(\mu_2, \sigma_2^2)^{\otimes 24} : (\mu_1, \mu_2, \sigma_1^2, \sigma_2^2) \in \mathbb{R}^2 \times \mathbb{R}_+^2\} \qquad (2.4)$$

with the unknown parameter $\theta = (\mu_1, \mu_2, \sigma_1^2, \sigma_2^2)$.

□

Example 2.13 (Soybeans) The growth of a soybean plant can be described by a classical simple linear regression model, with equidistant design points

$$X_i = a + bi + \varepsilon_i, \qquad i = 1, \dots, 7. \qquad (2.5)$$

For the errors ε_i it is assumed

$$E_\theta \varepsilon_i = 0, \quad \text{Var}_\theta \varepsilon_i = \sigma^2, \quad \text{and} \quad E_\theta \varepsilon_i \varepsilon_j = 0 \quad \text{for} \quad i \neq j.$$

Therefore the model for $\mathbf{X}$ which takes values in the sample space $\mathcal{X} = \mathbb{R}^7$ can be written in the form

$$\mathcal{P} = \{P_\theta : E_\theta X_i = a + bi, \quad \text{Var}_\theta X_i = \sigma^2, \, i = 1, \dots, 7,$$
$$\theta = (a, b, \sigma^2, \kappa) \in \Theta \subseteq \mathbb{R}^2 \times \mathbb{R}_+ \times \mathcal{K}\},$$

where κ characterizes the distribution of ε_i up to the first and second moments. Usually one is interested in the parameter (a, b, σ^2). Since (a, b, σ^2) is independent of κ, very frequently this (infinite) parameter κ is omitted in the description of the model. Nevertheless it is a semiparametric model. $\mathcal{P}$ becomes a parametric model, if the type of the distribution is suitably restricted, for example: If we assume a normal distribution for the errors.   □

Example 2.14 (Genotypes) The Hardy–Weinberg model states that the genotypes AA, Aa and aa occur with the following probabilities:

$$\mathsf{P}_\theta(\mathtt{aa}) = \theta^2, \quad \mathsf{P}_\theta(\mathtt{Aa}) = 2\theta(1-\theta), \quad \mathsf{P}_\theta(\mathtt{AA}) = (1-\theta)^2,$$

where θ is an unknown parameter in $\Theta = (0,1)$.   □

Parametrization. Let us say some more words about the parametrization. The parametrization is not unique. Instead of $\mathcal{P} = \{\mathsf{P}_\theta : \theta \in \Theta\}$ one can choose $\mathcal{P} = \{\mathsf{P}_\xi : \xi \in \Xi\}$ where $\theta = h^{-1}(\xi)$ for some one-to-one function h. A classical example is the exponential distribution, a typical life time distribution. We can parameterize it with respect to the expected life time, say μ, or with respect to the failure rate, mostly denoted by λ, where $\mu = 1/\lambda$.

Although parametrization is not unique it must be so that the parameters are identifiable, that is, if the parameters are different then also the distributions are different. Throughout the book we will assume the following condition:

Reg 0 (Identifiability) For $\theta_1, \theta_2 \in \Theta$ with $\theta_1 \neq \theta_2$ the distributions P_{θ_1} and P_{θ_2} are different.

The postulation of a **statistical model** is essential for the further investigation and to derive conclusions. Roughly speaking, all models are wrong. But some of them are useful—namely when properties characterizing the underlying phenomena are detected by the model. A very precise and complicated model which describes all details is not necessarily suitable. Because of its complexity it can happen that a mathematical investigation is too difficult or unfeasible.

> Mathematical statements derived from statistical methods **are exact under the chosen model**, but their validity in practice depends on how well the model reflects the problem.

2.3 Statistic

We can see in the flower example above that the interesting information of the experiment is given by the number of flowers of each color. This is a function of the random sample which essentially compresses the data. In the following we will consider arbitrary functions of the data. These functions are called statistics.

Definition 2.2 A statistic T is a function of the sample

$$T : \mathbf{x} \in \mathcal{X} \to T(\mathbf{x}) = t \in \mathcal{T}$$

where $\mathcal{T}$ is a suitable set. With the r.v. $\mathbf{X}$ the function

$$T(\mathbf{X}) : \omega \in \Omega \to T(\mathbf{x}) = t \in \mathcal{T}$$

is a random variable. The distribution of T is given by

$$\mathsf{P}_\theta^T(B) = \mathsf{P}_\theta(\{\mathbf{x} : T(\mathbf{x}) \in B\}).$$

Note that from the theoretical point of view statistics cannot be arbitrary functions. They have to be measurable. But in this textbook we disregard all problems of measure theory. Let us say, if we have continuous functions with only a countable set of discontinuity points, then we are on the safe side. The property "measurable" assures the existence of the measure P_θ^T.

Example 2.15 (Ballpoint pens) The differences between the days do not seem essential. A reasonable statistic is given by $T(\mathbf{X}) = \sum_{i=1}^7 X_i$ and its observed value is $T(\mathbf{x}) = 8 + 5 + 9 + 4 + 6 + 8 + 10 = 50$. $T(\mathbf{X})$ is a random variable over $\mathcal{T} = \{0, \ldots, 7 \times 400\}$. Under the binomial model (2.2) the statistical model for $T(\mathbf{X})$ is

$$\mathcal{P} = \{\mathrm{Bin}(7 \times 400, \theta) : \theta \in (0, 1)\}.$$

$\square$

Example 2.16 (Dolphins) Since the first four data are taken from younger dolphins, we omit these values and consider the reduced sample: $T(\mathbf{X}) = (X_5, \ldots, X_{28})$ with $\mathcal{T} = \mathbb{R}^{24}$. The observed value is $T(\mathbf{x}) = (101.00, \ldots, 318.00)$.

We can assume that the reduced sample consists of i.i.d. random variables. Moreover under the assumption of normality the statistical model for $T(\mathbf{X})$ is given by

$$\mathcal{P} = \{ \mathsf{N}(\mu, \sigma^2)^{\otimes 24} : (\mu, \sigma^2) \in \mathbb{R} \times \mathbb{R}_+ \}.$$

$\square$

Example 2.17 (Flowers) The most interesting statistics here are the numbers of flowers of the different colors. The number of white, violet and pink flowers is given by

$$n_j = \sum_{i=1}^{n} I_j(x_i), \qquad j = 1, 2, 3.$$

The n_j's with $\sum_{j=1}^{3} n_j = n$ are realizations of the r.v.'s $N_j = \sum_{i=1}^{n} I_j(X_i)$. The distribution of these counts (N_1, N_2, N_3) depends on the sampling strategy.

Fix the number n of collected flowers in advance and choose randomly the n flowers and determine the colors, then the vector (N_1, N_2) has a multinomial distribution $\mathsf{Mult}(n, \theta)$:

$$\mathsf{P}_\theta(\mathbf{n}) = p(\mathbf{n}; \theta) = \frac{n!}{n_1! \, n_2! \, n_3!} p_1^{n_1} p_2^{n_2} p_3^{n_3}, \quad \theta = (p_1, p_2), \ \mathbf{n} = (n_1, n_2) \quad (2.6)$$

with $n_3 = n - n_1 - n_2$, $p_3 = 1 - p_1 - p_2$ and $0 < p_j < 1$ for $j = 1, 2, 3$. $\square$

2.4 Exponential Families

Exponential families play an important role in statistics. If one can show that the distribution of a sample belongs to an exponential family the data can be reduced to a statistic of smaller dimension.

We consider a class of probability measures $\mathcal{P} = \{ \mathsf{P}_\theta : \theta \in \Theta \}$ and assume that for each P_θ there exists the probability function $p(\cdot; \theta)$.

Definition 2.3 A class of probability measures $\mathcal{P} = \{P_\theta : \theta \in \Theta\}$ is called an **exponential family**, if there exist a number $k \in \mathbb{N}$, real-valued functions $\zeta_1, \ldots, \zeta_k$ on Θ, real-valued statistics $T_1, \ldots, T_k$ and a function h on $\mathcal{X}$ such that the probability function has the form

$$p(x; \theta) = A(\theta) \exp \left(\sum_{j=1}^{k} \zeta_j(\theta) T_j(x) \right) h(x). \qquad (2.7)$$

By definition one can assume that $h(x) \geq 0$ and $A(\theta) > 0$. The quantity $A(\theta)$ is the normalizing factor. The statistical properties are determined by the exponential expression

$$\exp \left(\sum_{j=1}^{k} \zeta_j(\theta) T_j(x) \right).$$

Although $\zeta = (\zeta_1, \ldots, \zeta_k)$, $T = (T_1, \ldots, T_k)$ and k are not uniquely determined we call (2.7) a **k-parameter exponential family** or a k-dimensional exponential family. We will see that optimal statistical procedures will only depend on the k-dimensional statistic T. Therefore one tries to choose k minimal. An exponential family with minimal number of summands is called a strictly k-parameter exponential family.

Special case 2.1 (Normal distribution) Let $\mathcal{P}$ be the class of all normal distributions with parameter $\theta = (\mu, \sigma^2) \in \mathbb{R} \times \mathbb{R}_+$. Then the density has the form

$$f(x; \theta) = \frac{1}{\sqrt{2\pi\sigma^2}} \exp \left(-\frac{(x-\mu)^2}{2\sigma^2} \right),$$

which can be decomposed into

$$f(x; \theta) = \frac{1}{\sqrt{2\pi\sigma^2}} \exp \left(-\frac{\mu^2}{2\sigma^2} \right) \exp \left(-\frac{x^2}{2\sigma^2} + \frac{x\mu}{\sigma^2} \right).$$

Thus the normal distribution is a two-parameter exponential family with

$$A(\theta) = \frac{1}{\sqrt{2\pi\sigma^2}} \exp \left(-\frac{\mu^2}{2\sigma^2} \right),$$

$$\zeta_1(\theta) = -\frac{1}{2\sigma^2}, \qquad T_1(x) = x^2$$

and

$$\zeta_2(\theta) = \frac{\mu}{\sigma^2} \quad \text{and} \quad T_2(x) = x.$$

The covariance matrix of the statistic (T_1, T_2) has the form

$$\begin{pmatrix} 2\sigma^4 + 4\sigma^2\mu^2 & 2\sigma^2\mu \\ 2\sigma^2\mu & \sigma^2 \end{pmatrix},$$

and is positive definite for all $\mu \in \mathbb{R}$ and $\sigma^2 \in \mathbb{R}_+$.   □

Special case 2.2 (Binomial distribution) Let $\mathcal{P}$ be the class of all binomial distributions with (known) parameter n and (unknown) $\theta \in (0, 1)$. Then the probability function is

$$p(x; \theta) = \binom{n}{x}\theta^x(1 - \theta)^{n-x} = (1 - \theta)^n \exp\left[\ln\left(\frac{\theta}{1 - \theta}\right)x\right]\binom{n}{x}.$$

Thus, this class is an exponential family with $A(\theta) = (1 - \theta)^n$, $h(x) = \binom{n}{x}$, $T(x) = x$ and $\zeta(\theta) = \ln\left(\frac{\theta}{1-\theta}\right)$.   □

Remark 2.1 Two probability measures P and Q are called equivalent if $\mathsf{P}(N) = 0$ iff $\mathsf{Q}(N) = 0$. If $\mathcal{P} = \{\mathsf{P}_\theta : \theta \in \Theta\}$ forms an exponential family, then all P_θ's are equivalent: Suppose $\mathsf{P}_{\theta_1}(N) = 0$. Let $\mathbb{1}_N$ be the indicator function of the set N. Since

$$\mathsf{P}_{\theta_1}(N) = A(\theta_1)\int \exp(\sum_{j=1}^{k}\zeta_j(\theta_1)T_j(x))h(x)\,\mathbb{1}_N(x)dx = 0$$

it follows that $h(x)\,\mathbb{1}_N(x) = 0$ for almost all x. But this implies that

$$\mathsf{P}_\theta(N) = A(\theta)\int \exp(\sum_{j=1}^{k}\zeta_j(\theta)T_j(x))h(x)\,\mathbb{1}_N(x)dx = 0$$

for arbitrary $\theta \in \Theta$.

Example 2.18 (Counterexample) Let $\mathcal{P}$ be the class of all two-parameter exponential distributions $\mathsf{E}(1, \theta)$, i.e., distributions with density

$$f(x; \theta) = \exp[-(x - \theta)]\,\mathbb{1}_{[\theta,\infty)}(x) \qquad \text{with} \quad \theta \in \mathbb{R}. \tag{2.8}$$

This family is not an exponential family. The factor $\mathbb{1}_{[\theta,\infty)}(x)$ is not of exponential form and cannot be transformed in such a form.

From the Remark 2.1 it follows that all probability measures $P_\theta \in \mathcal{P}$ are pairwise equivalent, but this is not the case for the probability measures defined by (2.8). Choose two parameters $\theta_1 < \theta_2$. Then $N = (-\infty, (\theta_1 + \theta_2)/2]$ is a P_{θ_2}-null set, but $P_{\theta_1}(N) > 0$. This is in contradiction to the assumption that $\mathcal{P}$ is an exponential family. □

To give sufficient conditions for a family to be strictly k-parametric we introduce the following notation: Define

$$\mathcal{A} = \{x : p(x; \theta) > 0\}.$$

Because of Remark 2.1 $\mathcal{A}$ is independent of θ. The functions $T_1, \ldots, T_k$ are called $\mathcal{P}$-affine independent, if for $c_j \in \mathbb{R}$ and $c_0 \in \mathbb{R}$

$$\sum_{j=1}^{k} c_j T_j(x) = c_0 \qquad \text{for all } x \in \mathcal{A} \qquad \text{implies} \qquad c_j = 0, \quad \text{for } j = 0, \ldots, k.$$

Without proof we give the following theorem:

Theorem 2.1 *Let $\mathcal{P}$ be an exponential family. Then*

1. *The family $\mathcal{P}$ is strictly k-dimensional if in (2.7) the functions $1, \zeta_1, \ldots, \zeta_k$ are linearly independent and the statistics $T_1, \ldots, T_k$ are $\mathcal{P}$-affine independent.*
2. *The functions $T_1, \ldots, T_k$ are $\mathcal{P}$-affine independent, if the covariance matrix $\mathrm{Cov}_\theta T$ is positive definite for all $\theta \in \Theta$.*

The proof can be found in Witting (1987).

Example 2.19 (Flowers) The probability function of X is

$$p(x; \theta) = p_1^{I_1(x)} p_2^{I_2(x)} p_3^{I_3(x)} \tag{2.9}$$

with

$$I_j(x) = \begin{cases} 1 & \text{for } x = j \\ 0 & \text{else} \end{cases}$$

for $j = 1, 2, 3$. We rewrite (2.9) as

$$p(x; \theta) = \exp(I_1(x) \ln(p_1) + I_2(x) \ln(p_2) + I_3(x) \ln(p_3)).$$

Thus X belongs to a three-parameter exponential family.

OBS! Because of

$$I_1(x) + I_2(x) + I_3(x) = 1$$

this family is not strictly three-dimensional! We reformulate

$$p(x; \theta) = \exp(I_1(x) \ln(p_1) + I_2(x) \ln(p_2) + (1 - I_1(x) - I_2(x)) \ln(p_3))$$
$$= p_3 \exp(I_1(x)(\ln(p_1) - \ln(p_3)) + I_2(x)(\ln(p_2) - \ln(p_3))).$$

Hence $p(x; \theta)$ equals

$$(1 - (p_1 + p_2)) \exp \left(I_1(x) \ln(\frac{p_1}{1 - (p_1 + p_2)}) + I_2(x) \ln(\frac{p_2}{1 - (p_1 + p_2)}) \right).$$

Thus X belongs to a strictly two-parameter exponential family with

$$A(\theta) = 1 - (p_1 + p_2), \quad h(x) = 1,$$

$$\zeta_1(\theta) = \ln \left(\frac{p_1}{1 - (p_1 + p_2)} \right), \quad \zeta_2(\theta) = \ln \left(\frac{p_2}{1 - (p_1 + p_2)} \right)$$

and

$$T_1(x) = I_1(x), \quad T_2(x) = I_2(x).$$

$\square$

Let us formulate the following important statement about the distribution of a sample of independent r.v.'s distributed according to a distribution from an exponential family:

Theorem 2.2

a) If $X_1, \ldots, X_n$ is a sample of independent r.v.'s with distributions belonging to an exponential family, then the joint distribution of the vector $\mathbf{X} = (X_1, \ldots, X_n)$ is an element of an exponential family.

b) If $X_1, \ldots, X_n$ is a sample of i.i.d. r.v.'s with a distribution of the form (2.7) with functions ζ_j and $T = (T_1, \ldots, T_k)$, then the distribution of $\mathbf{X}$ belongs to an exponential family with functions ζ_j, $j = 1, \ldots, k$ and $T_{(n)}(\mathbf{x}) = \sum_{i=1}^{n} T(x_i)$.

PROOF: a) Since the distribution of X_i belongs to an exponential family the probability function of the sample is given by

$$p(\mathbf{x}; \theta) = \prod_{i=1}^{n} A_i(\theta) \exp \left[\sum_{j=1}^{k_i} \zeta_{ij}(\theta) T_{ij}(x_i) \right] h_i(x_i)$$

$$= \tilde{A}(\theta) \exp \left[\sum_{i=1}^{n} \sum_{j=1}^{k_i} \zeta_{ij}(\theta) T_{ij}(x_i) \right] \tilde{h}(\mathbf{x}) \qquad (2.10)$$

with $\tilde{A}(\theta) = \prod_{i=1}^{n} A_i(\theta)$ and $\tilde{h}(\mathbf{x}) = \prod_{i=1}^{n} h_i(x_i)$. Thus, again, it has the form of an exponential family.

b) In the case that the X_i's are identically distributed we have for the last equation in (2.10)

$$p(\mathbf{x}; \theta) = \tilde{A}(\theta) \exp \left[\sum_{j=1}^{k} \zeta_j(\theta) \sum_{i=1}^{n} T_j(x_i) \right] \tilde{h}(\mathbf{x}).$$

Therefore, the distribution of $\mathbf{X} = (X_1, \ldots, X_n)$ belongs to an exponential family with the functions ζ_j and the statistic $T_{(n)}(\mathbf{x}) = \sum_{i=1}^{n} T(x_i)$.

$\square$

Example 2.20 *(**Linear structural model**) Consider a simple linear model where the observations (x_i, y_i), $i = 1, \ldots, n$ fulfill the following relation

$$y_i = \beta \xi_i + \varepsilon_i \quad \text{and} \quad x_i = \xi_i + \delta_i.$$

Here ξ_i, ε_i and δ_i are realizations of independent standard normal random variables. Then we have for $Z_i^\mathsf{T} = (X_i, Y_i)$

$$Z_i \sim \mathsf{N}_2(0, \Sigma(\beta)), \quad \text{with } \Sigma(\beta) = \begin{pmatrix} 2 & \beta \\ \beta & \beta^2 + 1 \end{pmatrix}.$$

Note for $z^\mathsf{T} = (x, y)$

$$z^\mathsf{T} \Sigma(\beta)^{-1} z = \frac{1}{\beta^2 + 2} \left((\beta^2 + 1) x^2 - 2\beta xy + 2y^2 \right).$$

Thus the joint distribution of the pair (X_i, Y_i) belongs to an exponential family with

$$T(z) = (x^2, xy, y^2) \quad \text{and} \quad \zeta(\beta) = \left(\frac{\beta^2 + 1}{\beta^2 + 2}, \frac{-2\beta}{\beta^2 + 2}, \frac{2}{\beta^2 + 2} \right).$$

This family is not a strictly 3-dimensional exponential family, because

$$\sum_{j=1}^{3} a_j \zeta_j(\beta) = a_0, \quad \text{for all } a \in \mathbb{R} \text{ and } a_0 = a_1 = a, a_2 = 0, a_3 = \frac{a}{2}. \qquad \square$$

Example 2.21 (Genotypes) Consider an i.i.d. sample $\mathbf{X}$ of size n where each component is a copy of a r.v. X distributed according to the Hardy–Weinberg model introduced in Example 2.14. Then

$$p(x; \theta) = p_1^{I_1(x)} p_2^{I_2(x)} p_3^{I_3(x)}$$

with

$$I_1(x) = \begin{cases} 1 & \text{for } x = \text{aa} \\ 0 & \text{else} \end{cases} \qquad I_2(x) = \begin{cases} 1 & \text{for } x = \text{Aa} \\ 0 & \text{else} \end{cases}$$

$$I_3(x) = \begin{cases} 1 & \text{for } x = \text{AA} \\ 0 & \text{else} \end{cases}$$

and

$$p_1 = \theta^2, \; p_2 = 2\theta(1-\theta), \; p_3 = (1-\theta)^2.$$

Note

$$I_1(x) + I_2(x) + I_3(x) = 1$$

and $p_1 + p_2 + p_3 = 1$. As in Example 2.19 we get

$$\begin{aligned}
p(x; \theta) &= p_3 \exp(I_1(x) \ln(p_1/p_3) + I_2(x) \ln(p_2/p_3)) \\
&= (1-\theta)^2 \exp\left(I_1(x) 2\ln\left(\frac{\theta}{1-\theta}\right) + I_2(x) \ln\left(\frac{2\theta}{1-\theta}\right) \right) \\
&= (1-\theta)^2 \exp\left((2I_1(x) + I_2(x)) \ln\left(\frac{\theta}{1-\theta}\right) \right) 2^{I_2(x)}.
\end{aligned}$$

Thus X belongs to a strictly one-parameter exponential family with

$$A(\theta) = (1-\theta)^2, \quad h(x) = 2^{I_2(x)}$$

and

$$\zeta_1(\theta) = \ln\left(\frac{\theta}{1-\theta}\right) \quad \text{and} \quad T(x) = 2I_1(x) + I_2(x).$$

The distribution of the sample $\mathbf{X}$ of size n belongs to a one-parameter exponential family with the same function ζ_1 and the statistic

$$T_{(n)}(\mathbf{x}) = 2\sum_{i=1}^{n} I_1(x_i) + \sum_{i=1}^{n} I_2(x_i).$$

$\square$

To get a further characterization of a k-parameter exponential family, let us introduce the notion of the natural parameter space. The quantity $A(\theta)$ is only a normalizing factor and depends on the parameter θ only via $\zeta(\theta)$. The class can also be parameterized by $\zeta := \zeta(\theta)$. The parameter ζ is called natural parameter

$$\mathcal{P} = \{P_\zeta : \zeta \in \mathcal{Z}\}, \qquad \mathcal{Z} := \zeta(\Theta).$$

The probability function can be written (up to sets of measure zero) in this new parametrization as

$$p(x; \zeta) = C(\zeta) \exp \left(\sum_{j=1}^{k} \zeta_j T_j(x) \right) h(x) \qquad (2.11)$$

with

$$C(\zeta) = \left(\int \exp \left[\sum_{j=1}^{k} \zeta_j T_j(x) \right] h(x) dx \right)^{-1}.$$

The set

$$\mathcal{Z}^* = \{\zeta : 0 < \int \exp \left(\sum_{j=1}^{k} \zeta_j T_j(x) \right) h(x) dx < \infty, \ \zeta \in \mathbb{R}^k\}$$

is called the **natural parameter space**. This means that the natural parameter space consists of all points for which (2.11) is a probability function. In general we have $\zeta(\Theta) \subset \mathcal{Z}^*$. The natural parameter space is characterized by the following theorem.

Theorem 2.3 *The natural parameter space of a strictly k-parameter exponential family is convex and contains a nonempty k-dimensional interval.*

The proof is given in Witting (1987).

Special case 2.3 (Bernoulli distribution) Let $\mathcal{P}$ be the class of all Bernoulli distributions with parameter $\theta \in (0, 1)$. Then the probability mass function is

$$p(x; \theta) = \theta^x (1 - \theta)^{1-x} = (1 - \theta) \exp \left[\ln \left(\frac{\theta}{1 - \theta} \right) x \right].$$

Thus, this class is an exponential family with

$$A(\theta) = (1 - \theta), \quad T(x) = x \quad \text{and} \quad \zeta(\theta) = \ln \left(\frac{\theta}{1 - \theta} \right).$$

The natural parameter $\zeta(\theta)$ is also called **log odds**. It is the logarithm of

$$\frac{\mathsf{P}(X=1)}{\mathsf{P}(X=0)} = \frac{\mathsf{P}(X=1)}{1-\mathsf{P}(X=1)}.$$

We have

$$p(x;\zeta) = C(\zeta)\exp(\zeta x) \qquad \text{with} \quad C(\zeta) = 1/(1+\exp(\zeta)).$$

The natural parameter space is $\mathbb{R}$. $\square$

Exercise 2.1 *Let X_1, X_2, X_3 be i.i.d. as $\mathsf{Bin}(1,\theta)$, $\theta \in (0,1)$. Define $Y_1 = X_1 + X_2$ and $Y_2 = X_2 + X_3$. Show that the joint distributions of (Y_1, Y_2) form an exponential family. What can you say about the natural parameter space?

In the next sections about optimality of statistical procedures we will make use of the following analytic properties of exponential families:

Theorem 2.4 *Let $\mathcal{P}$ be a k-parameter exponential family with probability function of the form (2.11) and the natural parameter space $\mathcal{Z}^*$. For all interior points ζ of the natural parameter space $\mathcal{Z}^*$ we have*

a) All moments of the statistic T (with respect to P_ζ^T) exist.

b) Let $\mathcal{Z}$ be an open subset of the natural parameter space $\mathcal{Z}^*$ and the function ϕ be integrable with respect to P_ζ for all $\zeta \in \mathcal{Z}$. The function $\beta(\zeta) = \mathsf{E}_\zeta\phi$ is arbitrarily often differentiable, and

$$\frac{\partial \mathsf{E}_\zeta \phi}{\partial \zeta_j} = \int \phi(x)T_j(x)d\mathsf{P}_\zeta(x) - \mathsf{E}_\zeta\phi \int T_j(x)d\mathsf{P}_\zeta(x) = \mathsf{Cov}_\zeta(\phi,T).$$

The proof is given in Witting (1987).

Exercise 2.2 * Show that for any θ, an interior point of the natural parameter space, the expectations and the covariances of the statistics T_j in the exponential family (2.11) are given by

$$\mathsf{E}_\zeta T_j = -\frac{\partial \ln C(\zeta)}{\partial \zeta_j}, \qquad j = 1,\ldots,k$$

$$\mathsf{Cov}_\zeta(T_i, T_j) = -\frac{\partial^2 \ln C(\zeta)}{\partial \zeta_i \partial \zeta_j}, \qquad i,j = 1,\ldots,k.$$

2.5 List of Problems

1. Formulate a statistical model for the Example 2.9 (Ballpoint pens) where the probability for defective pens during the weekend production is higher than during the working days production.

2. Formulate a statistical model for the Example 2.2 (Flowers) where the probability for the colors depends on the soil and on the weather. Assume that the flowers are collected from five different areas on two different days.

3. Consider a population which consists of two parts. Both parts can be described by normal distributions, one with $N(\mu_1, \sigma_1^2)$ the other with $N(\mu_2, \sigma_2^2)$. The parameters of the normal distributions are unknown. Also the proportion of the two parts is not known. Suppose an element is drawn at random from the population. Write down a statistical model.

4. Check whether the family of all uniform distributions over the interval $[0, \theta]$, $\theta > 0$ forms an exponential family.

5. Do the following distributions belong to an exponential family?
   a) $N(0, \sigma^2)$, b) $N(1, \sigma^2)$
   c) $N(\mu, \sigma^2)$, $\theta = (\mu, \sigma^2)$, d) $N(\mu, \mu)$, $\mu > 0$

6. Do the following distributions belong to an exponential family?
   a) $Poi(\lambda)$ b) $Geo(p)$ c) $Rayleigh(\alpha)$

7. Construct a distribution which doesn't belong to an exponential family.

8. Let $X_{11}, \ldots, X_{1n_1}, X_{21}, \ldots, X_{2n_2}$ be independent random variables with $X_{ij} \sim Exp(\lambda_i)$, $j = 1, \ldots, n_i$, $i = 1, 2$ and $\theta = (\lambda_1, \lambda_2)$. Show that the (joint) distribution of $\mathbf{X} = (X_{11}, \ldots, X_{1n_1}, X_{21}, \ldots, X_{2n_2})$ belongs to a k-parameter exponential family. Determine k.

9. Let $\mathcal{P}$ be the class of the multinomial distributions with (known) parameter n and unknown parameter $\theta = (\pi_1, \ldots, \pi_m)$, where $\sum_{j=1}^{m} \pi_j = 1$ and $\pi_j \in (0, 1)$. Show that $\mathcal{P}$ is a k-parameter exponential family. Determine k.

2.6 Further Reading

In the book of Davison (2003) the process of finding an appropriate model is discussed. Starting from describing the variation in the data he shows how the knowledge of this variation can be transformed into statements about a model. Davison presents a lot of examples from different fields of application of statistics—so the reader can reinforce the material presented in the Chapter 2.

The approach presented in this textbook is based on a statistical model $\mathcal{P} = \{P_\theta : \theta \in \Theta\}$, where the underlying parameter θ is treated as an unknown fixed constant. The information contained in the data $\mathbf{x}$ is used to make inferences about θ. Another approach—the Bayesian approach—includes nonsample information about θ. It is assumed in addition that θ is a random variable with a known so-called prior distribution. Now, instead of P_θ the conditional distribution of **X** given θ is considered. Then the information about θ is summarized in the posterior distribution of θ given the data **x**. Discussions about both approaches, the first one is also called the frequentist, repeated sampling or classical approach, are found, for example, in Cox (2006) or in Pawitan (2001).

Both approaches can be considered under a decision-theoretic point of view. That is, the consequences of decisions are quantified by a loss function. The textbook of Robert (2001) contains a decision-theoretic foundation of the Bayesian analysis, inference procedures and computational methods. In Berger (1985) the frequentist and the Bayesian decision principles are presented and discussed.

In Section 2.4 exponential families are introduced. Properties which are the basis for deriving results in the following chapters are presented. For a general treatment of exponential families we refer to Barndorff-Nielsen (1978) and Brown (1986).

Finally we recommend the following books on the history of statistics: Peters (1987), Hald (1998) and Stigler (2000).

Chapter 3

Inference Principles

3.1 Likelihood Function

Just for illustration let us start with the example of the Lion's appetite which is taken from Dudewicz and Mishra (1988, p.348). Of course this has no real meaning.

Example 3.1 (Lion's appetite) Suppose that the appetite of a lion has three different stages:

$$\theta \in \Theta = \{\text{hungry}, \text{moderate}, \text{lethargic}\} = \{\theta_1, \theta_2, \theta_3\}.$$

Each night the lion eats x people with a probability $P_\theta(x)$ given by the following table:

x	0	1	2	3	4
θ_1	0	0.05	0.05	0.8	0.1
θ_2	0.05	0.05	0.8	0.1	0
θ_3	0.9	0.05	0.05	0	0

If we observe $x = 4$ we conclude that the lion was hungry. If we observe $x = 0$ the most likely conclusion is that our lion was lethargic. In the case $x = 1$ it is difficult to decide which stage he was in. But for $x = 3$ we would conclude that the lion was hungry; for $x = 2$ we suspect that he was in a moderate mood. □

Figure 3.1: Lion's appetite.

Suppose we have stated a statistical model $\mathcal{P} = \{\mathsf{P}_\theta : \theta \in \Theta\}$ for the random variable $\mathbf{X}$. Given $\mathbf{x}$ we know the value of the probability function $p(\mathbf{x}; \theta)$ apart from θ. From $\mathbf{x}$ we will conclude in a deductive way which value of the parameter is more likely. So it is useful to consider $p(\mathbf{x}; \theta)$ for fixed $\mathbf{x}$ as a function of θ. This idea leads to the following definition:

Definition 3.1 (Likelihood function) For a fixed observation $\mathbf{x}$ of a random variable $\mathbf{X}$ with a probability function $p(\cdot; \theta)$ the likelihood function $L(\cdot; \mathbf{x}) : \Theta \to \mathbb{R}_+$ is defined by

$$L(\theta; \mathbf{x}) = p(\mathbf{x}; \theta).$$

If $\mathbf{X} = (X_1, \ldots, X_n)$ is a sample of independent r.v.'s, then

$$L(\theta; \mathbf{x}) = \prod_{i=1}^{n} \mathsf{P}_{i,\theta}(x_i) \quad \text{in the discrete case} \qquad (3.1)$$

and

$$L(\theta; \mathbf{x}) = \prod_{i=1}^{n} f_i(x_i; \theta) \quad \text{in the continuous case,} \qquad (3.2)$$

where X_i is distributed according to $\mathsf{P}_{i,\theta}$ and $f_i(\cdot; \theta)$, respectively.

Example 3.2 (Lion's appetite) For $x = 4$ the likelihood function is given by

	θ_1	θ_2	θ_3
$L(\theta; x)$	0.1	0	0

□

The **maximum likelihood principle** was formulated in Lindgren (1962, p. 225) as follows:

"A statistical inference or procedure should be consistent with the assumption that the best explanation of a set of data $\mathbf{x}$ is provided by $\widehat{\theta}$ a value of θ that maximizes $L(\theta; \mathbf{x})$."

Example 3.3 (Lion's appetite) For $x = 1$ the likelihood function is:

$L(\theta; x)$	θ_1	θ_2	θ_3
	0.05	0.05	0.05

That means the observation $x = 1$ cannot tell us any preference over the three possible stages. □

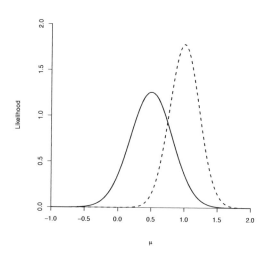

Figure 3.2: Likelihood functions for two samples from normal distributions. The solid line shows the likelihood based on a sample of size $n = 10$ and $\bar{x} = 0.5$, the dashed line that of a sample of size $n = 20$ with $\bar{x} = 1$.

Special case 3.1 (Normal distribution, variance known)
Let $X_1, \ldots, X_n$ be i.i.d. r.v.'s according to $\mathsf{N}(\theta, 1)$. Then

$$L(\theta; \mathbf{x}) = \prod_{i=1}^{n} \frac{1}{\sqrt{2\pi}} \exp\left(-\frac{1}{2}(x_i - \theta)^2\right) = (2\pi)^{-\frac{n}{2}} \exp\left(-\frac{1}{2}\sum_{i=1}^{n}(x_i - \theta)^2\right)$$

$$\propto \exp\left(-\frac{1}{2}\sum_{i=1}^{n}(x_i - \theta)^2\right).$$

Since $\sum_{i=1}^{n}(x_i - \bar{x})(\bar{x} - \theta) = 0$ we obtain

$$\sum_{i=1}^{n}(x_i - \theta)^2 = \sum_{i=1}^{n}(x_i - \bar{x} + \bar{x} - \theta)^2 = \sum_{i=1}^{n}(x_i - \bar{x})^2 + \sum_{i=1}^{n}(\bar{x} - \theta)^2.$$

Hence

$$L(\theta; \mathbf{x}) \quad \propto \quad \exp\left(-\frac{1}{2}\left(\sum_{i=1}^{n}(x_i - \overline{x})^2 + \sum_{i=1}^{n}(\overline{x} - \theta)^2\right)\right)$$

$$\propto \quad \exp\left(-\frac{n}{2}(\overline{x} - \theta)^2\right). \tag{3.3}$$

The likelihood function is proportional to the density function of the normal distribution with expectation $\overline{\mathbf{x}}$ and variance $\frac{1}{n}$. Note that the likelihood function is a function of the unknown parameter and is not a density! It has the form of the Gaussian bell curve. We can use our knowledge of the normal distribution to discuss its form; compare Figure 3.2. The maximum of $L(\theta; \mathbf{x})$ is reached at the sample mean $\overline{\mathbf{x}}$. The larger the sample sizes the more concentrated at $\overline{\mathbf{x}}$ the function is. $\square$

 Example 3.4 (Flowers) Let $X_1, \ldots, X_n$ be i.i.d. r.v.'s according to a three-point distribution introduced in Example 2.10 on page 10. The likelihood function is given by

$$L(p_1, p_2; \mathbf{x}) \quad = \quad \prod_{i=1}^{n} p_1^{I_1(x_i)} p_2^{I_2(x_i)}(1 - (p_1 + p_2))^{I_3(x_i)}$$

$$= \quad p_1^{n_1} p_2^{n_2}(1 - (p_1 + p_2))^{n - n_1 - n_2}, \tag{3.4}$$

where $n_j = \sum_{i=1}^{n} I_j(x_i)$ is the number of flowers of the respective color and $n = n_1 + n_2 + n_3$. $\square$

OBS! In Example 3.4 we see that the likelihood function can be continuous even for discrete distributions.

Example 3.5 (Bernoulli distribution) We are interested in a random experiment with two outcomes (success and failure). The probability of success, say θ, is unknown. Let us repeat this experiment ten times. The result is $\mathbf{x} = (0, 1, 0, 0, 0, 0, 1, 0, 1, 0)$, where $x_i = 1$ stands for success and $x_i = 0$ for failure. Since the x_i's are values of Bernoulli variables the likelihood function is

$$L(\theta; \mathbf{x}) = \prod_{i=1}^{10} \theta^{x_i}(1 - \theta)^{1 - x_i} = \theta^3(1 - \theta)^7.$$

The maximum of the likelihood function is taken at $\theta = \frac{3}{10}$. This follows from

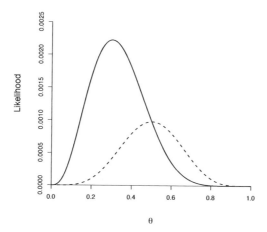

Figure 3.3: Likelihood functions in Bernoulli experiments. The dashed curve shows the likelihood function for data x_i with $\sum_{i=1}^{n} x_i = 5$. The solid line is the likelihood for three successes, i.e., $\sum_{i=1}^{n} x_i = 3$.

$$\frac{\partial L(\theta; \mathbf{x})}{\partial \theta} = 3\theta^2(1-\theta)^7 - 7\theta^3(1-\theta)^6 = \theta^2(1-\theta)^6(3(1-\theta) - 7\theta)$$

and

$$\frac{\partial L(\theta; \mathbf{x})}{\partial \theta} = 0 \qquad \text{iff} \qquad \theta = \frac{3}{10}.$$

Suppose in another series of ten experiments five successes are observed. Here the maximum is taken at $\theta = \frac{1}{2}$. Both likelihood functions are plotted in Figure 3.3. These functions show where θ is likely to fall after observing $\mathbf{x}$. □

It is convenient to use instead of the likelihood function the **log-likelihood function**

$$l(\theta; \mathbf{x}) = \ln L(\theta; \mathbf{x}) = \ln p(\mathbf{x}; \theta).$$

Since the ln-transformation is monotone the maximum likelihood principle is still the same.

The likelihood function in Definition 3.1 depends on the data $\mathbf{x}$. To draw conclusions about the underlying statistical model we have to take into account all possible samples. Thus, for fixed $\theta \in \Theta$ we consider $L(\theta; \mathbf{X})$ and $l(\theta; \mathbf{X})$ as random variables.

Let us consider an example where we characterize the log-likelihood function as random:

Example 3.6 (Normal distribution) Let $X \sim \mathsf{N}(\theta, 1)$ and $Y \sim \mathsf{N}(2\theta, 1)$ be independent, then

$$l(\theta; X) = -\frac{1}{2}\ln(2\pi) - \frac{1}{2}(X - \theta)^2$$

where the r.v. $-2l(\theta; X) - \ln(2\pi) = (X - \theta)^2$ is distributed according to a χ^2-distribution with one degree of freedom. Since

$$l(\theta; Y) = -\frac{1}{2}\ln(2\pi) - \frac{1}{2}(Y - 2\theta)^2$$

we get for the vector (X, Y)

$$
\begin{aligned}
l(\theta; X, Y) &= l(\theta; X) + l(\theta; Y) \\
&= -\ln(2\pi) - \frac{1}{2}(X - \theta)^2 - \frac{1}{2}(Y - 2\theta)^2
\end{aligned}
$$

and $-2l(\theta; X, Y) - \ln(2\pi) \sim \chi_2^2$ is distributed according to a χ^2-distribution with two degrees of freedom. □

Let us recall once more the **maximum likelihood principle**: The likelihood function should be large at the parameter, which fits best to the data $\mathbf{x}$. This is reflected by the following inequality. It says that the expected value of the log-likelihood function is maximal at the underlying parameter, say θ_0, or in other words: "In average the log-likelihood of the underlying distribution P_{θ_0} is larger than the log-likelihood of another distribution P_{θ}."

Lemma 3.1

$$\mathsf{E}_{\theta_0} l(\theta; \mathbf{X}) \le \mathsf{E}_{\theta_0} l(\theta_0; \mathbf{X}) \qquad \text{for all } \theta_0, \theta \in \Theta.$$

PROOF: The proof is based on Jensen's inequality which says: If Y is a real-valued r.v. and h is a convex function then $\mathsf{E}h(Y) \ge h(\mathsf{E}Y)$. Since the function $-\ln$ is convex we obtain

$$\mathsf{E}_{\theta_0}\ln p(\mathbf{X}; \theta_0) - \mathsf{E}_{\theta_0}\ln p(\mathbf{X}; \theta) = -\mathsf{E}_{\theta_0}\ln\left(\frac{p(\mathbf{X}; \theta)}{p(\mathbf{X}; \theta_0)}\right) \ge -\ln\left(\mathsf{E}_{\theta_0}\frac{p(\mathbf{X}; \theta)}{p(\mathbf{X}; \theta_0)}\right).$$

Further (for simplicity we write it for continuous distributions only)

$$\ln\left(\mathsf{E}_{\theta_0}\frac{p(\mathbf{X}; \theta)}{p(\mathbf{X}; \theta_0)}\right) = \ln\int\frac{p(\mathbf{x}; \theta)}{p(\mathbf{x}; \theta_0)}p(\mathbf{x}; \theta_0)d\mathbf{x} = \ln\int p(\mathbf{x}; \theta)d\mathbf{x} = \ln 1 = 0.$$

□

Example 3.7 (Lion's appetite) Taking the logarithm in the
table given in Example 3.1 we obtain the log-likelihood

x	0	1	2	3	4
θ_1	$-\infty$	-2.996	-2.996	-0.223	-2.303
θ_2	-2.996	-2.996	-0.223	-2.303	$-\infty$
θ_3	-0.105	-2.996	-2.996	$-\infty$	$-\infty$

Using this table we can derive the expectations. For example:

$$\mathsf{E}_{\theta_1} l(\theta_2; X) = l(\theta_2; 1) \cdot 0.05 + l(\theta_2; 2) \cdot 0.05 + l(\theta_2; 3) \cdot 0.8 + l(\theta_2; 4) \cdot 0.1$$
$$= -2.996 \cdot 0.05 - 2.223 \cdot 0.05 - 2.303 \cdot 0.8 - \infty \cdot 0.1 = -\infty.$$

Similarly computing the other expectations, we obtain the following matrix
of expectations $\mathsf{E}_\theta l(\theta'; X)$:

	θ'_1	θ'_2	θ'_3
θ_1	-0.703	$-\infty$	$-\infty$
θ_2	$-\infty$	-0.542	$-\infty$
θ_3	$-\infty$	-2.961	-0.394

The largest values are in the diagonal, i.e., for $\theta_i = \theta'_i$.   □

In Chapter 4 we will use the maximum likelihood principle for estimation
of unknown parameters. Roughly speaking, the maximum of the likelihood
function provides a point estimate. However, the likelihood function represents
more information than a single value. From the curvature at the maximum
we can also draw conclusions about the plausibility of the parameter. This
concept of measuring information contained in the data via the likelihood
function leads to the Fisher information which will be studied in the next
section. In Chapter 5 we will use the likelihood function again. There we
compare likelihood functions at different parameter values to decide between
two hypotheses.

3.2   Fisher Information

The likelihood approach can be used to measure the information of a statistical
model. Such an information measure should be nonrandom, nonnegative and
additive for independent experiments. Moreover, it should not decrease with
increasing sample size.

To define the desired information measure we need some regularity conditions.

Reg 1 The distributions $\{P_\theta : \theta \in \Theta\}$ have a common support, so that without loss of generality the set

$$\mathcal{A} = \{\mathbf{x} : p(\mathbf{x}; \theta) > 0\}$$

is independent of θ.

Remark 3.1 Distributions belonging to an exponential family satisfy condition **Reg 1**; compare Remark 2.1.

Example 3.8 (Counterexamples) Uniform distributions on the interval $[0, \theta]$, i.e., $\{U[0, \theta] : \theta \in (0, \infty)\}$ and exponential distributions with densities of the form

$$\{f(x; a, b) = \frac{1}{b} \exp(-(x - a)/b)\, \mathbb{1}_{[a,\infty)}(x) : a, b \in (0, \infty)\},$$

have supports depending on the parameter. That is, for these classes of distributions **Reg 1** is not satisfied.   □

3.2.1 The One-Dimensional Parameter Case

We start our considerations with scalar parameters; moreover we assume:

Reg 2 The parameter space $\Theta \subseteq \mathbb{R}$ is an open interval (finite or infinite).

We will consider models where the likelihood function is smooth. Thus we will assume:

Reg 3 For any $\mathbf{x} \in \mathcal{A}$ and all $\theta \in \Theta$, the derivative $\frac{\partial p(\mathbf{x};\theta)}{\partial \theta}$ exists and is finite.

Remark 3.2 The derivative of the likelihood function (w.r.t. its argument θ) is denoted by prime; for the derivative of $p(\mathbf{x}; \theta)$ w.r.t. θ we use $\frac{\partial p(\mathbf{x};\theta)}{\partial \theta}$.

Definition 3.2 (Score function) Suppose conditions **Reg 1**–**Reg 3** are satisfied. For every $\mathbf{x} \in \mathcal{A}$ we define the score function as derivative of the log-likelihood function, i.e.,

$$V(\theta; \mathbf{x}) = l'(\theta; \mathbf{x}) = \frac{\partial \ln L(\theta; \mathbf{x})}{\partial \theta}.$$

Let us consider the score function in more detail. Since

$$l'(\theta; \mathbf{x}) = \frac{L'(\theta; \mathbf{x})}{L(\theta; \mathbf{x})} = \lim_{\tau \to 0} \frac{\frac{1}{\tau}(L(\theta + \tau; \mathbf{x}) - L(\theta; \mathbf{x}))}{L(\theta; \mathbf{x})}$$

we can interpret the score function as the relative rate at which the likelihood function $L(.; \mathbf{x})$ changes.

The regularity conditions **Reg 1–Reg 3** imply that integration (or summation) and differentiation can be reversed, i.e., for all $\theta \in \Theta$

$$\frac{\partial}{\partial \theta} \int_A f(\mathbf{x}; \theta) d\mathbf{x} = \int_A \frac{\partial}{\partial \theta} f(\mathbf{x}; \theta) d\mathbf{x} \qquad \text{for continuous distributions}$$

and

$$\frac{\partial}{\partial \theta} \sum_{\mathbf{x} \in A} \mathsf{P}_\theta(\mathbf{x}) = \sum_{\mathbf{x} \in A} \frac{\partial}{\partial \theta} \mathsf{P}_\theta(\mathbf{x}) \qquad \text{for discrete distributions.}$$

Using this we can prove the first result. At first glance it may be surprising. Suppose the underlying distribution is P_{θ_0}. Under the regularity assumptions we can write the derivative of the expected value of the log-likelihood function $C(\theta) = \mathsf{E}_{\theta_0} l(\theta; \mathbf{X})$ as $\mathsf{E}_{\theta_0} V(\theta; \mathbf{X})$. Recalling the maximum likelihood principle and Lemma 3.2 we expect that the function $C(\theta)$ has its maximum at θ_0; thus the derivative of $C(\theta)$ at θ_0 is zero.

Theorem 3.1 *Under the regularity conditions* **Reg 1–Reg 3** *we have*

$$\mathsf{E}_\theta V(\theta; \mathbf{X}) = 0 \qquad \text{for all } \theta \in \Theta.$$

PROOF: We prove the statement for continuous distributions. The discrete case is left as an exercise. Since integration and differentiation can be reversed we have

$$\mathsf{E}_\theta V(\theta; \mathbf{X}) = \int_A l'(\theta; \mathbf{x}) f(\mathbf{x}; \theta) d\mathbf{x} = \int_A \frac{\frac{\partial}{\partial \theta} f(\mathbf{x}; \theta)}{f(\mathbf{x}; \theta)} f(\mathbf{x}; \theta) d\mathbf{x}$$

$$= \int_A \frac{\partial}{\partial \theta} f(\mathbf{x}; \theta) d\mathbf{x} = \frac{\partial}{\partial \theta} \int_A f(\mathbf{x}; \theta) d\mathbf{x} = 0.$$

□

Exercise 3.1 Give the proof for discrete distributions.

Example 3.9 (Normal distribution) Let us demonstrate Theorem 3.1 with help of Example 3.6 on page 30. The score function for X is given by

$$V(\theta; X) = l'(\theta; X) = \frac{\partial}{\partial \theta}\left(-\frac{1}{2}\ln(2\pi) - \frac{1}{2}(X - \theta)^2\right) = X - \theta.$$

Thus $V(\theta; X) \sim N(0, 1)$, in particular $E_\theta V(\theta; X) = 0$.   □

To define an information measure let us come back to the interpretation of the score function as the rate at which the likelihood function changes. If the underlying parameter is θ_0 and the value $V^2(\theta_0; x)$ is large then we can easier distinguish θ_0 from the neighboring values θ. Thus $V^2(\theta; x)$ provides information about θ. Since the score function is random we will take the expectation of $V^2(\theta; X)$ w.r.t. P_θ. In this sense, the larger $E_\theta V^2(\theta; X)$ is, the more informative is X. Since the score function has expectation zero this is equivalent to the variance and we define:

Definition 3.3 (Fisher information) Suppose the conditions **Reg 1–Reg 3** are fulfilled. The Fisher information is defined by

$$I_X(\theta) = \text{Var}_\theta V(\theta; X).$$

OBS! Note that the "X" in $I_X(\theta)$ is only a symbol. It does not mean that the Fisher information is random. It indicates that the Fisher information corresponds to the r.v. X.

Special case 3.2 (Normal distribution, known variance) Let $X_1, \ldots, X_n$ be i.i.d. r.v.'s according to $N(\mu, \sigma^2)$, where σ^2 is known. Then

$$L(\mu; x) \propto \exp\left(-\frac{1}{2\sigma^2}\sum_{i=1}^n (x_i - \mu)^2\right),$$

$$l(\mu; x) = -\left(\frac{1}{2\sigma^2}\sum_{i=1}^n (x_i - \mu)^2\right) + \text{const.}$$

and

$$V(\mu; x) = l'(\mu; x) = \frac{1}{\sigma^2}\sum_{i=1}^n (x_i - \mu).$$

Thus,

$$I_X(\mu) = \frac{n}{\sigma^2}.$$

This gives a reasonable interpretation. The data contain more information about μ if the variance is smaller. Moreover, with increasing sample size the information increases.   □

Exercise 3.2 Is it possible to compute the Fisher information for the lion's appetite in Example 3.1?

If we additionally assume that the likelihood function is twice differentiable we can derive an equivalent expression for the Fisher information.

Reg 4 For all $\mathbf{x} \in \mathcal{A}$ and all $\theta \in \Theta$ the log-likelihood function is twice differentiable and for all $\theta \in \Theta$

$$\frac{\partial^2}{\partial \theta^2} \int_{\mathcal{A}} f(\mathbf{x}; \theta) d\mathbf{x} = \int_{\mathcal{A}} \frac{\partial^2}{\partial \theta^2} f(\mathbf{x}; \theta) d\mathbf{x} \qquad \text{for continuous distributions}$$

and

$$\frac{\partial^2}{\partial \theta^2} \sum_{\mathbf{x} \in \mathcal{A}} \mathsf{P}_\theta(\mathbf{x}) = \sum_{\mathbf{x} \in \mathcal{A}} \frac{\partial^2}{\partial \theta^2} \mathsf{P}_\theta(\mathbf{x}) \qquad \text{for discrete distributions.}$$

Theorem 3.2 *Suppose the regularity conditions* **Reg 1**–**Reg 4**. *Then*

$$I_{\mathbf{X}}(\theta) = -\mathsf{E}_\theta l''(\theta; \mathbf{X}) = -\mathsf{E}_\theta \frac{\partial^2}{\partial \theta^2} \ln p(\mathbf{X}; \theta).$$

PROOF: We consider continuous distributions. The rules of differentiation imply

$$
\begin{aligned}
l''(\theta; \mathbf{x}) &= \frac{\partial^2}{\partial \theta^2} \ln f(\mathbf{x}; \theta) = \frac{\partial}{\partial \theta} \frac{\frac{\partial}{\partial \theta} f(\mathbf{x}; \theta)}{f(\mathbf{x}; \theta)} \\
&= \frac{\frac{\partial^2}{\partial \theta^2} f(\mathbf{x}; \theta) f(\mathbf{x}; \theta) - \left(\frac{\partial}{\partial \theta} f(\mathbf{x}; \theta)\right)^2}{f(\mathbf{x}; \theta)^2} \\
&= \frac{\frac{\partial^2}{\partial \theta^2} f(\mathbf{x}; \theta)}{f(\mathbf{x}; \theta)} - \left(\frac{\frac{\partial}{\partial \theta} f(\mathbf{x}; \theta)}{f(\mathbf{x}; \theta)}\right)^2.
\end{aligned}
$$

By **Reg 4** we obtain

$$\mathsf{E}_\theta \frac{\frac{\partial^2}{\partial \theta^2} f(\mathbf{X}; \theta)}{f(\mathbf{X}; \theta)} = \int_{\mathcal{A}} \frac{\frac{\partial^2}{\partial \theta^2} f(\mathbf{x}; \theta)}{f(\mathbf{x}; \theta)} f(\mathbf{x}; \theta) \, d\mathbf{x}$$

$$= \int_{\mathcal{A}} \frac{\partial^2}{\partial \theta^2} f(\mathbf{x}; \theta) \, d\mathbf{x} = \frac{\partial^2}{\partial \theta^2} \int_{\mathcal{A}} f(\mathbf{x}; \theta) \, d\mathbf{x} = 0.$$

Thus

$$-\mathsf{E}_\theta l''(\theta; \mathbf{X}) = \mathsf{E}_\theta \left(\frac{\frac{\partial}{\partial \theta} f(\mathbf{X}; \theta)}{f(\mathbf{X}; \theta)} \right)^2 = \mathsf{E}_\theta V^2(\theta; \mathbf{X}).$$

□

Theorem 3.2 implies an equivalent definition of the Fisher information as expectation of the negative second derivative of the log-likelihood function. It is convenient to denote this negative derivative by the term "observed information."

Definition 3.4 (Observed information) The quantity

$$J(\theta; \mathbf{x}) = -\frac{\partial^2 \ln L(\theta; \mathbf{x})}{\partial \theta^2}$$

is called observed information.

Let us now discuss the statement of Theorem 3.2. We have

$$I_{\mathbf{X}}(\theta) = \mathsf{E}_\theta J(\theta; \mathbf{X}).$$

The curvature of the log-likelihood function at a maximum $\hat{\theta}$ is characterized by the second derivative. This derivative is negative and a large value of $J(\hat{\theta}; \mathbf{x})$ is associated with a strong peak around $\hat{\theta}$. That means that the data give a high prevalence to $\hat{\theta}$ and it is less plausible that the data come from a distribution P_θ where θ is far away from $\hat{\theta}$.

Example 3.10 In Example 3.5 on page 28 we have

$$V(\theta; \mathbf{x}) = \frac{\sum_{i=1}^{10} x_i}{\theta} - \frac{10 - \sum_{i=1}^{10} x_i}{1 - \theta},$$

$$J(\theta; \mathbf{x}) = \frac{\sum_{i=1}^{10} x_i}{\theta^2} + \frac{10 - \sum_{i=1}^{10} x_i}{(1 - \theta)^2}.$$

For the first sample with 3 successes we obtain $J(\hat{\theta}; \mathbf{x}) = 47.619$, for the second sample with 5 successes we get $J(\hat{\theta}; \mathbf{x}) = 40$. The observed information of the first sample is larger. In Figure 3.3 we see that the solid curve corresponding to the first sample has a stronger peak.   □

Special case 3.3 (Binomial distribution) The log-likelihood and the score function for the Binomial distribution $\mathsf{Bin}(n, \theta)$ with $\theta \in (0, 1)$ are given by

$$l(\theta; x) = \ln \binom{n}{x} + (n - x)\ln(1 - \theta) + x\ln\theta \quad \text{and} \quad V(\theta; x) = -\frac{n - x}{1 - \theta} + \frac{x}{\theta}.$$

The derivative of V is given by

$$V'(\theta; x) = -\frac{n - x}{(1 - \theta)^2} - \frac{x}{\theta^2}.$$

With $\mathsf{E}_\theta X = n\theta$ we obtain for $I_X(\theta) = -\mathsf{E}_\theta V'(\theta; X)$

$$I_X(\theta) = \frac{n - \mathsf{E}_\theta X}{(1 - \theta)^2} + \frac{\mathsf{E}_\theta X}{\theta^2} = \frac{n(1 - \theta)}{(1 - \theta)^2} + \frac{n\theta}{\theta^2} = \frac{n}{(1 - \theta)\theta}.$$

□

Exercise 3.3 Plot the Fisher information (as a function of θ) from Special case 3.3. Discuss the information for rare events, i.e., for small θ.

Theorem 3.3 *Let X and Y be independent r.v.'s. If $I_X(\theta)$ and $I_Y(\theta)$ are the Fisher information about θ contained in X and Y, respectively, then the Fisher information contained in (X, Y) is given by*

$$I_{(X,Y)}(\theta) = I_X(\theta) + I_Y(\theta).$$

PROOF: If $p(\cdot; \theta)$ and $q(\cdot; \theta)$ are the probability functions of X and Y, respectively, then the log-likelihood function is

$$l(\theta; (X, Y)) = \ln p(X; \theta) + \ln q(Y; \theta)$$

and the score function is given by

$$V(\theta; (X, Y)) = \frac{\partial}{\partial\theta} \ln p(X; \theta) + \frac{\partial}{\partial\theta} \ln q(Y; \theta) = V(\theta; X) + V(\theta; Y).$$

Because of the independence of X and Y

$$I_{(X,Y)}(\theta) = \mathsf{Var}_\theta V(\theta; (X, Y)) = \mathsf{Var}_\theta (V(\theta; X) + V(\theta; Y))$$
$$= \mathsf{Var}_\theta V(\theta; X) + \mathsf{Var}_\theta V(\theta; Y) = I_X(\theta) + I_Y(\theta).$$

□

Corollary 3.1 *Let* $\mathbf{X} = (X_1, \ldots, X_n)$ *be i.i.d. copies of a r.v.* X *with Fisher information* $I_X(\theta)$. *Then the information contained in the sample is*

$$I_{\mathbf{X}}(\theta) = nI_X(\theta).$$

Sometimes $I_X(\theta)$ *is also denoted by* $i(\theta)$.

Summarizing we can say:

> If the number of independent observations increases the Fisher information is increasing as well.

Special case 3.4 (Normal distribution, known mean)
Let $X_1, \ldots, X_n$ be a sample of i.i.d. r.v.'s from $\mathsf{N}(0, \theta)$. By Corollary 3.1 it is enough to carry out the computations for a single $X \sim \mathsf{N}(0, \theta)$. The log-likelihood function is

$$l(\theta; X) = -\frac{1}{2}\ln(2\pi) - \frac{1}{2}\ln(\theta) - \frac{1}{2\theta}X^2$$

and the score function

$$V(\theta; X) = l'(\theta; X) = -\frac{1}{2\theta} + \frac{1}{2\theta^2}X^2.$$

Let us verify that $\mathsf{E}_\theta V(\theta; X) = 0$: Since $X \sim \mathsf{N}(0, \theta)$ we have

$$\mathsf{E}_\theta\left(-\frac{1}{2\theta} + \frac{X^2}{2\theta^2}\right) = \left(-\frac{1}{2\theta} + \frac{\theta}{2\theta^2}\right) = 0.$$

Further $\mathsf{Var}_\theta X^2 = \mathsf{E}_\theta X^4 - \theta^2 = 2\theta^2$ and

$$I_X(\theta) = \mathsf{Var}_\theta V(\theta; X) = \frac{1}{4\theta^4}\mathsf{Var}_\theta X^2 = \frac{2\theta^2}{4\theta^4} = \frac{1}{2\theta^2}.$$

The information contained in a sample is $I_{\mathbf{X}}(\theta) = \frac{n}{2\theta^2}$. Here we have a similar interpretation to the one in Special case 3.2: The information is larger when the variance of the distribution is smaller.   □

In Chapter 2 on page 12 we mentioned that a statistical model can be parameterized in different ways. The Fisher information depends on the particular parametrization.

Theorem 3.4 (Dependence on parametrization) *Let X be a r.v. with distribution P_θ. Suppose that another parametrization is given by $\theta = h(\xi)$ where h is differentiable. Then the information contained in X about ξ is given by*

$$I_X^*(\xi) = I_X(h(\xi))[h'(\xi)]^2.$$

Here $I_X(\theta)$ is the information that $X \sim \mathsf{P}_\theta$ contains about θ and $I_X^*(\xi)$ is the information that $X \sim \mathsf{P}_{h(\xi)}$ contains about ξ.

PROOF: The log-likelihood function w.r.t. $\mathsf{P}_{h(\xi)}$ is $l^*(\xi; x) = \ln p(x; h(\xi))$; thus the score function is

$$V^*(\xi; x) = \frac{\partial}{\partial \xi} \ln p(x; h(\xi)) = \frac{\partial}{\partial \theta} \ln p(x; \theta)\, |_{\theta = h(\xi)}\, h'(\xi)$$

and

$$\mathsf{Var}_\xi V^*(\xi; X) = \mathsf{Var}_\xi\big(V(h(\xi); X) h'(\xi)\big) = I_X(h(\xi))[h'(\xi)]^2.$$

$\square$

Example 3.11 (Standard deviation) Consider $X \sim \mathsf{N}(\mu, \sigma^2)$. The parameter of interest is σ. From Special case 3.4 we know $I_X(\sigma^2) = \frac{1}{2\sigma^4}$. By Theorem 3.4 we have for $\sigma^2 = h(\sigma)$ with $h'(\sigma) = 2\sigma$ and

$$I_X^*(\sigma) = I_X(\sigma^2)[h'(\sigma)]^2 = \frac{1}{2\sigma^4} \cdot 4\sigma^2 = \frac{2}{\sigma^2}.$$

The direct way to compute $I_X^*(\sigma)$ is

$$l^*(\sigma; x) = -\frac{1}{2}(\ln(2\pi)) - \ln \sigma - \frac{1}{2\sigma^2}(x - \mu)$$

and

$$V^*(\sigma; x) = -\frac{1}{\sigma} + \frac{1}{\sigma^3}(x - \mu),$$

thus

$$I_X^*(\sigma) = \mathsf{Var}_\sigma V^*(\sigma; X) = \frac{1}{\sigma^6}(3\sigma^4 - \sigma^4) = \frac{2}{\sigma^2}.$$

$\square$

Exercise 3.4 Derive the Fisher information for the natural parameter of the Poisson distribution a) directly b) using the transformation formula given in Theorem 3.4.

Example 3.12 (Dolphins) The statistical model (2.4) with arbitrary known parameters $\mu_2, \sigma_1^2, \sigma_2^2$ and with $m = 4$ and $n = 28$ is supposed. The parameter of interest is the mean of the younger dolphins $\theta = \mu_1$.

$$\mathcal{P} = \{N(\theta, \sigma_1^2)^{\otimes m} \otimes N(\mu_2, \sigma_2^2)^{\otimes(n-m)} : \theta \in \mathbb{R}\}. \tag{3.5}$$

We cannot use formula (3.2) since the population is inhomogeneous and we have no i.i.d. sample. Let us calculate the Fisher information for θ step by step. The likelihood function $L(\cdot; \mathbf{X})$ is

$$\left(\frac{1}{2\pi\sigma_1^2}\right)^{\frac{m}{2}} \left(\frac{1}{2\pi\sigma_2^2}\right)^{\frac{n-m}{2}} \exp\left(-\frac{1}{2\sigma_1^2} \sum_{i=1}^{m} (X_i - \theta)^2 - \frac{1}{2\sigma_2^2} \sum_{i=m+1}^{n} (X_i - \mu_2)^2\right)$$

and the log-likelihood $l(\cdot; \mathbf{X})$ is

$$-\frac{4}{2} \ln(2\pi\sigma_1^2) - \frac{1}{2\sigma_1^2} \sum_{i=1}^{4} (X_i - \theta)^2 - \frac{24}{2} \ln(2\pi\sigma_2^2) - \frac{1}{2\sigma_2^2} \sum_{i=5}^{28} (X_i - \mu_2)^2.$$

The score function contains only the observations related to the younger dolphins:

$$V(\theta; \mathbf{X}) = l'(\theta; \mathbf{X}) = \frac{\partial}{\partial\theta} \left(-\frac{1}{2\sigma_1^2} \sum_{i=1}^{4} (X_i - \theta)^2\right) = \frac{1}{\sigma_1^2} \sum_{i=1}^{4} (X_i - \theta).$$

The second derivative of the log-likelihood function is $-\frac{4}{\sigma_1^2}$. Thus

$$I_{\mathbf{X}}(\theta) = \frac{4}{\sigma_1^2}.$$

The 24 observations of the older dolphins give no information on $\theta = \mu_1$. Only the observations of new youngsters would increase the information about θ.□

Example 3.13 (Dolphins) Let us continue Example 3.12 but with additional information. The situation becomes quite different, if we assume that the mean of the adult dolphins is 20 times as much as that of the youngsters: $\mu_2 = 20\,\mu_1$. This means instead of the statistical model (3.5) we have

$$\mathcal{P} = \{N(\theta, \sigma_1^2)^{\otimes m} \otimes N(20\,\theta, \sigma_2^2)^{\otimes(n-m)} : \theta \in \mathbb{R}\}. \tag{3.6}$$

Let us calculate the Fisher information for this case. We have now that $L(\theta; \mathbf{X})$ is proportional to

$$\exp\left(-\frac{1}{2\sigma_1^2}\sum_{i=1}^{m}(X_i - \theta)^2 - \frac{1}{2\sigma_2^2}\sum_{i=m+1}^{n}(X_i - 20\,\theta)^2\right) \tag{3.7}$$

and $l(\theta; \mathbf{X})$ equals

$$-\frac{1}{2\sigma_1^2}\sum_{i=1}^{m}(X_i - \theta)^2 - \frac{1}{2\sigma_2^2}\sum_{i=m+1}^{n}(X_i - 20\,\theta)^2 + \text{const.}$$

and

$$V(\theta; \mathbf{X}) = \frac{1}{\sigma_1^2}\sum_{i=1}^{m}(X_i - \theta) + \frac{20}{\sigma_2^2}\sum_{i=m+1}^{n}(X_i - 20\,\theta).$$

For $m = 4$ and $n = 28$ the Fisher information is

$$I_{\mathbf{X}}(\theta) = \frac{4}{\sigma_1^2} + \frac{9600}{\sigma_2^2}.$$

Even for a very large variance σ_2^2 the Fisher information of model (3.6) with the additional assumption, $\mu_2 = 20\,\mu_1$, is higher than in model (3.5), because in model (3.6) all observations contribute information on the unknown parameter of interest. For $\sigma_1^2 = \sigma_2^2$ the information is 2400 times larger.   □

After this textbook example we come to a more serious application:

Special case 3.5 (Censored distribution) Suppose we study the lifetime of some technical item described by a r.v. Y. The experiment is started at time t_0 and is finished at time t^*. Some of the items fail in this time interval; some are still working at t^*. Thus the observations are realizations of the r.v. (X_i, Δ_i) with

$$X_i = \begin{cases} Y_i & \text{if } Y_i \leq t^* \\ t^* & \text{if } Y_i > t^* \end{cases} \qquad \Delta_i = \begin{cases} 1 & \text{if } Y_i \leq t^* \\ 0 & \text{if } Y_i > t^* \end{cases}.$$

We assume that the $Y_1, \ldots, Y_n$ are i.i.d. according to $\mathsf{Exp}\,(\lambda)$. The distribution function of X_i can be computed by $F(t) = \mathsf{P}_\lambda(X_i \leq t) = H(t, 1) + H(t, 0)$ (shown in Figure 3.4), where

$$H(t, \delta) = \mathsf{P}_\lambda(X_i \leq t, \Delta_i = \delta) = \begin{cases} \mathsf{P}_\lambda(Y_i \leq t, Y_i \leq t^*) & \delta = 1 \\ \mathsf{P}_\lambda(t^* \leq t, Y_i > t^*) & \delta = 0 \end{cases}.$$

For $t < t^*$

$$H(t, \delta) = \begin{cases} 1 - \exp(-\lambda t) & \delta = 1 \\ 0 & \delta = 0 \end{cases}$$

and for $t \geq t^*$

$$H(t, \delta) = \begin{cases} 1 - \exp(-\lambda t^*) & \delta = 1 \\ \exp(-\lambda t^*) & \delta = 0 \end{cases}.$$

From the subdistribution function H we conclude that the contribution of an uncensored observation (x_i, δ_i) with $\delta_i = 1$ to the likelihood function is $(\lambda \exp(-\lambda x_i))^{\delta_i}$. A censored observation $(x_i, \delta_i) = (t^*, 0)$ contributes the "jump" $1 - F(t^*; \lambda)$. (Compare in Figure 3.4.) Thus the likelihood function has the form

$$\begin{aligned} L(\lambda; \mathbf{x}, \delta) &= \prod_{i=1}^{n} (\lambda \exp(-\lambda x_i))^{\delta_i} \exp(-\lambda x_i (1 - \delta_i)) \\ &= \lambda^r \exp(-\lambda S_{\text{total}}), \end{aligned}$$

where $r = \sum_{i=1}^{n} \delta_i$ is the observed number of uncensored data and $S_{\text{total}} = \sum_{i=1}^{n} x_i$ is the total time on the test for all objects under study. Now, let us derive the Fisher information. Differentiating the log-likelihood yields

$$l(\lambda; \mathbf{x}, \delta) = r \log \lambda - \lambda S_{\text{total}} \qquad \frac{\partial^2 l(\lambda; \mathbf{x}, \delta)}{\partial \lambda^2} = -\frac{r}{\lambda^2}.$$

Thus we have to compute the expectation of $R = \sum_{i=1}^{n} \Delta_i$:

$$\mathsf{E}_\lambda R = n \mathsf{E}_\lambda \Delta_1 = n \mathsf{P}_\lambda((0, t^*]) = n(1 - \exp(-\lambda t^*)).$$

The Fisher information is

$$I^{\text{cens}}(\lambda) = n \left(1 - \exp(-\lambda t^*)\right)/\lambda^2.$$

Let us discuss this result to emphasize the meaning of "information."

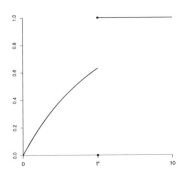

Figure 3.4: Distribution of the r.v. X_i in the exponential model under censoring.

- Suppose we observe for a very long time, that is the censoring point is very large. Mathematically this is described by the limit $t^* \to \infty$. Then $I^{\text{cens}}(\lambda) \to I(\lambda) = \frac{n}{\lambda^2}$, which is the Fisher information of the exponential distribution for data without censoring. That is, for large t^* we do not lose much information compared with the case without censoring.

- Suppose we have money only for a short study. For $t^* \to 0$ we have $I^{\text{cens}}(\lambda) \to 0$. That is, for very small t^* the observations contain little information about the parameter λ.

- Suppose that the adviser of this experiment has some experience and chooses t^* around twice the expected lifetime. We describe this as follows: The expectation of Y_i is $1/\lambda$; we set $t^* = 2/\lambda$. Then we get $I^{\text{cens}}(\lambda) = n\,(1-\exp(-2))/\lambda^2$ and the information in the censored sample is reduced by the factor $1 - \exp(-2) = 0.864$.

$\square$

The following theorem confirms the role of an information measure. If one takes an arbitrary statistic T, then the information about θ contained in $T(X)$ is less or equals $I_{\mathbf{X}}(\theta)$.

Theorem 3.5 *Let $\mathbf{X}$ be a r.v. with $\mathbf{X} \sim P_\theta$ and T any statistic. Suppose the regularity conditions Reg 1–Reg 3. Then*

$$I_{T(\mathbf{X})}(\theta) \leq I_{\mathbf{X}}(\theta),$$

where $I_{T(\mathbf{X})}(\theta)$ is computed with respect to the distribution of $T(\mathbf{X})$.

PROOF: We will prove this only in the discrete case. Let $p^T(\cdot;\theta)$ be the probability function of T. First we show that the score functions V and $\tilde{V}$ defined by $V(\theta;\mathbf{x}) = \frac{\partial}{\partial \theta} \ln p(\mathbf{x};\theta)$ and $\tilde{V}(\theta;t) = \frac{\partial}{\partial \theta} \ln p^T(t;\theta)$ satisfy

$$\mathsf{E}_\theta\left(V(\theta;\mathbf{X}) \mid T(\mathbf{X}) = t\right) = \tilde{V}(\theta;t). \tag{3.8}$$

We have

$$p^T(t;\theta) = \mathsf{P}_\theta\left(\{\mathbf{x} : T(\mathbf{x}) = t\}\right) = \sum_{\mathbf{x}:T(\mathbf{x})=t} p(\mathbf{x};\theta)$$

and the conditional probability

$$h(\mathbf{x} \mid t;\theta) = \frac{\mathsf{P}_\theta(\{\mathbf{x}\} \cap \{\mathbf{x}' : T(\mathbf{x}') = t\})}{\mathsf{P}_\theta\left(\{\mathbf{x}' : T(\mathbf{x}') = t\}\right)}$$

$$= \begin{cases} \frac{p(\mathbf{x};\theta)}{p^T(t;\theta)} & \text{for } \mathbf{x} \text{ with } T(\mathbf{x}) = t \\ 0 & \text{otherwise} \end{cases}.$$

Thus we get (3.8) by the following chain of equalities:

$$\mathsf{E}_\theta\left(V(\theta;\mathbf{X})\mid T(\mathbf{X})=t\right)$$

$$= \mathsf{E}_\theta\left(\frac{\partial p(\mathbf{X};\theta)}{\partial\theta}\frac{1}{p(\mathbf{X};\theta)}\mid T(\mathbf{X})=t\right) = \sum_{\mathbf{x}}\frac{\partial p(\mathbf{x};\theta)}{\partial\theta}\frac{1}{p(\mathbf{x};\theta)}h(\mathbf{x}\mid t;\theta)$$

$$= \frac{1}{p^T(t;\theta)}\sum_{\mathbf{x}:T(\mathbf{x})=t}\frac{\partial p(\mathbf{x};\theta)}{\partial\theta} = \frac{1}{p^T(t;\theta)}\frac{\partial}{\partial\theta}\sum_{\mathbf{x}:T(\mathbf{x})=t}p(\mathbf{x};\theta)$$

$$= \frac{1}{p^T(t;\theta)}\frac{\partial}{\partial\theta}p^T(t;\theta) = \frac{\partial}{\partial\theta}\ln p^T(t;\theta) = \tilde{V}(\theta;t).$$

Consider now $I_{T(\mathbf{X})}(\theta) = \mathsf{E}_\theta[\tilde{V}(\theta;T)]^2$. Using (3.8) and the rules for the conditional mean $\mathsf{E}X = \mathsf{E}(\mathsf{E}(X\mid T))$ we obtain

$$\mathsf{E}_\theta[\tilde{V}(\theta;T)]^2 = \mathsf{E}_\theta\left(\tilde{V}(\theta;T)[\mathsf{E}_\theta(V(\theta;\mathbf{X})|T)]\right)$$

$$= \mathsf{E}_\theta\mathsf{E}_\theta\left(\tilde{V}(\theta;T)V(\theta;\mathbf{X})\mid T\right) = \mathsf{E}_\theta\left(\tilde{V}(\theta;T)V(\theta;\mathbf{X})\right).$$

Thus

$$0 \leq \mathsf{E}_\theta\left(\tilde{V}(\theta;T) - V(\theta;\mathbf{X})\right)^2 = \mathsf{E}_\theta\left(V(\theta;\mathbf{X})\right)^2 - \mathsf{E}_\theta\left(\tilde{V}(\theta;T)\right)^2$$

$$= I_{\mathbf{X}}(\theta) - I_{T(\mathbf{X})}(\theta).$$

$\square$

Example 3.14 (Dolphins) Continuation of Example 3.13 on page 40. Consider the statistics

$$T_1(\mathbf{X}) = \frac{1}{m}\sum_{i=1}^{m}X_i \qquad \text{and}$$

$$T_2(\mathbf{X}) = a^{-1}\left(\frac{1}{\sigma_1^2}\sum_{i=1}^{m}X_i + \frac{20}{\sigma_2^2}\sum_{i=m+1}^{n}X_i\right) \quad \text{with } a = \frac{m}{\sigma_1^2} + \frac{20^2\,(n-m)}{\sigma_2^2} = I_{\mathbf{X}}(\theta).$$

It holds

$$T_1(\mathbf{X}) \sim \mathsf{N}(\theta,\frac{\sigma_1^2}{m}) \quad \text{and} \quad T_2(\mathbf{X}) \sim \mathsf{N}(\theta,a^{-1}).$$

Analogous to Special case 3.2 we obtain

$$I_{T_1(\mathbf{X})}(\theta) = \frac{m}{\sigma_1^2} < I_{\mathbf{X}}(\theta) \quad \text{and} \quad I_{T_2(\mathbf{X})}(\theta) = I_{\mathbf{X}}(\theta).$$

Later in Example 3.22 on page 57 we will see that the statistic T_2 is called sufficient. $\square$

3.2.2   The Multivariate Case

Now we consider the case where the parameter θ is k-dimensional, i.e., $\theta \in \mathbb{R}^k$. We assume the regularity condition **Reg 1**; the remaining conditions are modified as follows:

Reg' 2 The parameter space $\Theta \subseteq \mathbb{R}^k$ is an open set.

Reg' 3 For all $\mathbf{x} \in \mathcal{A}$ the likelihood function has finite partial derivatives.

The derivative of the log-likelihood function is now a vector.

Definition 3.5 (Score function) For all $\mathbf{x} \in \mathcal{A}$ the vector of partial derivatives of the log-likelihood function

$$V(\theta; \mathbf{x}) = \nabla_\theta l(\theta; \mathbf{x}) = (\frac{\partial}{\partial \theta_1} l(\theta; \mathbf{x}), \ldots, \frac{\partial}{\partial \theta_k} l(\theta; \mathbf{x}))^\mathsf{T}$$

is called score function or score vector.

As in the one-dimensional case we have $\mathsf{E}_\theta V(\theta; \mathbf{X}) = 0$. We define:

Definition 3.6 (Fisher information matrix) The $k \times k$ matrix

$$I_{\mathbf{X}}(\theta) = \mathsf{Cov}_\theta V(\theta; \mathbf{X})$$

is called Fisher information. The elements have the form

$$I_{\mathbf{X}}(\theta)_{jr} = \mathsf{E}_\theta(\frac{\partial}{\partial \theta_j} l(\theta; \mathbf{X}) \frac{\partial}{\partial \theta_r} l(\theta; \mathbf{X})).$$

With the additional assumption

Reg' 4 For all $\mathbf{x} \in \mathcal{A}$ the likelihood function has second partial derivatives and for all $\theta \in \Theta$ and $j, r = 1, \ldots, k$

$$\frac{\partial^2}{\partial \theta_j \partial \theta_r} \int_{\mathcal{A}} f(\mathbf{x}; \theta) d\mathbf{x} = \int_{\mathcal{A}} \frac{\partial^2}{\partial \theta_j \partial \theta_r} f(\mathbf{x}; \theta) d\mathbf{x} \quad \text{for continuous distributions,}$$

$$\frac{\partial^2}{\partial \theta_j \partial \theta_r} \sum_{\mathbf{x} \in \mathcal{A}} \mathsf{P}_\theta(\mathbf{x}) = \sum_{\mathbf{x} \in \mathcal{A}} \frac{\partial^2}{\partial \theta_j \partial \theta_r} \mathsf{P}_\theta(\mathbf{x}) \qquad \text{for discrete distributions}$$

we can formulate the following theorem.

Theorem 3.6 *Suppose* **Reg 1, Reg' 2–Reg' 4**. *Let* $J(\theta; \mathbf{X})$ *be the matrix of the negative second derivatives of the log-likelihood function, that is* $J(\theta; \mathbf{X})$ *has the elements*

$$J(\theta; \mathbf{X})_{jr} = -\frac{\partial^2 l(\theta; \mathbf{X})}{\partial \theta_j \partial \theta_r} \qquad j, r = 1, \ldots, k.$$

Then the Fisher information is

$$I_{\mathbf{X}}(\theta) = \mathsf{E}_\theta J(\theta; \mathbf{X}).$$

The $k \times k$ matrix $J(\theta; \mathbf{X})$ is called the observed Fisher information matrix.

Special case 3.6 (Normal distribution) Let X be $\mathsf{N}\left(\mu, \sigma^2\right)$-distributed with $\theta = (\mu, \sigma^2)$. Then from

$$l(\theta; X) = -\frac{1}{2}\ln(2\pi\sigma^2) - \frac{1}{2\sigma^2}(X - \mu)^2$$

we obtain for the score vector

$$V(\mu, \sigma^2; X) = (\frac{1}{\sigma^2}(X - \mu), -\frac{1}{2\sigma^2} + \frac{1}{2\sigma^4}(X - \mu)^2)^\mathsf{T}.$$

We obtain:

$$I_X(\mu, \sigma^2)_{11} = \mathsf{Var}_{\mu,\sigma^2} \frac{1}{\sigma^2}(X - \mu) = \frac{1}{\sigma^2},$$

$$I_X(\mu, \sigma^2)_{12} = \mathsf{Cov}_{\mu,\sigma^2}(\frac{1}{\sigma^2}(X - \mu), \frac{1}{2\sigma^4}(X - \mu)^2)) = 0,$$

$$I_X(\mu, \sigma^2)_{22} = \mathsf{Var}_{\mu,\sigma^2} \frac{1}{2\sigma^4}(X - \mu)^2 = \frac{1}{4\sigma^8}(\mathsf{E}_{\mu,\sigma^2}(X - \mu)^4 - \sigma^4) = \frac{1}{2\sigma^4}.$$

Summarizing

$$I_X\left(\mu, \sigma^2\right) = \begin{pmatrix} \frac{1}{\sigma^2} & 0 \\ 0 & \frac{1}{2\sigma^4} \end{pmatrix}.$$

Since Theorem 3.3 and Corollary 3.1 apply also in the multidimensional case we have: For a sample of n i.i.d. r.v.'s from $\mathsf{N}(\mu, \sigma^2)$ the Fisher information matrix is

$$I_{\mathbf{X}}\left(\mu, \sigma^2\right) = \frac{n}{\sigma^2}\begin{pmatrix} 1 & 0 \\ 0 & \frac{1}{2\sigma^2} \end{pmatrix}. \qquad (3.9)$$

$\square$

Let us calculate the Fisher information matrix for the multivariate normal distribution.

Special case 3.7 (Multivariate normal distribution) The sample $\mathbf{X} = (X_1, \ldots, X_n)$ consists of independent random vectors distributed according to $N_p\left(\mu, \sigma^2 \Sigma\right)$. The $p \times p$ matrix Σ is assumed to be nonsingular and known. Note that the mean μ is now a $p \times 1$ vector. The unknown parameter $\theta^\mathsf{T} = (\mu^\mathsf{T}, \sigma^2)$ is a $(p+1) \times 1$ vector. The log-likelihood function of a single r.v. $X \sim N_p(\mu, \sigma^2 \Sigma)$ is given by

$$l(\mu, \sigma^2; X) = -\frac{p}{2} \ln(2\pi\sigma^2) - \frac{1}{2} \ln\left(\det\left(\Sigma\right)\right) - \frac{1}{2\sigma^2}(X - \mu)^\mathsf{T} \Sigma^{-1}(X - \mu).$$

For the computation of the score vector and the information matrix we make use of the following rule of differentiation: Let a and b be $p \times 1$ vectors and A a symmetric $p \times p$ matrix, then

$$\frac{\partial}{\partial b} a^\mathsf{T} b = a, \qquad \frac{\partial}{\partial b} b^\mathsf{T} Ab = 2Ab \quad \text{and} \quad \frac{\partial^2}{\partial b \partial b^\mathsf{T}} b^\mathsf{T} Ab = 2A.$$

Thus the score vector is given by

$$V(\theta; X) = \left(\frac{1}{\sigma^2} \Sigma^{-1}(X - \mu), -\frac{p}{2\sigma^2} + \frac{1}{2\sigma^4}(X - \mu)^\mathsf{T} \Sigma^{-1}(X - \mu) \right)^\mathsf{T}$$

and the elements of the observed information are

$$J(\mu, \sigma^2; X)_{11} = \frac{1}{\sigma^2} \Sigma^{-1},$$

$$J(\mu, \sigma^2; X)_{12} = \frac{1}{\sigma^4} \Sigma^{-1}(X - \mu),$$

$$J(\mu, \sigma^2; X)_{22} = -\frac{p}{2\sigma^4} + \frac{1}{\sigma^6}(X - \mu)^\mathsf{T} \Sigma^{-1}(X - \mu).$$

Now, take $I_X(\mu, \sigma) = \mathsf{E}_{(\mu, \sigma^2)} J(\mu, \sigma^2; X)$.

The computation of the first two expectations is obvious; for the third we obtain with the identity $\mathsf{E}(Z^\mathsf{T} AZ) = \text{trace}(\mathsf{E}(ZZ^\mathsf{T})A)$ for a $p \times 1$ vector Z and a $p \times p$ matrix A

$$\mathsf{E}_{\mu, \sigma^2} \left(-\frac{p}{2\sigma^4} + \frac{1}{\sigma^6}(X - \mu)^\mathsf{T} \Sigma^{-1}(X - \mu) \right)$$

$$= -\frac{p}{2\sigma^4} + \frac{1}{\sigma^6} \sigma^2 \, \text{trace}(I_{p \times p}) = \frac{p}{2\sigma^4}.$$

Thus

$$I_X(\mu, \sigma^2) = \frac{1}{\sigma^2} \begin{pmatrix} \Sigma^{-1} & 0 \\ 0 & \frac{p}{2\sigma^2} \end{pmatrix},$$

and the information contained in the sample $\mathbf{X}$ is $I_{\mathbf{X}}(\mu, \sigma^2) = nI_X(\mu, \sigma^2)$. $\square$

Example 3.15 (Dolphins) We suppose the model:

$$\mathcal{P} = \{N(\mu_1, \sigma_1^2)^{\otimes m} \otimes N(\mu_2, \sigma_2^2)^{\otimes (n-m)} : (\mu_1, \mu_2) \in \mathbb{R}^2\}.$$

Let the variances be known. The parameter of interest is $\theta = (\mu_1, \mu_2)$. The information about μ_1 comes from the first m observations. From Special case 3.2 we know it is $\frac{m}{\sigma_1^2}$. The remaining observations contain information about μ_2 and which is equal to $\frac{n-m}{\sigma_2^2}$. Since information is additive we obtain

$$I_{\mathbf{X}}(\mu_1, \mu_2) = m I_{X_1}(\theta) + (n-m) I_{X_{m+1}}(\theta),$$

where

$$I_{X_1}(\theta) = \begin{pmatrix} \frac{1}{\sigma_1^2} & 0 \\ 0 & 0 \end{pmatrix} \quad \text{and} \quad I_{X_{m+1}}(\theta) = \begin{pmatrix} 0 & 0 \\ 0 & \frac{1}{\sigma_2^2} \end{pmatrix}.$$

Thus for the sample $\mathbf{X}$ we have $I_{\mathbf{X}}(\theta) = \begin{pmatrix} \frac{m}{\sigma_1^2} & 0 \\ 0 & \frac{n-m}{\sigma_2^2} \end{pmatrix}.$ □

Exercise 3.5 (Dolphins) Calculate the Fisher information matrix in the dolphin example on page 11 under the model (2.4), where all parameters are unknown: $\theta = (\mu_1, \mu_2, \sigma_1^2, \sigma_2^2)$.

Example 3.16 (Flowers) Recall Equation (3.4) on page 28. Then the log-likelihood is

$$l(\theta; \mathbf{x}) = n_1 \ln(p_1) + n_2 \ln(p_2) + (n - n_1 - n_2) \ln(1 - (p_1 + p_2)).$$

The score vector is

$$V(\theta; \mathbf{x}) = \left(\frac{n_1}{p_1} - \frac{n - n_1 - n_2}{1 - p_1 - p_2}, \frac{n_2}{p_2} - \frac{n - n_1 - n_2}{1 - p_1 - p_2} \right)^{\mathsf{T}}.$$

Let us check up Theorem 3.1 and calculate the expectation. Since the N_j's are sums of independent Bernoulli variables $I_j(X_i)$ we have $N_j \sim \text{Bin}(n, p_j)$ and $\mathsf{E}_\theta N_j = np_j$ and $\text{Var}_\theta N_j = np_j(1 - p_j)$. We obtain

$$\mathsf{E}_\theta V(\theta; \mathbf{X}) = \left(\frac{np_1}{p_1} - \frac{n(1 - p_1 - p_2)}{1 - p_1 - p_2}, \frac{np_2}{p_2} - \frac{n(1 - p_1 - p_2)}{1 - p_1 - p_2} \right)^{\mathsf{T}} = (0, 0)^{\mathsf{T}}.$$

The observed information matrix is given by

$$J(\theta; \mathbf{x})_{jj} = \frac{n_j}{p_j^2} + \frac{n - n_1 - n_2}{(1 - p_1 - p_2)^2} \quad \text{and} \quad J(\theta; \mathbf{x})_{12} = \frac{n - n_1 - n_2}{(1 - p_1 - p_2)^2}.$$

Thus with

$$E_\theta J(\theta; \mathbf{X})_{jj} = \frac{np_j}{p_j^2} + \frac{n(1 - p_1 - p_2)}{(1 - p_1 - p_2)^2} \quad \text{and} \quad J(\theta; \mathbf{X})_{12} = \frac{n(1 - p_1 - p_2)}{(1 - p_1 - p_2)^2}$$

and $p_3 = 1 - (p_1 + p_2)$ we get

$$I_{\mathbf{X}}(\theta) = n \begin{pmatrix} \frac{1}{p_1} + \frac{1}{p_3} & \frac{1}{p_3} \\ \frac{1}{p_3} & \frac{1}{p_2} + \frac{1}{p_3} \end{pmatrix}. \tag{3.10}$$

□

Exercise 3.6 Derive the Fisher information for the multinomial distribution describing an experiment with m $(m > 3)$ outcomes.

Example 3.17 (Identifiability) Let $X_1, \ldots, X_n$ be an i.i.d. sample from $N(\mu, \lambda\sigma^2)$. Also if μ is known, the unknown parameter $\theta = (\lambda, \sigma^2) \in \mathbb{R}_+ \times \mathbb{R}_+$ is not identifiable. (Compare condition **Reg 0.**) For example, choose $\theta_1 = (2, 6)$ and $\theta_2 = (3, 4)$, then $\theta_1 \neq \theta_2$ but $P_{\theta_1} = P_{\theta_2}$.
Suppose in addition we have a sample $Y_1, \ldots, Y_m$ of i.i.d. r.v. from $N(\mu, \sigma^2)$. For simplicity we assume that μ is known. Then the model is

$$\mathcal{P} = \{N(\mu, \lambda\sigma^2)^{\otimes n} \otimes N(\mu, \sigma^2)^{\otimes m} : \theta = (\lambda, \sigma^2) \in \mathbb{R}_+ \times \mathbb{R}_+\}.$$

Now the parameter is identifiable. For different values of θ the probability measures in $\mathcal{P}$ are different.
The information in $\mathbf{Y}$ about σ^2 is $\frac{m}{2\sigma^4}$ (compare Special case 3.4 on page 38). For the information matrix w.r.t. θ we obtain

$$I_{\mathbf{Y}}(\lambda, \sigma^2) = \begin{pmatrix} 0 & 0 \\ 0 & \frac{m}{2\sigma^4} \end{pmatrix}.$$

The information about θ contained in the sample $\mathbf{X}$ is computed as follows:

$$l(\lambda, \sigma^2; \mathbf{x}) = -\frac{n}{2} \left(\ln \lambda + \ln \sigma^2 \right) - \frac{1}{2\lambda\sigma^2} \sum_{i=1}^{n} (x_i - \mu)^2,$$

$$V(\lambda, \sigma^2; \mathbf{x}) = \left(-\frac{n}{2\lambda} + \frac{1}{2\lambda^2\sigma^2} \sum_{i=1}^{n} (x_i - \mu)^2, \ -\frac{n}{2\sigma^2} \frac{1}{2\lambda\sigma^4} \sum_{i=1}^{n} (x_i - \mu)^2 \right)^{\mathsf{T}},$$

and

$$\mathsf{Var}_\theta\left(-\frac{n}{2\lambda} + \frac{1}{2\lambda^2\sigma^2}\sum_{i=1}^{n}(X_i - \mu)^2\right) = \frac{2\lambda^2\sigma^4 n}{4\lambda^4\sigma^4} = \frac{n}{2\lambda^2},$$

$$\mathsf{Var}_\theta\left(-\frac{n}{2\sigma^2} + \frac{1}{2\lambda\sigma^4}\sum_{i=1}^{n}(X_i - \mu)^2\right) = \frac{2\lambda^2\sigma^4 n}{4\lambda^2\sigma^8} = \frac{n}{2\sigma^4},$$

$$\mathsf{Cov}_\theta\left(\frac{1}{2\lambda^2\sigma^2}\sum_{i=1}^{n}(X_i - \mu)^2, \frac{1}{2\lambda\sigma^4}\sum_{i=1}^{n}(X_i - \mu)^2\right) = \frac{2\lambda^2\sigma^4 n}{4\lambda^3\sigma^6} = \frac{n}{2\lambda\sigma^2}.$$

Thus

$$I_{\mathbf{X}}(\lambda, \sigma^2) = \begin{pmatrix} \frac{n}{2\lambda^2} & \frac{n}{2\lambda\sigma^2} \\ \frac{n}{2\lambda\sigma^2} & \frac{n}{2\sigma^4} \end{pmatrix}.$$

Note that $\det(I_{\mathbf{X}}(\lambda, \sigma^2)) = 0$ and $\det(I_{\mathbf{Y}}(\lambda, \sigma^2)) = 0$. This can be explained as follows: The parameter θ cannot be identified by $\mathbf{X}$ alone, and $\mathbf{Y}$ contains zero information about λ. However, for

$$I_{(\mathbf{X},\mathbf{Y})}(\lambda, \sigma^2) = I_{\mathbf{X}}(\theta) + I_{\mathbf{Y}}(\theta) = \begin{pmatrix} \frac{n}{2\lambda^2} & \frac{n}{2\lambda\sigma^2} \\ \frac{n}{2\lambda\sigma^2} & \frac{n+m}{2\sigma^4} \end{pmatrix}$$

we have $\det(I_{(\mathbf{X},\mathbf{Y})}(\theta)) = \frac{nm}{4\lambda^2\sigma^4}$. Thus, the Fisher information is nonsingular. Note, this is already true for $m = 1$.   □

3.3   Sufficiency

In this section we answer the following questions. Do we have to store each value of the sample? The likelihood function of the normal distribution in (3.3) depends only on the sample mean. If we trust in the likelihood function, we only need to know the sample size and the sample mean. How can we save data storage without losing information? Which statistics of the sample contain all information?

Any statistic T

$$T : \mathbf{x} \in \mathcal{X} \to T(\mathbf{x}) = t \in \mathcal{T}$$

generates a partition of the sample space

$$\mathcal{X} = \bigcup_{T(\mathbf{x})=t} \mathcal{X}_t \quad \text{with} \quad \mathcal{X}_t = \{\mathbf{x} : T(\mathbf{x}) = t\}.$$

Example 3.18 (Dice) Roll a dice twice. The sample space $\mathcal{X} = \{(x_1, x_2) : x_i \in \{1, \dots, 6\}\}$ consists of the following 36 elements:

$$
\begin{array}{cccccc}
(1,1) & (1,2) & (1,3) & (1,4) & (1,5) & (1,6) \\
(2,1) & (2,2) & (2,3) & (2,3) & (2,4) & (2,5) \\
(3,1) & (3,2) & (3,3) & (3,4) & (3,5) & (3,6) \\
(4,1) & (4,2) & (4,3) & (4,4) & (4,5) & (4,6) \\
(5,1) & (5,2) & (5,3) & (5,4) & (5,5) & (5,6) \\
(6,1) & (6,2) & (6,3) & (6,4) & (6,5) & (6,6).
\end{array}
$$

Consider as a statistic the total number of dots $T(x_1, x_2) = x_1 + x_2$. The possible values of $T(x_1, x_2)$ are $k = 2, \dots, 12$. The statistic T leads to the following partition of $\mathcal{X}$ into 11 disjoint sets $\mathcal{X}_k = \{(x_1, x_2) : x_1 + x_2 = k\}$, that is

$$\mathcal{X}_2 = \{(1,1)\}, \mathcal{X}_3 = \{(1,2),(2,1)\}, \mathcal{X}_4 = \{(1,3),(2,2),(3,1)\},$$
$$\mathcal{X}_5 = \{(1,4),(2,3),(3,2),(4,1)\}, \mathcal{X}_6 = \{(1,5),(2,4),(3,3),(4,2),(5,1)\},$$
$$\mathcal{X}_7 = \{(1,6),(2,5),(3,4),(4,3),(5,2),(6,1)\},$$
$$\mathcal{X}_8 = \{(2,6),(3,5),(4,4),(5,3),(6,2)\}, \mathcal{X}_9 = \{(3,6),(4,5),(5,4),(6,3)\},$$
$$\mathcal{X}_{10} = \{(4,6),(5,5),(6,4)\}, \mathcal{X}_{11} = \{(5,6),(6,5)\}, \mathcal{X}_{12} = \{(6,6)\}.$$

One can identify each statistic with the respective decomposition of the sample space. That is, if one knows to which $\mathcal{X}_k$ the observation $\mathbf{x}$ belongs then one knows the value of $T(\mathbf{x})$. □

Taking any partition (statistic) incurs loss of information. But some information may be irrelevant for inference on the parameter of interest. A statistic which produces a reduction without loss of relevant information will later be defined as a sufficient statistic.

Sufficiency principle

> Summarizing the data set does not necessarily lead to the loss of anything contained in them that could help us to choose a probability distribution among distributions belonging to the model.

Let us illustrate this principle with the following example.

Special case 3.8 (Bernoulli) Let $\mathbf{X} = (X_1, \dots, X_n)$ be a sample of independent Bernoulli variables. Then the probability function is

$$p(\mathbf{x}; \theta) = \prod_{i=1}^{n} \theta^{x_i} (1 - \theta)^{1 - x_i} = \theta^k (1 - \theta)^{n-k},$$

where $k = \sum_{i=1}^{n} x_i$. Consider

$$l(\theta; \mathbf{X}) = T \ln \theta + (n - T) \ln(1 - \theta),$$

where

$$T = \sum_{i=1}^{n} X_i \qquad \text{and} \quad T \sim \text{Bin}(n, \theta),$$

with $p^T(k; \theta) = \binom{n}{k} \theta^k (1 - \theta)^{n-k}$, $k = 0, \ldots, n$.

The statistic T generates a partition of the sample space $\mathcal{X} = \bigcup_{k=0}^{n} \mathcal{X}_k$, with $\mathcal{X}_k = \{(x_1, \ldots, x_n) : k = \sum_{i=1}^{n} x_i\}$. Note that each set $\mathcal{X}_k$ has $\binom{n}{k}$ elements. As example consider the case $n = 3$ where the partition of $\mathcal{X}$ consists of the following four sets:

$$\mathcal{X}_0 = \{(0,0,0)\}, \mathcal{X}_1 = \{(1,0,0), (0,1,0), (0,0,1)\},$$
$$\mathcal{X}_2 = \{(1,1,0), (0,1,1), (1,0,1)\}, \mathcal{X}_3 = \{(1,1,1)\}.$$

The conditional distribution of $\mathbf{X}$ given $T = k$ is a distribution on $\mathcal{X}_k$ only. Thus for $x \in \mathcal{X}_k = \{x' : \sum_i x_i' = k\}$

$$\frac{\mathsf{P}_\theta(\{\mathbf{x}\} \cap \mathcal{X}_k)}{\mathsf{P}_\theta^T(\{k\})} = \frac{p(\mathbf{x}; \theta)}{p^T(k; \theta)} = \frac{\theta^k (1 - \theta)^{n-k}}{\binom{n}{k} \theta^k (1 - \theta)^{n-k}} = \frac{1}{\binom{n}{k}}.$$

This means that after knowing $T(\mathbf{x}) = k$, the remaining random effect concerns only the positions of the zeros and ones. Each order has the same probability, in other words we have a discrete uniform distribution. The positions of ones and zeros do not provide any information about θ.   □

Definition 3.7 (Sufficient statistic) A statistic T is said to be **sufficient** for the statistical model $\{\mathsf{P}_\theta : \theta \in \Theta\}$ of $\mathbf{X}$ if the conditional distribution of $\mathbf{X}$ given $T = t$ is independent of θ for all t.

For sufficient statistics T we have:

$$\mathbf{X} \sim \mathsf{P}_\theta \qquad T \sim \mathsf{P}_\theta^T \qquad \text{but} \quad \mathbf{X} \mid T \sim \mathsf{P}.$$

A sufficient statistic contains all information about θ included in $\mathbf{X}$. One also says T is sufficient for θ or T is sufficient for $\mathbf{X}$.

Example 3.19 (Flowers) Let us check whether it is enough to know numbers of white, violet and pink flowers to make statistical inference about the unknown parameter $\theta = (p_1, p_2)$. The distribution of $\mathbf{X}$ is given by

$$p(\mathbf{x}; \theta) = p_1^{n_1} p_2^{n_2} p_3^{n_3}$$

with $n_j = I_j(x_i)$ and $n_3 = n - n_1 - n_2$ and $p_3 = 1 - (p_1 + p_2)$. The vector $T(\mathbf{X}) = (N_1, N_2)$ has a multinomial distribution (see (5.2)) with probability function

$$p^T(\mathbf{n}; \theta) = \binom{n}{n_1, n_2, n_3} p_1^{n_1} p_2^{n_2} p_3^{n_3}.$$

The conditional distribution of $\mathbf{X}$ given $T(\mathbf{x}) = (n_1, n_2)$ is zero if $\sum_i I_1(x_i) \neq n_1$ or $\sum_i I_2(x_i) \neq n_2$; otherwise it is equal to

$$\frac{p(\mathbf{x}; \theta)}{p^T(\mathbf{n}; \theta)} = \frac{p_1^{n_1} p_2^{n_2} p_3^{n_3}}{\binom{n}{n_1, n_2, n_3} p_1^{n_1} p_2^{n_2} p_3^{n_3}} = \frac{1}{\binom{n}{n_1, n_2, n_3}},$$

which is independent of the unknown parameter $\theta = (p_1, p_2)$. Hence the statistic $T(\mathbf{X}) = (N_1, N_2)$ is sufficient. The remaining random effect is the position of each color in the sample, but that contains no information on the probability of its occurrence. □

Example 3.20 (Dolphins) Suppose model (3.5) which is

$$\mathcal{P} = \{N(\theta, \sigma_1^2)^{\otimes m} \otimes N(\mu_2, \sigma_2^2)^{\otimes(n-m)} : \theta \in \mathbb{R}\}.$$

The parameter of interest is $\theta = \mu_1$, the mean related to the m younger dolphins. Consider $T(\mathbf{X}) = (X_1, \ldots, X_m)$. Since the two parts of the sample are independent, the conditional distribution of $\mathbf{X}$ given T is independent of θ. Thus $T(\mathbf{X}) = (X_1, \ldots, X_m)$ is sufficient. The second part of the sample contains no information about θ. □

Exercise 3.7 Suppose model (3.6) considered in Example 3.13.

$$\mathcal{P} = \{N(\theta, \sigma_1^2)^{\otimes m} \otimes N(20\,\theta, \sigma_2^2)^{\otimes(n-m)} : \theta \in \mathbb{R}\},$$

where σ_j^2, $j = 1, 2$ are known. Show that $T(\mathbf{X}) = (X_1, \ldots, X_m)$ is not sufficient for θ.

The definition of sufficiency given in Definition 3.7 is connected with the concept of conditional probability. The definition of the conditional probability is not elementary in many interesting cases. Fortunately, there is a criterion for sufficiency which is much simpler. The following theorem gives an easy way to check if a particular statistic is sufficient.

Theorem 3.7 (Factorization criterion) *Let* $\mathcal{P} = \{P_\theta : \theta \in \Theta\}$ *be a statistical model with probability function* $p(\cdot; \theta)$. *A statistic* T *is sufficient for* $\mathcal{P}$ *if and only if there exist nonnegative functions* $g(\cdot; \theta)$ *and* h *such that the probability functions* $p(\cdot; \theta)$ *satisfy*

$$p(\mathbf{x}; \theta) = g(T(\mathbf{x}); \theta)h(\mathbf{x}). \tag{3.11}$$

PROOF: We will prove the factorization criterion for discrete distributions. This gives an intuitive idea; moreover it does not require knowledge of measure theory. For a proof of the continuous case we refer to Liese and Miescke (2008, page 182).

Suppose (3.11) holds. We show that the conditional probability of $\mathbf{X}$ given $T = t$ does not depend on θ. For $\mathbf{x}$ with $T(\mathbf{x}) \neq t$ this probability is zero. Otherwise it is equal to

$$\frac{P_\theta(\{\mathbf{x}\} \cap \mathcal{X}_t)}{p^T(t; \theta)} = \frac{p(\mathbf{x}; \theta)}{p^T(t; \theta)},$$

where $\mathcal{X}_t = \{\mathbf{x} : T(\mathbf{x}) = t\}$. Since

$$p^T(t; \theta) = P_\theta(\mathcal{X}_t) = \sum_{\mathbf{x} \in \mathcal{X}_t} p(\mathbf{x}; \theta),$$

by (3.11)

$$\frac{p(\mathbf{x}; \theta)}{p^T(t; \theta)} = \frac{p(\mathbf{x}; \theta)}{\sum\limits_{\mathbf{x} \in \mathcal{X}_t} p(\mathbf{x}; \theta)} = \frac{g(t; \theta)h(\mathbf{x})}{\sum\limits_{\mathbf{x} \in \mathcal{X}_t} g(t; \theta)h(\mathbf{x})} = \frac{h(\mathbf{x})}{\sum\limits_{\mathbf{x} \in \mathcal{X}_t} h(\mathbf{x})}$$

is independent of θ, thus T is sufficient.

Now suppose T is sufficient. Since $\mathcal{X} = \cup_t \mathcal{X}_t$ there exists a t^* with $\mathbf{x} \in \mathcal{X}_{t^*}$ and

$$p(\mathbf{x}; \theta) = P_\theta(\{\mathbf{x}\} \cap \mathcal{X}_{t^*}) = \frac{P_\theta(\{\mathbf{x}\} \cap \mathcal{X}_{t^*})}{P_\theta(\mathcal{X}_{t^*})} P_\theta(\mathcal{X}_{t^*}).$$

The first factor is the conditional probability of $\mathbf{X}$ given $T(\mathbf{x}) = t^*$. Since T is sufficient it does not depend on θ and will form the factor $h(\mathbf{x})$. The probability $P_\theta(T(\mathbf{x}) = t^*)$ depends on θ, but only via $T(\mathbf{x})$; this is $g(T(\mathbf{x}); \theta)$. □

It is much simpler to establish sufficiency of a statistic with the help of Theorem 3.7 than by the original Definition 3.7. See the following example.

Example 3.21 (Flowers) For $T(\mathbf{x}) = (n_1, n_2)$ the probability function

$$p(\mathbf{x}; \theta) = p_1^{n_1} p_2^{n_2} p_3^{n_3}$$

can be decomposed into

$$g(T(\mathbf{x}); p_1, p_2) = p_1^{n_1} p_2^{n_2} (1 - p_1 - p_2)^{n - n_1 - n_2} \quad \text{and} \quad h(\mathbf{x}) = 1.$$

Thus (N_1, N_2) is a sufficient statistic. By the same argument we get the sufficiency of (N_1, N_3) and (N_2, N_3). □

Corollary 3.2

a) *A statistic T is sufficient for $\theta \in \Theta$ if and only if*

$$L(\theta; \mathbf{x}) \propto g(T(\mathbf{x}), \theta) \qquad \text{for all } \mathbf{x}.$$

b) *Let T be a sufficient statistic and suppose that the Fisher information can be computed for $\{\mathsf{P}_\theta : \theta \in \Theta\}$ and for $\{\mathsf{P}_\theta^T : \theta \in \Theta\}$. Then*

$$I_{\mathbf{X}}(\theta) = I_{T(\mathbf{X})}(\theta).$$

PROOF: By Theorem 3.7

$$L(\theta; \mathbf{x}) = p(\mathbf{x}; \theta) = g(T(\mathbf{x}); \theta) h(\mathbf{x}) \propto g(T(\mathbf{x}); \theta).$$

Let the likelihood function for the model $\{\mathsf{P}_\theta^T : \theta \in \Theta\}$ be $\widetilde{L}(\theta; t) = p^T(t; \theta)$. For discrete random variables we have

$$p^T(t; \theta) = \sum_{\mathbf{x} \in \mathcal{X}_t} p(x; \theta) = g(t; \theta) \sum_{\mathbf{x} \in \mathcal{X}_t} h(x),$$

and for continuous random variables it holds $p^T(t; \theta) = g(t; \theta) \int_{\mathcal{X}_t} h(x) dx$. Hence $\widetilde{L}(\theta; t) \propto g(t; \theta)$ and the derivatives of $L(\theta; \mathbf{x})$ and $\widetilde{L}(\theta; t)$ w.r.t. θ are the same; thus the score functions coincide and so $I_{\mathbf{X}}(\theta)$ and $I_{T(\mathbf{X})}(\theta)$ are the same.

□

Corollary 3.3 *Suppose $\mathbf{X} = (X_1, \ldots, X_n)$ is a sample of i.i.d. r. v.'s with distribution F. Then the order statistic $(X_{[1]}, \ldots, X_{[n]})$ is sufficient for F.*

PROOF: The probability function is

$$p(\mathbf{x}; F) = \prod_{i=1}^{n} p(x_i; F) = \prod_{i=1}^{n} p(x_{[i]}; F).$$

The desired statement follows from Theorem 3.7.

□

Remark 3.3 Corollary 3.3 says that independent observations can always be permutated without losing information. This statement holds for parametric models as well as for nonparametric models. However there are also parametric models where the order statistic is the only sufficient statistic. Compare Special case 3.14.

Special case 3.9 (Geometric distribution) Let $\mathbf{X} = (X_1, \ldots, X_n)$ be i.i.d. according to $\mathsf{Geo}(\theta)$. The probability function of the sample is

$$p(\mathbf{x}; \theta) = \prod_{i=1}^{n} \theta(1-\theta)^{x_i} = \theta^n (1-\theta)^{\sum_{i=1}^{n} x_i} \qquad x_i \in \{0, 1, \ldots\}.$$

Applying Theorem 3.7 with

$$g(T(\mathbf{x}), \theta) = \theta^n (1-\theta)^{\sum_{i=1}^{n} x_i} \qquad \text{and} \quad h(\mathbf{x}) = 1$$

we obtain that $\sum_{i=1}^{n} X_i$ is a sufficient statistic. □

Special case 3.10 (Likelihood ratio) Suppose that the parameter set consists only of two values: $\Theta = \{\theta_0, \theta_1\}$. Let us define the likelihood ratio statistic

$$\Lambda(\mathbf{x}) = \frac{L(\theta_0; \mathbf{x})}{L(\theta_1; \mathbf{x})} = \frac{p(\mathbf{x}; \theta_0)}{p(\mathbf{x}; \theta_1)}$$

and the functions

$$g(\lambda; \theta) = \begin{cases} \sqrt{\lambda} & \text{for} \quad \theta = \theta_0 \\ \frac{1}{\sqrt{\lambda}} & \text{for} \quad \theta = \theta_1 \end{cases} \qquad \text{and} \quad h(\mathbf{x}) = \sqrt{p_{\theta_0}(\mathbf{x}) p_{\theta_1}(\mathbf{x})}.$$

Then we have

$$p(\mathbf{x}; \theta) = \begin{cases} \sqrt{\Lambda(\mathbf{x})}\, h(\mathbf{x}) & \text{for} \quad \theta = \theta_0 \\ \frac{1}{\sqrt{\Lambda(\mathbf{x})}}\, h(\mathbf{x}) & \text{for} \quad \theta = \theta_1 \end{cases} = g(\Lambda(\mathbf{x}); \theta) h(\mathbf{x})$$

and Theorem 3.7 implies that the likelihood ratio is a sufficient statistic for $\Theta = \{\theta_0, \theta_1\}$. □

Special case 3.11 (Uniform distribution) Let $X_1, \ldots, X_n$ be i.i.d. sample from $\mathsf{U}[a, b]$ with $a < b$. The unknown parameter is $\theta = (a, b)$. The joint density is given by

$$f(\mathbf{x}; a, b) = \prod_{i=1}^{n} \frac{1}{b-a} \mathbb{1}_{[a,b]}(x_i) = \frac{1}{(b-a)^n} \prod_{i=1}^{n} \mathbb{1}_{[a,b]}(x_i).$$

The following transformations hold

$$\prod_{i=1}^{n} \mathbb{1}_{[a,b]}(x_i) = 1 \quad \Leftrightarrow \quad a \le x_i \le b \quad \text{for all } i$$

$$\Leftrightarrow \quad a \le x_{\min} \le x_{\max} \le b.$$

Thus with $x_{[1]} = x_{\min}$ and $x_{[n]} = x_{\max}$ we can write

$$f(\mathbf{x}; a, b) = \frac{1}{(b-a)^n} \, \mathbb{1}_{[a,\infty)}(x_{[1]}) \, \mathbb{1}_{(-\infty,b]}(x_{[n]})$$

and by Theorem 3.7 we get that T with $T(\mathbf{x}) = \left(x_{[1]}, x_{[n]}\right)$ is sufficient. □

Special case 3.12 (Uniform distribution) Let $X_1, \dots, X_n$ be i.i.d. r.v.'s from $U[0, \theta]$, that is we modify Special case 3.11 and assume that $a = 0$. Now

$$f(\mathbf{x}; \theta) = \frac{1}{\theta^n} \, \mathbb{1}_{(-\infty,\theta]}(\max_i x_i).$$

Applying Theorem 3.7 with

$$h(\mathbf{x}) = 1 \quad \text{and} \quad g(T(\mathbf{x}), \theta) = \frac{1}{\theta^n} \, \mathbb{1}_{(-\infty,\theta]}(x_{[n]}),$$

we obtain that $X_{[n]}$ is a sufficient statistic. □

Exercise 3.8 Let $X_1, \dots, X_n$ be an i.i.d. sample from $U[a, b]$.
a) Assume $b = 1$ and show that $X_{\min}$ is a sufficient statistic for the parameter $a < 1$.
b) Assume $a = \theta - \frac{1}{2}$ and $b = \theta + \frac{1}{2}$. Verify that $T = (X_{\min}, X_{\max})$ is sufficient.

Example 3.22 (Dolphins) (Continuation of Example 3.13 on page 40.) We examine the model

$$\mathcal{P} = \{N(\theta, \sigma_1^2)^{\otimes m} \otimes N(20\,\theta, \sigma_2^2)^{\otimes(n-m)} : \theta \in \mathbb{R}\}.$$

The likelihood function is (compare also (3.7))

$$L(\theta; \mathbf{x}) \quad \propto \quad \exp\left(\left(\frac{1}{\sigma_1^2} \sum_{i=1}^{m} x_i + \frac{20}{\sigma_2^2} \sum_{i=m+1}^{n} x_i \right)\theta - \left(\frac{m}{2\sigma_1^2} + \frac{20\,(n-m)}{\sigma_2^2} \right)\theta^2 \right).$$

From Corollary 3.2 it follows that

$$T(\mathbf{X}) = \frac{1}{\sigma_1^2} \sum_{i=1}^{m} X_i + \frac{20}{\sigma_2^2} \sum_{i=m+1}^{n} X_i$$

is a sufficient statistic. The statistic $\sum_{i=1}^{m} X_i$ is not sufficient!
OBS! The sample mean $\frac{1}{n} \sum_{i=1}^{n} X_i$ is not sufficient in a model where the variances satisfy $20\sigma_1^2 \neq \sigma_2^2$!   □

3.3.1 Minimal Sufficiency

In the above section we learned that it is possible to reduce the data without loss of information about the unknown parameter. Now the question arises: Is it possible to reduce the data further? For instance, in Corollary 3.3 we got that the order statistic is sufficient for each i.i.d. sample. In Example 3.11 we obtained that for the uniform distribution a statistic including only the minimum and the maximum of the sample is already sufficient. Is it possible to take only the range? By the way, the answer in this special case is no. But how can we show that?

Let us describe the characterization "no further reduction possible without losing information" from the viewpoint of the partition of the sample space. A statistic is minimal sufficient if it generates the coarsest sufficient partition. To construct such a partition we define the following equivalence relation: Two elements $\mathbf{x}, \mathbf{y} \in \mathcal{X}$ are equivalent if there exists a $k(\mathbf{x}, \mathbf{y})$, which does not depend on θ, such that $p(\mathbf{x}; \theta) = k(\mathbf{x}, \mathbf{y}) p(\mathbf{y}; \theta)$. We write

$$\mathbf{x} \simeq \mathbf{y} \text{ iff } \quad p(\mathbf{x}; \theta) = k(\mathbf{x}, \mathbf{y}) p(\mathbf{y}; \theta).$$

This equivalence relation generates a partition of the sample space $\mathcal{X}$ with

$$\mathcal{D}(\mathbf{x}) = \{\mathbf{y} : \mathbf{x} \simeq \mathbf{y}\} \quad \text{and} \quad \mathcal{D}_0 = \{\mathbf{x} : p(\mathbf{x}; \theta) = 0\}.$$

For each class $\mathcal{D}(\mathbf{x})$ choose a representative, say $\tilde{\mathbf{x}}$. Now, let us define the following statistic $G : \mathcal{X} \to \mathcal{X}$. For all $\mathbf{y} \in \mathcal{D}(\mathbf{x})$ set $G(\mathbf{y}) = \tilde{\mathbf{x}}$. In other words, G is constant on each class of the partition. Let us show, (i) that G is sufficient and (ii) that a partition generated by another sufficient statistic is not coarser than the partition based on the considered equivalence relation:
(i) Let $\mathbf{x}$ arbitrarily be fixed. Then $\mathbf{x} \simeq \tilde{\mathbf{x}}$ and therefore

$$p(\mathbf{x}; \theta) = k(\mathbf{x}, \tilde{\mathbf{x}}) p(\tilde{\mathbf{x}}; \theta) = k(\mathbf{x}, G(\mathbf{x})) p(G(\mathbf{x}); \theta),$$

and by Theorem 3.7 it follows that G is sufficient.
(ii) Let S be an arbitrary sufficient statistic with values in $\mathcal{S}$. Denote the partition generated by S by

$$\mathcal{E}_s = \{\mathbf{y} : S(\mathbf{y}) = s\}.$$

Consider an arbitrary point $\mathbf{x}$ belonging to $\mathcal{E}_s$ and to $\mathcal{D}(\mathbf{x})$. We show that then $\mathcal{E}_s \subseteq \mathcal{D}(\mathbf{x})$, that is, the partition $\{\mathcal{E}_s\}_{s \in \mathcal{S}}$ is finer (or equal) than the partition in equivalence classes.

For $\mathbf{y} \in \mathcal{E}_s$ we obtain with $S(\mathbf{x}) = S(\mathbf{y})$ by the factorization theorem

$$p(\mathbf{x}; \theta) = g(S(\mathbf{x}; \theta))h(\mathbf{x}) = g(S(\mathbf{y}; \theta))h(\mathbf{x})$$

and

$$p(\mathbf{y}; \theta) = g(S(\mathbf{y}; \theta))h(\mathbf{y}),$$

where $h(\mathbf{y}) \neq 0$. For $h(\mathbf{y}) = 0$, we have $p(\mathbf{y}; \theta) = 0$ for all θ and thus $\mathsf{P}_\theta(h(\mathbf{y}) = 0) = 0$ for all θ. Hence

$$p(\mathbf{x}; \theta) = g(S(\mathbf{y}; \theta))h(\mathbf{y})\frac{h(\mathbf{x})}{h(\mathbf{y})} = p(\mathbf{y}; \theta)\frac{h(\mathbf{x})}{h(\mathbf{y})}.$$

The ratio $h(\mathbf{x})/h(\mathbf{y})$ is independent of θ, thus $\mathbf{y} \simeq \mathbf{x}$ and $\mathcal{E}_s \subseteq \mathcal{D}(\mathbf{x})$.

Every statistic T which is a function of G generates a partition which is not finer than the equivalence relation partition; every *sufficient* statistic T which is a function of G generates the same partition. Thus every sufficient statistic which is a function of G is minimal sufficient.

Note, the construction of a minimal sufficient statistic T is not unique, but the sets $\mathcal{D}(\mathbf{x})$ are unique—in other words, if T_1 and T_2 are minimal sufficient statistics then they generate the same partition.

Example 3.23 (Flowers) Continuation of Example 3.21 on page 55. The statistic $T(\mathbf{X}) = (N_1, N_2, N_3)$ is sufficient but not minimal sufficient, because the proper reduction (N_1, N_2) is sufficient too. □

Without using the idea of the partition of the sample space by a statistic we use the following definition:

Definition 3.8 (Minimal sufficiency) A statistic T is **minimal sufficient** iff T is a function of any other sufficient statistic.

That is for all sufficient statistics S there exists a function H, such that $T = H(S)$.

Both approaches to minimal sufficiency are aligned by the following criterion: Let $\mathcal{K}$ be the set of all pairs $(\mathbf{x}, \mathbf{y})$ for which there is a $k(\mathbf{x}, \mathbf{y}) > 0$ such that

$$L(\theta; \mathbf{x}) = k(\mathbf{x}, \mathbf{y})L(\theta; \mathbf{y}) \qquad \text{for all } \theta \in \Theta.$$

Theorem 3.8 *Let T be a sufficient statistic for $\mathcal{P} = \{\mathsf{P}_\theta : \theta \in \Theta\}$. If for all $(\mathbf{x}, \mathbf{y}) \in \mathcal{K}$ the statistic T satisfies $T(\mathbf{x}) = T(\mathbf{y})$ then T is minimal sufficient.*

PROOF: Let S be an arbitrary sufficient statistic and the pair $(\mathbf{x}, \mathbf{y})$ such that $S(\mathbf{x}) = S(\mathbf{y})$. By the factorization theorem we have

$$L(\theta; \mathbf{x}) = g(S(\mathbf{x}); \theta)h(\mathbf{x}) \quad \text{and} \quad L(\theta; \mathbf{y}) = g(S(\mathbf{y}); \theta)h(\mathbf{y})$$

and therefore

$$L(\theta; \mathbf{x}) = \frac{h(\mathbf{x})}{h(\mathbf{y})} L(\theta, \mathbf{y}).$$

Thus $(\mathbf{x}, \mathbf{y}) \in \mathcal{K}$ and $T(\mathbf{y}) = T(\mathbf{x})$. But this means T is a function of S.

$\square$

We illustrate this concept with the help of the following example.

Example 3.24 (Bernoulli distribution) For simplicity let us take $n = 3$ independent variables X_1, X_2, and X_3 distributed according to a Bernoulli distribution. As shown in Special case 3.8 the statistic $T(\mathbf{x}) = x_1 + x_2 + x_3$ leads to a partition of the sample space into four sets $\mathcal{X}_k$. To demonstrate what is behind the criterion take another sufficient statistic, say $S(\mathbf{x}) = (x_1, x_2 + x_3)$. Let us write down the partition generated by S:

$$\mathcal{X}_{(0,0)} = \{(0,0,0)\}, \quad \mathcal{X}_{(0,2)} = \{(0,1,1)\}, \quad \mathcal{X}_{(0,1)} = \{(0,1,0), (0,0,1)\},$$

$$\mathcal{X}_{(1,0)} = \{(1,0,0)\}, \quad \mathcal{X}_{(1,2)} = \{(1,1,1)\}, \quad \mathcal{X}_{(1,1)} = \{(1,1,0), (1,0,1)\}.$$

This partition consists of six elements; it is finer than that generated by T. Let us apply the Theorem 3.8. From

$$\frac{L(\theta; \mathbf{x})}{L(\theta; \mathbf{y})} = \frac{\theta^{\sum_{i=1}^{3} x_i}(1 - \theta)^{3 - \sum_{i=1}^{3} x_i}}{\theta^{\sum_{i=1}^{3} y_i}(1 - \theta)^{3 - \sum_{i=1}^{3} y_i}}$$

it follows that points $(\mathbf{x}, \mathbf{y})$ are in $\mathcal{K}$ iff $\sum_i x_i = \sum_i y_i$. For the statistic T this is satisfied only if $T(\mathbf{x}) = T(\mathbf{y})$. Consider S and take $\mathbf{x} = (1,0,0)$ and $\mathbf{y} = (0,1,0)$. These values are in $\mathcal{K}$ but $S(\mathbf{x}) \neq S(\mathbf{y})$, i.e., they are in different sets of the partition generated by S. That is, S is not minimal sufficient. $\square$

Summarizing we have:

- A statistic T generates a partition of the sample space.

- A partition generated by a sufficient statistic has the property that if we know into which $\mathcal{X}_t = \{x : T(x) = t\}$ the data fall, we can determine the likelihood function (up to a factor not depending on the parameter).

> The sample itself gives the finest partition; a minimal sufficient statistic leads to the coarsest partition of the sample space, which still contains all essential information.

Special case 3.13 (Normal distribution) Consider a sample $\mathbf{X}$ of independent $N(\mu, \sigma^2)$-distributed r.v.'s. For the parameter $\theta = (\mu, \sigma^2) \in \mathbb{R} \times \mathbb{R}_+$ the ratio of the likelihood function at $\mathbf{x}$ and $\mathbf{y}$ is

$$\frac{L(\theta; \mathbf{x})}{L(\theta; \mathbf{y})} = \frac{(\frac{1}{2\pi\sigma^2})^{\frac{n}{2}} \exp\left(-\frac{1}{2\sigma^2} \sum_{i=1}^n (x_i - \mu)^2\right)}{(\frac{1}{2\pi\sigma^2})^{\frac{n}{2}} \exp\left(-\frac{1}{2\sigma^2} \sum_{i=1}^n (y_i - \mu)^2\right)}$$

$$= \exp\left(-\frac{1}{2\sigma^2} \sum_{i=1}^n (x_i - \mu)^2 + \frac{1}{2\sigma^2} \sum_{i=1}^n (y_i - \mu)^2\right)$$

$$= \exp\left(-\frac{1}{2\sigma^2}\left(\sum_{i=1}^n x_i^2 - \sum_{i=1}^n y_i^2 - 2\mu\left(\sum_{i=1}^n x_i - \sum_{i=1}^n y_i\right)\right)\right).$$

This ratio is independent of θ if $\sum_{i=1}^n x_i = \sum_{i=1}^n y_i$ and $\sum_{i=1}^n x_i^2 = \sum_{i=1}^n y_i^2$. Thus

$$T(\mathbf{x}) = (T_1(\mathbf{x}), T_2(\mathbf{x})) = \left(\sum_{i=1}^n x_i, \sum_{i=1}^n x_i^2\right)$$

forms a minimal sufficient statistic. Note that the sample mean $\bar{x} = \frac{1}{n}T_1(\mathbf{x})$ and the sample variance

$$s^2 = \frac{1}{n-1} \sum_{i=1}^n (x_i - \bar{x})^2 = \frac{1}{n-1}\left(\sum_{i=1}^n x_i^2 - n\bar{x}^2\right)$$

$$= \frac{1}{n-1}\left(T_2(\mathbf{x}) - \frac{1}{n}T_1(\mathbf{x})^2\right)$$

are in one-to-one correspondence with $T(\mathbf{x})$. Hence the pair $(\overline{X}, S^2)$ is minimal sufficient for the normal distribution, where both parameters are unknown.□

Example 3.25 (Pendulum) Applying the concept of sufficiency to Example 2.3 on page 6 we see that it is not necessary to save the whole data set 2.1. It is sufficient to know the sample mean and the sample variance $\left(1.998, (0.013)^2\right)$. But that is exactly what we have to keep, keeping less means a loss of information!   □

Exercise 3.9 Consider **X** an i.i.d. sample from $N(\theta, 1)$, $\theta \in \mathbb{R}$. Show that the sample mean is minimal sufficient.

Exercise 3.10 Consider **X** an i.i.d. sample from $N(0, \sigma^2)$, $\theta = \sigma^2 \in \mathbb{R}_+$. Is the sample variance minimal sufficient?

Special case 3.14 (Cauchy distribution) Let $X_1, \ldots, X_n$ be i.i.d. r.v.'s according to a Cauchy distribution with location parameter $\theta \in \mathbb{R}$ and density

$$f(x; \theta) = \frac{1}{\pi(1 + (x - \theta)^2)}.$$

The ratio of the likelihood functions at two points of the sample space is given by

$$\frac{L(\theta; \mathbf{x})}{L(\theta; \mathbf{y})} = \frac{\prod_{i=1}^{n}(1 + (y_i - \theta)^2)}{\prod_{i=1}^{n}(1 + (x_i - \theta)^2)} = \frac{\mathrm{pol}_{\mathrm{num}}(\theta)}{\mathrm{pol}_{\mathrm{den}}(\theta)},$$

where $\mathrm{pol}_{\mathrm{num}}(\theta)$ and $\mathrm{pol}_{\mathrm{den}}(\theta)$ are polynomials in θ. These polynomials are of degree $2n$ and their leading coefficient is 1. If their ratio is independent of θ, then $\mathrm{pol}_{\mathrm{num}}(\theta) = \mathrm{pol}_{\mathrm{den}}(\theta)$. This means that their $2n$ zeros agree. The $2n$ zeros of $\mathrm{pol}_{\mathrm{num}}(\theta)$ are $y_j \pm \imath$ for $j = 1, \ldots, n$, and those of $\mathrm{pol}_{\mathrm{den}}(\theta)$ are $x_j \pm \imath$ for $j = 1, \ldots, n$. (Here $\imath$ denotes the imaginary unit.) Thus

$$\{y_j \pm \imath : j = 1, \ldots, n\} = \{x_j \pm \imath : j = 1, \ldots, n\},$$

hence

$$\{y_j : j = 1, ..., n\} = \{x_j : j = 1, ..., n\}.$$

Consequently, the ratio of the likelihood functions at **x** and **y** is independent of θ iff

$$(y_{[1]}, \ldots, y_{[n]}) = (x_{[1]}, \ldots, x_{[n]}).$$

We obtain that the order statistic

$$T(\mathbf{x}) = (x_{[1]}, \ldots, x_{[n]})$$

is minimal sufficient. From Corollary 3.3 we know that the order statistic is sufficient for all i.i.d. samples. But in the Cauchy family no more reduction of the data is possible without losing information. □

Example 3.26 (Patients data) Let us apply the logistic regression model to patients data. Suppose for n patients with mean cholesterol levels $z_1, \ldots, z_n$ we have registered whether a heart disease (HD) occurs or not. We define

$$x_i = \begin{cases} 1 & \text{patient } i \text{ had a HD} \\ 0 & \text{patient } i \text{ had no HD} \end{cases}.$$

Suppose the z_i are fixed and the x_i's are realizations of Bernoulli variables with probabilities $\pi(z_i)$. A common approach to model the "probability of illness as a function of risk factors" by a parametric model is to assume that the natural parameter of the Bernoulli distribution is linear in the z_i. That is

$$\ln \frac{\pi(z_i)}{1 - \pi(z_i)} = \beta_0 + \beta_1 z_i. \tag{3.12}$$

The unknown parameter is $\theta = (\beta_0, \beta_1)$. Let us derive a minimal sufficient statistic for θ. Equivalently to the logit transformation (3.12) we have

$$\pi(z_i) = \frac{\exp(\beta_0 + \beta_1 z_i)}{1 + \exp(\beta_0 + \beta_1 z_i)}.$$

The likelihood function is

$$\prod_{i=1}^{n} \pi(z_i)^{x_i} (1 - \pi(z_i))^{1-x_i} = \frac{\exp(\sum_{i=1}^{n}(\beta_0 x_i + \beta_1 z_i x_i))}{\prod_{i=1}^{n}(1 + \exp(\beta_0 + \beta_1 z_i))}.$$

Let us apply Theorem 3.8: The ratio of the likelihood at $\mathbf{x} = (x_1, \ldots, x_n)$ and $\mathbf{y} = (y_1, \ldots, y_n)$ is

$$\frac{\exp(\sum_{i=1}^{n}(\beta_0 x_i + \beta_1 z_i x_i))}{\exp(\sum_{i=1}^{n}(\beta_0 y_i + \beta_1 z_i y_i))}.$$

This ratio is independent of θ iff $(\sum_{i=1}^{n} x_i, \sum_{i=1}^{n} z_i x_i) = (\sum_{i=1}^{n} y_i, \sum_{i=1}^{n} z_i y_i)$. Thus the statistic T defined by

$$T(\mathbf{x}) = \left(\sum_{i=1}^{n} x_i, \sum_{i=1}^{n} z_i x_i\right)$$

is minimal sufficient. □

Sometimes the application of Theorem 3.8 requires long computations. If one has already shown that the considered family of distributions forms an exponential family we can use the following result: Recall that by (2.7) the distribution of X belongs to an exponential family iff

$$p(x; \theta) = A(\theta) \exp\left[\sum_{j=1}^{k} \zeta_j(\theta) T_j(x)\right] h(x),$$

and for a sample $\mathbf{X}$ of independent copies of X we have

$$p(\mathbf{x}; \theta) = \prod_{i=1}^{n} p(x_i; \theta) = A(\theta)^n \exp\left[\sum_{j=1}^{k} \zeta_j(\theta) \sum_{i=1}^{n} T_j(x_i)\right] \prod_{i=1}^{n} h(x_i).$$

Theorem 3.7 implies that

$$T_{(n)}(\mathbf{x}) = \left(\sum_{i=1}^{n} T_1(x_i), \ldots, \sum_{i=1}^{n} T_k(x_i)\right)$$

is sufficient.

Theorem 3.9 *For a sample of i.i.d. r.v.'s from a strictly k-parameter exponential family it holds:*

1. *The statistic*

$$T_{(n)}(\mathbf{x}) = \left(\sum_{i=1}^{n} T_1(x_i), \ldots, \sum_{i=1}^{n} T_k(x_i)\right) \tag{3.13}$$

   is minimal sufficient.
2. *The distribution of $T_{(n)}(\mathbf{X})$ belongs to a k-parameter exponential family.*

PROOF: We will apply Theorem 3.8. Consider the ratio of the likelihood functions at points $\mathbf{x}$ and $\mathbf{y}$

$$\frac{L(\theta; \mathbf{x})}{L(\theta; \mathbf{y})} = \frac{\prod_{i=1}^{n} h(x_i)}{\prod_{i=1}^{n} h(y_i)} \exp\left[\sum_{j=1}^{k} \zeta_j(\theta) \left(\sum_{i=1}^{n} T_j(x_i) - \sum_{i=1}^{n} T_j(y_i)\right)\right].$$

The ratio is independent of θ, i.e., $(\mathbf{x}, \mathbf{y}) \in \mathcal{K}$ iff

$$\sum_{i=1}^{n} T_j(x_i) = \sum_{i=1}^{n} T_j(y_i) \qquad \text{for all } j = 1, \ldots, k.$$

Hence the statistic (3.13) is minimal sufficient.

We will show the second statement for discrete distributions only. The probability function for $T_{(n)}(\mathbf{X})$ is

$$p^{T_{(n)}}(t;\theta) = \mathsf{P}_\theta(\{\mathbf{x} : \sum_{i=1}^{n} T_1(x_i) = t_1, \ldots, \sum_{i=1}^{n} T_k(x_i) = t_k\})$$

$$= \sum_{\mathbf{x}:T(\mathbf{x})=t} A(\theta)^n \exp\left[\sum_{j=1}^{k} \zeta_j(\theta) \sum_{i=1}^{n} T_j(x_i)\right] \prod_{i=1}^{n} h(x_i)$$

$$= A(\theta)^n \exp\left[\sum_{j=1}^{k} \zeta_j(\theta)t_j\right] \sum_{\mathbf{x}:T(\mathbf{x})=t} \prod_{i=1}^{n} h(x_i).$$

Set $a(\theta) = A(\theta)^n$ and $H(t) = \sum_{\mathbf{x}:T(\mathbf{x})=t} \prod_{i=1}^{n} h(x_i)$, then we have

$$p^{T_{(n)}}(t;\theta) = a(\theta)\exp\left[\sum_{j=1}^{k} \zeta_j(\theta)t_j\right] H(t),$$

which belongs to a k-parameter exponential family.

□

Let us formulate a non-i.i.d. version of Theorem 3.9. Applying Theorem 3.9 to the whole sample we get:

Corollary 3.4 *For a sample* $\mathbf{X} = (X_1, \ldots, X_n)$ *from a strictly k-parameter exponential family with*

$$p(\mathbf{x};\theta) = A(\theta)\exp\left[\sum_{j=1}^{k} \zeta_j(\theta)T_j(\mathbf{x})\right] h(\mathbf{x}).$$

The statistic
$$T(\mathbf{x}) = (T_1(\mathbf{x}), \ldots, T_k(\mathbf{x}))$$
is minimal sufficient and the distribution of T belongs to a k-parameter exponential family.

Special case 3.15 (Two-sample problem) Suppose that

$$\mathbf{Z} = (X_1, \ldots, X_{n_1}, Y_1, \ldots, Y_{n_2})$$

is a sample of independent r.v.'s of size $n = n_1 + n_2$. The X_i's are normally distributed with mean μ_1 and variance σ^2. The Y_j's are distributed according to $\mathsf{N}(\mu_2, \sigma^2)$. Thus

$$\mathcal{P} = \{\mathsf{N}(\mu_1, \sigma^2)^{\otimes n_1} \otimes \mathsf{N}(\mu_2, \sigma^2)^{\otimes n_2} : (\mu_1, \mu_2, \sigma^2) \in \mathbb{R} \times \mathbb{R} \times \mathbb{R}_+\}.$$

This family forms a 3-parameter exponential family with the sufficient statistic

$$T(\mathbf{Z}) = \left(\sum_{i=1}^{n_1} X_i, \sum_{j=1}^{n_2} Y_j, \sum_{i=1}^{n_1} X_i^2 + \sum_{j=1}^{n_2} Y_j^2 \right).$$

The distribution of each single sample belongs to a 2-parameter exponential family with

$$T_1(\mathbf{X}) = \left(\sum_{i=1}^{n_1} X_i, \sum_{i=1}^{n_1} X_i^2 \right) \quad \text{and} \quad T_2(\mathbf{Y}) = \left(\sum_{j=1}^{n_2} Y_j, \sum_{j=1}^{n_2} Y_j^2 \right).$$

□

Exercise 3.11 (Dolphins) Show for the two-sample problem considered in Example 3.15 that the statistic $T = (T_1, T_2)$ is sufficient but not minimal sufficient.

Exercise 3.12 (Dolphins) Consider Example 3.12 on page 40. Show that in the model (3.5) the statistic $\sum_{i=1}^{m} x_i$ is minimal sufficient.

Exercise 3.13 (Dolphins) Consider Example 3.22 on page 57. Show that the statistic $\frac{1}{\sigma_1^2} \sum_{i=1}^{m} x_i + \frac{20}{\sigma_2^2} \sum_{i=m+1}^{n} x_i$ is minimal sufficient.

Exercise 3.14 (Flowers) Show that the statistics $T_2(\mathbf{x}) = (n_2, n_3)$ and $T_3(\mathbf{x}) = (n_1, n_3)$ are minimal sufficient in the Example 3.4.

 Example 3.27 (Genotypes) Recall Example 2.21 on page 20. Introduce $\sum_{i=1}^{n} I_1(x_i) = n_{\mathrm{aa}}$ and $\sum_{i=1}^{n} I_2(x_i) = n_{\mathrm{Aa}}$, where n_{aa} is the number of genotypes aa and n_{Aa} is the number of genotypes Aa. Theorem 3.9 implies that

$$T(\mathbf{x}) = 2n_{\mathrm{aa}} + n_{\mathrm{Aa}}$$

is minimal sufficient. □

3.4   List of Problems

1. Derive the likelihood function for a sample of n i.i.d. random variables from:
   a) $N(0, \sigma^2)$,   b) $N(1, \sigma^2)$,   c) $N(\mu, \sigma^2)$ with $\theta = (\mu, \sigma^2)$,   d) $N(\mu, \mu)$ with $\mu > 0$.

2. Derive the likelihood function for a sample of n i.i.d. random variables from:
   a) $\text{Exp}(\lambda)$,   b) $\text{Poi}(\lambda)$,   c) $\text{Geo}(p)$.

3. Suppose that the underlying model consists of three discrete distributions: $\mathcal{P} = \{P_1, P_2, P_3\}$. These distributions are defined by

	Values				
	a	b	c	d	e
P_1	0.1	0.4	0.1	0.2	0.2
P_2	0	0.2	0	0.1	0.7
P_2	0.8	0.1	0.05	0.05	0

   a) Suppose $x = a$ is observed. Give the likelihood function. Which P_j maximizes the likelihood function?
   b) Give the likelihood function, if $x = b$ is observed. Compare it with the likelihood function if $x = d$ is observed.

4. Consider a random sample from $U[\theta - \frac{1}{2}, \theta + \frac{1}{2}]$. Find the likelihood function and determine all maxima.

5. Derive the minimal sufficient statistics for a sample of n i.i.d. r. v.'s from:
   a) $N(0, \sigma^2)$,   b) $N(1, \sigma^2)$,   c) $N(\mu^2, 1)$,   d) $N(\mu, \sigma^2)$ with $\theta = (\mu, \sigma^2)$,
   e) $N(\mu, \mu)$ with $\mu > 0$.

6. Given a sample of i.i.d. r.v.'s from $N(\mu, \sigma^2)$: a) Give an example of a statistic which is not sufficient for $\theta = (\mu, \sigma^2)$. b) Give an example of a sufficient but not minimal sufficient statistic.

7. Give the proof of Theorem 3.6 for continuous distributions.

8. Consider the Kungsängslilja example, where p_1 and p_2 are unknown parameters.
   a) Calculate the covariance between the number of white and pink flowers. Why are they correlated in this direction?
   b) Derive the 2×2 Fisher information matrix for the Kungsängslilja example.

9. Let $X_1, \ldots, X_n$ be i.i.d. r.v.'s with density

$$f(x; \theta) = \frac{1}{2\theta} \exp\left(-\frac{|x|}{\theta}\right)$$

   (Double exponential distribution).
   a) Show that $T = \sum_{i=1}^{n} |X_i|$ is a sufficient statistic.
   b) Compute the distribution of the X_i's given $T = t$.

10. Let (X, Y) be distributed according to the two-dimensional normal distribution $N_2(0, C)$, where

$$C = \begin{pmatrix} 1 & \rho \\ \rho & 1 \end{pmatrix} \qquad -1 < \rho < 1.$$

Compute the Fisher information $I_{(X,Y)}(\rho)$.

11. Estimate the parameter $\gamma = \vartheta^2$. You can choose between the following experiments:
    a) You observe i.i.d. r.v.'s $X_1, \ldots, X_n$ with $X_i \sim \mathrm{Bin}(1, \vartheta^2)$.
    b) You observe i.i.d. r.v.'s $Y_1, \ldots, Y_n$ with $Y_i \sim \mathrm{Bin}(1, \vartheta)$.
    Compare both experiments. Which experiment is more suitable (under which condition)?

12. Let X be Poisson-distributed with parameter μ. The r.v. Z has a so-called truncated Poisson distribution, i.e.,

$$\mathsf{P}_\mu(Z = k) = \mathsf{P}_\mu(X = k | X > 0).$$

Compute the Fisher informations $I_X(\mu)$ and $I_Z(\mu)$. Compare.

13. Given $X \sim \mathrm{Exp}(\lambda)$. The parameter of interest is

$$\gamma = \mathsf{P}(X > t_0),$$

where t_0 is a known time point. Compute the Fisher information w.r.t. γ.

14. The aim is to estimate the success probability θ of a certain event. You can draw a sample
    a) $X_1, \ldots, X_n$ with $X_i \sim \mathrm{Bin}(1, \theta)$
    b) $Y_1, \ldots, Y_n$ with $Y_i \sim \mathrm{Geo}(\theta)$.
    Which type of drawn sample is more informative?

3.5   Further Reading

A discussion about the likelihood principle and its generalization can be found in the monograph of Berger and Wolpert (1988).

In our textbook the likelihood is defined for parametric models. It is also possible to define a likelihood function for nonparametric models, the so-called empirical likelihood; see, for example, Owen (2001).

The proof of Lemma 3.1 led to the expression $\mathsf{E}_\theta \ln\left(\frac{p(X;\theta)}{p(X;\theta')}\right)$. This expectation is called Kullback–Leibler divergence or discrepancy. Although it is often called a distance metric between the two distributions P_θ and $\mathsf{P}_{\theta'}$ it is not a metric. Since different distributions can be compared w.r.t. the Kullback–Leibler divergence this distance is a useful tool in model selection. For a discussion of the Kullback–Leibler discrepancy and other divergence measures in statistics we refer to the book of Pardo (2006) and to the book of Konishi and Kitagawa (2008), where a generalized information criterion is introduced.

The Fisher information matrix can be used to define a Riemannian metric for a regular parametric model. In Kass and Paul (1997) statistical manifolds and their curvature properties are studied.

For advanced graduate students we recommend the book of Liese and Miescke (2008). It presents the main ideas of decision theory in a mathematically rigorous manner, while observing statistical relevance. All of the major topics are introduced at an elementary level, then developed incrementally to higher levels. The authors present the major results of classical finite sample size decision theory and the concepts of modern asymptotic decision theory.

Chapter 4

Estimation

Suppose that we have formulated a statistical model $\mathcal{P}$ for the phenomenon under consideration: $\mathbf{X} \sim \mathsf{P}_\theta \in \mathcal{P}$. We want to draw conclusions about the distribution P_θ on the basis of the data $\mathbf{x} \in \mathcal{X}$. If $\mathcal{P}$ is a parametric model, one is mostly interested in statements about the finite-dimensional parameter θ. If the underlying model is not parameterized by a finite-dimensional parameter, the quantity of interest can be the distribution function, some other function or a finite-dimensional parameter depending on the unknown distribution P_θ. To formalize this approach, we introduce a function g defined on the parameter space Θ and take values in a set Γ, i.e., $g : \Theta \to \Gamma$. Our aim is to derive conclusions about the value of g at θ, i.e., about the quantity $\gamma = g(\theta)$.

The first step of a statistical analysis is to specify a plausible value for γ or to determine a subset of Γ of plausible values for γ. The first task is the problem of **point estimation**, the second is that of deriving **confidence sets**. In this section we will present some basic ideas about point estimation procedures and we will discuss criteria for "good" estimation methods. Before we do this let us define what an estimator is:

Definition 4.1 (Estimator) A function $T: \mathcal{X} \to \Gamma$ is an **estimator**. It is used to estimate $\gamma = g(\theta)$. The value $T(\mathbf{x})$ is called the **estimate** of $g(\theta)$. It is the realization of the random variable $T(\mathbf{X})$.

In other words:

> An **estimator is a rule** of how to use the data to construct a plausible value for γ. The **estimate is the result** of applying this rule.

Remark 4.1 To be exact, we will always assume that an estimator T is a measurable function, i.e., the probability $\mathsf{P}_\theta^T(B) = \mathsf{P}_\theta(\{\mathbf{x} : T(\mathbf{x}) \in B\})$ is defined.

Remark 4.2 Usually the estimate as well as the estimator of the parameter γ are denoted by decorations: $\widehat{\gamma}, \widetilde{\gamma}, \widetilde{\widetilde{\gamma}}, \overline{\gamma}, \overline{\overline{\gamma}}$. The used method is often given as a subscript: $\widehat{\gamma}_{\mathrm{MLE}}, \widehat{\gamma}_{\mathrm{LSE}}$. This type of notation is very common and we will apply it too. Unfortunately this system does not differentiate between estimator and estimate.

Example 4.1 (Ballpoint pens) Consider Example 2.1 on page 5, the quality control for ballpoint pens. Here the parameter of interest is the probability of producing a defective pen, that means the success probability in the underlying binomial model, i.e., $g(\theta) = \theta$ and $\Gamma = (0,1)$. Of course, an intuitive estimate is the arithmetic mean of the relative frequency of defective pen items in our sample:

$$T(\mathbf{x}) = \frac{1}{7} \sum_{i=1}^{n} \frac{x_i}{n} = \frac{8+5+9+4+6+8+10}{7*400} = \frac{1}{56}.$$

But also

$$\widetilde{T}(\mathbf{x}) = \frac{x_1}{n} = \frac{8}{400} = \frac{1}{50}$$

is an estimate in the sense of Definition 4.1.   □

Example 4.2 (Pendulum) Consider Example 2.3 on page 6. Suppose that we are interested in the precision of the measurements. Then, the parameter of interest is the variance of the assumed normal distribution, i.e., $\gamma = g(\mu, \sigma^2) = \sigma^2$ and $\Gamma = \mathbb{R}_+$. Usually this parameter is estimated by the empirical variance defined by

$$T(\mathbf{x}) = \frac{1}{n-1} \sum_{i=1}^{n} (x_i - \overline{x})^2 = 0.0001.$$

But what can we say in favor of this choice?   □

Example 4.3 (Soybeans) In the simple linear regression model (2.5), where we do not assume a parametric form for the distribution of the errors, we wish to determine the values of the parameters of the linear relationship. Thus, we have $\gamma = g(a, b, \sigma^2, \kappa) = (a, b)$ with $\Gamma = \mathbb{R}^2$. We will see in Section 6.3 that the least squares method will provide good values for the coefficients a and b. But, under which assumptions?
Should we take another estimation method if the errors can be assumed to be normally distributed? □

4.1 Methods of Estimation

4.1.1 The Method of Moments

This method is old and simple, no special parametric form of the underlying distribution is required. It is suitable for parameters γ which depend on moments of the underlying distribution. The idea is to replace these moments by their empirical versions. To formalize this approach let $\mathbf{X} = (X_1, \ldots, X_n)$ be a sample of i.i.d. real-valued random variables with distribution function F, that is, we assume the model

$$\mathcal{P} = \{P_F^{\otimes n} : F \in \Theta\}, \quad \text{where} \quad \Theta = \{F : \int x^r dF < \infty\} \tag{4.1}$$

is the set of all distribution functions with finite r-th moment for some r. The moment of order j is given by

$$m_j = m_j(F) = \mathsf{E}X^j = \int x^j dF. \tag{4.2}$$

Recall: For a continuous r.v. X with density f the integral $\int x^j dF$ has the form $\int x^j dF = \int x^j f(x) dx$. If X is a discrete r.v. with $\mathsf{P}(X = z_m) = p_m$, $m \in \mathbb{N}$ then $\int x^j dF = \sum_m z_m^j p_m$.

The basic assumption is that the parameter of interest γ depends on the unknown parameter $\theta = F$ via the moments m_j and can be expressed as

$$\gamma = h(m_1(F), \ldots, m_r(F)), \tag{4.3}$$

where h is a known function. A simple estimation method consists in replacing the m_j's in (4.3) by their empirical versions. This leads to the following definition:

Definition 4.2 (Method of moments estimator) Suppose the statistical model (4.1). The moment estimate for the parameter γ given in (4.3) is defined by

$$\hat{\gamma}_{\text{MME}} = h(\hat{m}_1, \ldots, \hat{m}_r),$$

where $\hat{m}_j$ is the **empirical moment** (or sample moment) of order j:

$$\hat{m}_j = \frac{1}{n} \sum_{i=1}^{n} x_i^j. \tag{4.4}$$

Let us consider the background of this type of estimator: The distribution function is defined by $F(t) = \mathrm{P}_F((-\infty, t])$. For fixed t, a natural estimate for the probability of the set $(-\infty, t]$ is the relative frequency. Thus, we define

$$\hat{F}_n(t) \; = \; \hat{F}_n(t; \mathbf{x}) \; = \; \frac{1}{n} \sum_{i=1}^{n} \mathbb{1}_{(-\infty, t]}(x_i).$$

Here $\mathbb{1}_A$ denotes the indicator function of the set A; $\hat{F}_n$ is called the **empirical distribution function**. Considered as a function of t, it is a step function. Using the definition of an integral with respect to $\hat{F}_n$ we can write the empirical moment $\hat{m}_j$ given in (4.4) in the form

$$\hat{m}_j \; = \; m_j(\hat{F}_n) \; = \; \int x^j \, d\hat{F}_n,$$

and the analogy to (4.2) is evident.

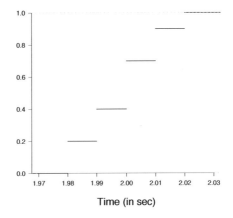

Figure 4.1: Empirical distribution function $\hat{F}_n$ of the Pendulum data.

In what follows we consider some important cases, not only for illustration, but also because they are interesting by themselves for applications. We assume model (4.1).

Special case 4.1 (Mean) Consider the case $r = 1$. Using the method of moments we estimate $\gamma = \int x \, dF$ by the sample mean.   □

Special case 4.2 (Standard deviation) The parameter of interest is the standard deviation σ, i.e., we suppose model (4.1) with $r = 2$. Since the variance is defined by $\sigma^2 = \int (x - m_1)^2 dF$ we have

$$\sigma = \sqrt{m_2 - m_1^2} = h(m_1, m_2).$$

Thus the estimate constructed by the method of moments is

$$\hat{\sigma}_{\text{MME}} = \sqrt{\hat{m}_2 - \hat{m}_1^2}.$$

Using

$$\hat{m}_2 - \hat{m}_1^2 = \frac{1}{n}\sum_{i=1}^{n} x_i^2 - \left(\frac{1}{n}\sum_{i=1}^{n} x_i\right)^2 = \frac{1}{n}\sum_{i=1}^{n}(x_i - \overline{x})^2,$$

we obtain

$$\hat{\sigma}_{\text{MME}} = \sqrt{\frac{1}{n}\sum_{i=1}^{n}(x_i - \overline{x})^2}.$$

OBS! The estimate $\hat{\sigma}_{\text{MME}}$ is not the sample standard deviation!   □

Special case 4.3 (Skewness) The skewness of a r.v. X with distribution function F is defined by

$$\gamma = \frac{\mathsf{E}_F(X - \mathsf{E}_F X)^3}{(\mathsf{Var}_F X)^{3/2}} = \frac{m_3 - 3m_2 m_1 + 2m_1^3}{(m_2 - m_1^2)^{3/2}},$$

and the moment estimate is given by

$$\hat{\gamma}_{\text{MME}} = \frac{\hat{m}_3 - 3\hat{m}_2 \hat{m}_1 + 2\hat{m}_1^3}{(\hat{m}_2 - \hat{m}_1^2)^{3/2}} = \frac{\frac{1}{n}\sum_{i=1}^{n}(x_i - \overline{x})^3}{(\frac{1}{n}\sum_{i=1}^{n}(x_i - \overline{x})^2)^{3/2}}.$$ □

Example 4.4 (Pendulum) Using the data given in Example 2.3 on page 6 we get

$$\hat{m}_1 = 1.998, \quad \hat{m}_2 = 3.99216 \quad \text{and} \quad \hat{m}_3 = 7.976959.$$

Using these values we obtain $\hat{\gamma}_{\mathrm{MME}}(\mathbf{x}) \approx 0.0739$. The value of $\hat{\gamma}_{\mathrm{MME}}$ is a good indicator for our assumption that the underlying distribution is a normal distribution, because the normal distribution is symmetric with $\gamma = 0$.   □

Special case 4.4 (Correlation) The estimation of the correlation is a slight extension of the method of moments introduced above. Suppose that we have a sample of i.i.d. bivariate r.v.'s (X_i, Y_i), $i = 1, \ldots, n$ which are copies of (X, Y) with joint distribution function F. Furthermore, we assume that the variances are positive and finite. The parameter of interest is

$$\rho(F) = \frac{\mathrm{Cov}_F(X, Y)}{\sqrt{\mathrm{Var}_F X}\sqrt{\mathrm{Var}_F Y}}$$

and it can be written in the form

$$\rho(F) = \frac{\mathsf{E}_F((X - \mathsf{E}_F X)(Y - \mathsf{E}_F Y))}{\sqrt{\mathsf{E}_F(X - \mathsf{E}_F X)^2}\sqrt{\mathsf{E}_F(Y - \mathsf{E}_F Y)^2}}.$$

Its empirical version is the well-known Pearson correlation coefficient r_{xy}. It is given by

$$r_{xy} = \hat{\rho}_{\mathrm{MME}} = \frac{\sum_{i=1}^{n}(x_i - \bar{x})(y_i - \bar{y})}{\sqrt{\sum_{i=1}^{n}(x_i - \bar{x})^2}\sqrt{\sum_{i=1}^{n}(y_i - \bar{y})^2}}.$$

                                                                                                □

Example 4.5 (Friesian cows) Consider example 2.7 on page 7. The computation of the empirical correlation leads to the value $\hat{\rho}_{\mathrm{MME}} = 0.9492$. Both the scatterplot in Figure 4.2 and $\hat{\rho}_{\mathrm{MME}}$ show a strong positive relationship between milk protein and milk production.   □

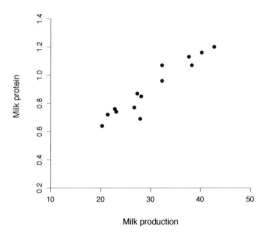

Figure 4.2: Scatterplot of milk protein and milk production of 12 cows.

Until now we have considered nonparametric models. Here are some examples of the application of the method of moments to the estimation in parametric models. Method of moments estimates can be used as initial values in numerical procedures for finding maximum likelihood estimates.

Special case 4.5 (Log normal) Let $X_1, \ldots, X_n$ be i.i.d. according to the log normal distribution with parameter $(\mu, \sigma^2) \in \mathbb{R}_+ \times \mathbb{R}_+$, i.e., X_i has the density

$$f(t; \mu, \sigma^2) = \frac{1}{\sigma t \sqrt{2\pi}} \exp\left(-\frac{(\ln t - \mu)^2}{2\sigma^2}\right) \mathbb{1}_{(0,\infty)}(t).$$

Since

$$m_1 = \exp(\mu + \sigma^2/2) \qquad \text{and} \qquad m_2 = \left(\exp(\mu + \sigma^2/2)\right)^2 \exp(\sigma^2)$$

we obtain

$$\hat{\mu}_{\text{MME}} = 2 \ln \hat{m}_1 - \frac{1}{2} \ln \hat{m}_2 \quad \text{and} \quad \hat{\sigma}^2_{\text{MME}} = \ln \hat{m}_2 - 2 \ln \hat{m}_1.$$

$\square$

The construction of estimators by the method of moments is not unique, as can been seen in the following case.

Special case 4.6 (Poisson distribution) Suppose that the observations are realizations of i.i.d. Poisson r.v.'s X_i, $i = 1, \ldots, n$, with parameter $\lambda > 0$. The parameter λ is the expectation of X_i; therefore, using the method of moments we choose as an estimate the sample mean, i.e., $\hat{\lambda}_{\mathrm{MME}} = \frac{1}{n} \sum_{i=1}^{n} x_i$. On the other hand, it is well known that the variance of X_i is also λ. Hence, another candidate for estimating this parameter is

$$\tilde{\lambda}_{\mathrm{MME}} = \hat{m}_2 - \hat{m}_1^2 = \frac{1}{n} \sum_{i=1}^{n} (x_i - \bar{x})^2.$$

In Special case 4.23 on page 110 we will see which choice is better.   □

Exercise 4.1 Derive the method of moment estimator for the natural parameter of a binomial distribution (log odds). Compute the corresponding estimate for the data of Example 2.1 (Ball point pens).

4.1.2 Maximum Likelihood Estimators

Maximum likelihood estimation is the most widely used and the most important method of estimation. It has an intuitive motivation. Moreover, maximum likelihood estimators usually have fairly good properties.

We assume a parametric statistical model, i.e., the distribution of the sample $\mathbf{X} = (X_1, \ldots, X_n)$ of independent r.v.'s belongs to

$$\mathcal{P} = \{\mathsf{P}_\theta = \mathsf{P}_{1,\theta} \otimes \cdots \otimes \mathsf{P}_{n,\theta} : \quad \theta \in \Theta \subseteq \mathbb{R}^k\}$$

for some k. Let us recall the likelihood function introduced in Section 3.1. Suppose that $\mathbf{X}$ has the probability function $p(\cdot; \theta)$, i.e.,

$$p(x_1, \ldots, x_n; \theta) = \prod_{i=1}^{n} \mathsf{P}_{i,\theta}(x_i) \quad \text{in the discrete case,}$$

or

$$p(x_1, \ldots, x_n; \theta) = \prod_{i=1}^{n} f_i(x_i; \theta) \quad \text{in the continuous case.}$$

For a fixed observation $\mathbf{x}$ the likelihood function $L(\cdot; \mathbf{x}) : \Theta \to [0, \infty)$ and the log-likelihood function are defined as functions of the parameter θ:

$$L(\theta; \mathbf{x}) = p(\mathbf{x}; \theta) \qquad l(\theta; \mathbf{x}) = \ln L(\theta; \mathbf{x}).$$

In Section 3.1 the maximum likelihood principle was formulated. It was suggested that the plausibility of a model should depend on its likelihood. We now define parameter estimates as those values that maximize the likelihood function. Thus a maximum likelihood estimate is a parameter value for which the data, which are actually obtained, have the highest probability.

Definition 4.3 (Maximum Likelihood Estimator) An estimator T is called maximum likelihood estimator **(MLE)** of θ, if

$$L(T(\mathbf{x}); \mathbf{x}) = \max_{\theta \in \Theta} L(\theta; \mathbf{x})$$

for all $\mathbf{x} \in \mathcal{X}$.

The MLE of θ is denoted by $\hat{\theta}_{\text{MLE}}$. For the estimation of a parameter γ the following result holds:

Theorem 4.1 *If* $\gamma = g(\theta)$ *and* g *is bijective, i.e.,* $\theta = g^{-1}(\gamma)$, *then* $\hat{\theta}$ *is a MLE for* θ *iff* $\hat{\gamma} = g(\hat{\theta})$ *is a MLE for* γ.

PROOF: The likelihood function is $L(\theta; \mathbf{x}) = p(\mathbf{x}; \theta)$. Since g is bijective the likelihood function with respect to γ is given by $\tilde{L}(\gamma; \mathbf{x}) = p(\mathbf{x}; g^{-1}(\gamma))$. Furthermore, we have

$$\tilde{L}(\hat{\gamma}; \mathbf{x}) \geq \tilde{L}(\gamma; \mathbf{x}) \quad \text{for all } \gamma$$
$$\Leftrightarrow p(\mathbf{x}; g^{-1}(\hat{\gamma})) \geq p(\mathbf{x}; g^{-1}(\gamma)) \quad \text{for all } \gamma$$
$$\Leftrightarrow p(\mathbf{x}; \hat{\theta}) \geq p(\mathbf{x}; \theta) \quad \text{for all } \theta$$
$$\Leftrightarrow L(\hat{\theta}; \mathbf{x}) \geq L(\theta; \mathbf{x}) \quad \text{for all } \theta.$$

$\square$

In cases where g is not bijective we will define:

Definition 4.4 (MLE for $g(\theta)$) The MLE of a parameter $\gamma = g(\theta)$ is defined by $\hat{\gamma}_{\text{MLE}} = g(\hat{\theta}_{\text{MLE}})$ where $\hat{\theta}_{\text{MLE}}$ is the MLE of θ.

Special case 4.7 (Binomial distribution) Let $x \in \mathcal{X} = \{0, \ldots, n\}$ be a realization of a $\text{Bin}(n, \theta)$-distributed r.v. X. The parameter set is $\Theta = (0, 1)$. The likelihood function and the log-likelihood function are given by

$$L(\theta; x) = \binom{n}{x} \theta^x (1 - \theta)^{n-x}$$

and

$$l(\theta; x) = \ln \binom{n}{x} + x \ln \theta + (n - x) \ln(1 - \theta).$$

Since $l(\cdot; x)$ is differentiable, $\hat{\theta}_{\mathrm{MLE}}$ is a solution of the equation

$$l'(\theta; x) = \frac{x}{\theta} - \frac{n - x}{1 - \theta} = 0.$$

If $x \neq 0$ and $x \neq n$ the solution exists, and we obtain $\widehat{\theta}_{\mathrm{MLE}}(x) = \frac{x}{n}$. Note that $l''(\hat{\theta}_{\mathrm{MLE}}; x) < 0$. $\qquad\square$

Example 4.6 (Bernoulli distribution) In a series of 10 trials $x = 3$ successes were observed. So the maximum likelihood estimate for the probability of success is $3/10$.

Suppose now that we have the result $\mathbf{x} = (1, 1, 1, 1, 1, 1, 1, 1, 1, 1)$, that is the binomial variable which counts the successes has the value $x = 10$. In the regular binomial model $\mathrm{Bin}(n, \theta)$ with $\theta \in (0, 1)$ the MLE does not exist. If we allow $\theta \in [0, 1]$, then obviously $\max\limits_{\theta \in [0,1]} L(\theta; x) = \theta^{10}$ is taken at $\hat{\theta}_{\mathrm{MLE}} = 1$. $\qquad\square$

Special case 4.8 (Normal distribution) Let $X_1, \ldots, X_n$ be independent $N(\mu, \sigma^2)$-distributed r.v.'s. The parameter of interest is $\theta = (\mu, \sigma^2)$ and the likelihood function is given by

$$L(\mu, \sigma^2; \mathbf{x}) \propto \frac{1}{(\sigma^2)^{n/2}} \exp\left(-\sum_{i=1}^{n} \frac{(x_i - \mu)^2}{2\sigma^2}\right).$$

Taking logarithm we get

$$l(\mu, \sigma^2; \mathbf{x}) = \ln L(\mu, \sigma^2; \mathbf{x}) = -\frac{n}{2} \ln(\sigma^2) - \sum_{i=1}^{n} \frac{(x_i - \mu)^2}{2\sigma^2} + \mathrm{const}.$$

Differentiating with respect to μ and σ^2 leads to the likelihood equations

$$\sum_{i=1}^{n} (x_i - \mu) = 0$$

$$-\frac{n}{2} \frac{1}{\sigma^2} + \sum_{i=1}^{n} \frac{(x_i - \mu)^2}{2\sigma^4} = 0.$$

The solution of this system of two equations is given by

$$\hat{\mu}_{\text{MLE}}(\mathbf{x}) = \frac{1}{n}\sum_{i=1}^{n} x_i = \overline{x} \quad \text{and} \quad \hat{\sigma}^2_{\text{MLE}}(\mathbf{x}) = \frac{1}{n}\sum_{i=1}^{n}(x_i - \overline{x})^2.$$

Note that in this case the MLE coincide with the method of moment estimators. □

Example 4.7 (Pendulum) Assume normal distribution for the times considered in Example 4.2. Computation of the MLE for the given data yields:

$$\hat{\mu}_{\text{MLE}}(\mathbf{x}) = 1.998, \quad \hat{\sigma}^2_{\text{MLE}}(\mathbf{x}) = 3.99216 - (1.998)^2 = 1.56 \times 10^{-4}.$$

□

In the following examples the likelihood function is not differentiable. Thus, its maximum has to be determined by other methods.

Special case 4.9 (Uniform distribution) Let $X_1, \ldots, X_n$ be a sample of i.i.d. r.v.'s according to the uniform distribution $U[0, \theta]$. That is

$$f(x; \theta) = \left\{ \begin{array}{ll} \frac{1}{\theta} & \text{for} \quad 0 \le x \le \theta \\ 0 & \text{else} \end{array} \right. = \frac{1}{\theta} \mathbb{1}_{[0,\theta]}(x).$$

Thus

$$L(\theta; \mathbf{x}) = \prod_{i=1}^{n} f(x_i; \theta) = \frac{1}{\theta^n} \prod_{i=1}^{n} \mathbb{1}_{[x_i,\infty)}(\theta).$$

Note

$$\prod_{i=1}^{n} \mathbb{1}_{[x_i,\infty)}(\theta) = 1 \quad \Leftrightarrow \quad \mathbb{1}_{[x_i,\infty)}(\theta) = 1 \ \text{ for all } i \quad \Leftrightarrow \quad x_i \le \theta \ \text{ for all } i$$

$$\Leftrightarrow \quad \max_{i} x_i \le \theta \quad \Leftrightarrow \quad \mathbb{1}_{[\max_{i} x_i,\infty)}(\theta) = 1.$$

Thus $L(\theta; \mathbf{x}) = \frac{1}{\theta^n} \mathbb{1}_{[\max_{i} x_i,\infty)}(\theta)$. The function $\frac{1}{\theta^n}$ is monotone decreasing, hence

$$L(\theta; \mathbf{x}) \le \frac{1}{\left(\max_{i} x_i\right)^n}$$

and the maximum likelihood estimator is $\hat{\theta}_{\text{MLE}} = \max_{i} X_i$. □

Exercise 4.2 (Shifted Gamma distribution) Let $X_1, \ldots, X_n$ be i.i.d. r.v.'s according to a $\mathsf{Gamma}(p, a, A)$-distribution with density

$$\frac{1}{\Gamma(p)} x^{p-1} \frac{1}{a^{p-1}} \exp\left(-\frac{x}{a}\right) \mathbb{1}_{[A, \infty]}.$$

The parameters p and a are known. Determine the maximum likelihood estimator for $\theta = A$.

Example 4.8 (Nonuniqueness of the MLE) Let X_1 and X_2 be independent r.v.'s with a Cauchy distribution. The likelihood function is

$$L(\theta; \mathbf{x}) = \prod_{i=1}^{2} \frac{1}{\pi(1 + (x_i - \theta)^2)}, \qquad \theta \in \mathbb{R}. \qquad (4.5)$$

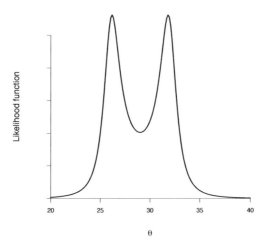

Figure 4.3: Likelihood function of data from a Cauchy distribution.

Figure 4.3 shows the likelihood function (4.5) for $\mathbf{x} = (26, 32)$. We see that there is no unique maximum.   □

Here is a classical example of nonexistence of a maximum likelihood estimator:

Example 4.9 (Nonexistence of MLE) Let $X_1, \ldots, X_n$ be i.i. $\text{Poi}(\lambda)$-distributed. The observations are the r.v.'s

$$Y_i = \begin{cases} 0 & X_i = 0 \\ 1 & X_i > 0 \end{cases}.$$

The problem is to estimate the parameter λ on the basis of $Y_1, \ldots, Y_n$. Since

$$\mathsf{P}_\lambda(Y_i = 0) = \exp(-\lambda) \quad \text{and} \quad \mathsf{P}_\lambda(Y_i = 1) = 1 - \exp(-\lambda),$$

we obtain

$$L(\lambda; \mathbf{y}) = (1 - \exp(-\lambda))^{\sum\limits_{i=1}^{n} y_i} \exp(-\lambda(n - \sum_{i=1}^{n} y_i)),$$

$$l(\lambda; \mathbf{y}) = \sum_{i=1}^{n} y_i \ln(1 - \exp(-\lambda)) - \lambda(n - \sum_{i=1}^{n} y_i). \tag{4.6}$$

Suppose $\sum\limits_{i=1}^{n} y_i \neq n$, then the likelihood equation $l'(\lambda; \mathbf{y}) = 0$ implies

$$\frac{\sum_{i=1}^{n} y_i}{1 - \exp(-\lambda)} \cdot \exp(-\lambda) = (n - \sum_{i=1}^{n} y_i),$$

and we obtain the MLE

$$\hat{\lambda}(\mathbf{y}) = -\ln(1 - \bar{y}).$$

For $\sum\limits_{i=1}^{n} y_i = n$ the log-likelihood function (4.6) is $n \ln(1 - \exp(-\lambda))$, which is monotone increasing in λ. Thus there does not exist a maximum. The probability that this case occurs is

$$\mathsf{P}_\lambda\left(\sum_{i=1}^{n} Y_i = n\right) = \mathsf{P}_\lambda(Y_1 = 1, \ldots, Y_n = 1) = (1 - \exp(-\lambda))^n \to 1$$

for $\lambda \to \infty$. This means that there are values of the parameter λ for which the probability that a MLE does not exist is near 1. On the other hand, for a fixed λ and $n \to \infty$ we have

$$\mathsf{P}_\lambda\left(\sum_{i=1}^{n} Y_i = n\right) = \mathsf{P}_\lambda(Y_1 = 1, \ldots, Y_n = 1) = (1 - \exp(-\lambda))^n \to 0.$$

$\square$

Example 4.10 (Taxi) A town has N taxis, numbered from 1 up to N. A passenger observes the numbers of n taxis. The problem is to estimate the unknown parameter $\theta = N$. Our model can be formulated as follows: We have a sample of i.i.d. r.v.'s X_i taking values in $\{1, \ldots, N\}$ with probability $1/N$. We assume that all taxis are working. The likelihood function is given by

$$L(N; \mathbf{x}) = \begin{cases} \frac{1}{N^n} & \text{if } x_i \leq N \quad \text{for all } i \\ 0 & \text{otherwise} \end{cases}.$$

The function $L(\cdot; \mathbf{x})$ is monotone decreasing; it takes its maximum at the maximum of the observed numbers. Thus, the maximum likelihood estimator is $\hat{N}_{\mathrm{MLE}} = x_{\max}$. In Figure 4.4 we see the function $L(\cdot; \mathbf{x})$ for $n = 10$ and $x_{\max} = 35$. □

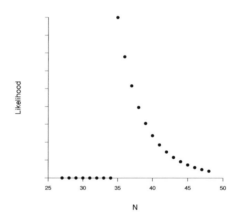

Figure 4.4: Likelihood function for the taxi example.

Special case 4.10 (Double exponential distribution) Suppose that X_i, $i = 1, \ldots, n$ are i.i.d. r.v.'s with density function

$$\frac{a}{2} \exp(-a \mid x - b \mid), \qquad a > 0, \, b \in \mathbb{R} \quad \text{and} \quad x \in \mathbb{R}.$$

The parameter a is assumed to be known and we are interested in an estimator of $\theta = b$. The likelihood and the log-likelihood function are given by

$$L(b; \mathbf{x}) \propto \exp\left(-a \sum_{i=1}^{n} \mid x_i - b \mid\right) \quad \text{and} \quad l(b; \mathbf{x}) = -a \sum_{i=1}^{n} \mid x_i - b \mid + \text{const.}$$

Thus to find $\hat{b}_{\mathrm{MLE}}(\mathbf{x})$ we have to minimize the expression

$$\sum_{i=1}^{n} |\, x_i - b \,| \,.$$

It is well known that (see Example 4.12) the median $b = x_{\mathrm{med}}$ is the solution of this minimization problem. For even sample size n the median is not uniquely determined. Usually one defines

$$x_{\mathrm{med}} = \begin{cases} x_{[\frac{n+1}{2}]} & \text{for} \quad n \quad \text{odd} \\ \frac{1}{2}\left(x_{[\frac{n}{2}]} + x_{[\frac{n}{2}]+1}\right) & \text{for} \quad n \quad \text{even} \end{cases} \,,$$

where $x_{[1]} \leq \cdots \leq x_{[n]}$ are the ordered observations. $\square$

In the following example we differentiate the log-likelihood function, but the resulting system of equations has no explicit solution.

Special case 4.11 (Weibull distribution) In survival analysis the data $x_1, \ldots, x_n$ are often assumed to be realizations of Weibull distributed r.v.'s, i.e., their density is

$$f(x; \theta) = \begin{cases} \dfrac{\alpha}{\beta}\left(\dfrac{x}{\beta}\right)^{\alpha-1} \exp\left(-\left(\dfrac{x}{\beta}\right)^{\alpha}\right) & \text{if } x > 0 \\ 0 & \text{otherwise} \end{cases} \,.$$

The parameter is $\theta = (\alpha, \beta) \in \mathbb{R}_+ \times \mathbb{R}_+$. The log likelihood function is given by

$$l(\alpha, \beta; \mathbf{x}) = n(\ln \alpha - \ln \beta) + (\alpha - 1) \sum_{i=1}^{n} \ln \frac{x_i}{\beta} - \sum_{i=1}^{n} \left(\frac{x_i}{\beta}\right)^{\alpha}.$$

Differentiation leads to the system of equations

$$-\frac{n\alpha}{\beta} + \frac{\alpha}{\beta} \sum_{i=1}^{n} \left(\frac{x_i}{\beta}\right)^{\alpha} = 0$$

$$\frac{n}{\alpha} + \sum_{i=1}^{n} \ln \frac{x_i}{\beta} - \sum_{i=1}^{n} \left(\frac{x_i}{\beta}\right)^{\alpha} \ln \left(\frac{x_i}{\beta}\right) = 0.$$

For the computation of the solution of this nonlinear system of equations iterative procedures are needed; for example, one can apply the Newton–Raphson algorithm. $\square$

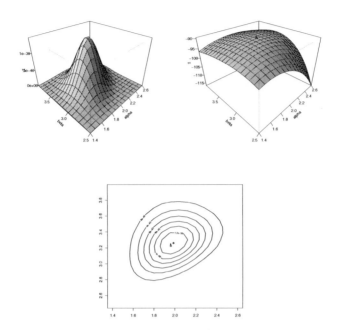

Figure 4.5: Likelihood, log likelihood function and contour plot for the Weibull distribution. Plots are based on simulated data for $n = 50$, $\alpha = 2$ and $\beta = 3$.

Summary: On existence and computation of the MLE

1. In many standard cases the likelihood function L is differentiable. In these cases we can maximize L by differentiating and equating the (partial) derivatives with zero. If L achieves its maximum at an interior point of Θ, then we have to find a solution $\widehat{\theta}_{\text{MLE}}$ of the system of likelihood equations

$$\frac{\partial L(\theta; \mathbf{x})}{\partial \theta_j} = 0, \qquad j = 1, \ldots, k. \tag{4.7}$$

Since the logarithm function is increasing, maximizing the log-likelihood function $l(\cdot; \mathbf{x}) = \ln L(\cdot; \mathbf{x})$ is equivalent to maximize $L(\cdot; \mathbf{x})$, and $l(\cdot; \mathbf{x})$ is often easier to deal with when differentiating.

2. Before we solve (4.7) we have to answer the questions:
   - Do the likelihood equations have a solution?
   - If they have a solution, is this solution unique?

Of course, it is easy to answer these questions when the likelihood equation can be solved explicitly, as in Special cases 4.7 and 4.8. However, in more complicated cases, where the solution can be computed only by numerical methods, more advanced methods are necessary. An example of such a case is the Weibull distribution (see Special case 4.11).

3. Moreover, one has to show that the solution of (4.7) is a local maximum. If the matrix of the second derivatives (the Hessian matrix) is negative definite at $\hat{\theta}$, then $\hat{\theta}$ is the maximum likelihood estimate.

4. **OBS!** Be careful! If the likelihood function is not differentiable, you must find another way to determine the MLE. Compare Special case 4.9 and Example 4.10.

4.1.3 M-Estimators

In this section we briefly consider the so-called M-estimators. This type of estimator was introduced by Huber (1964) to reduce the effect of outliers in the estimation of a parameter or, in other words, to construct **robust estimators**. M-estimators are solutions of equations corresponding to a minimization problem. For simplicity let us start with the consideration of one-dimensional parameters.

Recall that the arithmetic mean $\overline{X}$ of a sample of i.i.d. r.v.'s $X_1, \ldots, X_n$ minimizes the sum of squared differences

$$\overline{x} = \arg\min_{\theta} \sum_{i=1}^{n} (x_i - \theta)^2$$

and $\overline{x}$ is the solution of

$$\sum_{i=1}^{n} (x_i - \theta) = 0. \tag{4.8}$$

The median, an estimator which is robust against outliers, minimizes the sum of absolute deviations:

$$x_{\text{med}} = \arg\min_{\theta} \sum_{i=1}^{n} |x_i - \theta|$$

and x_{med} is the solution of

$$\sum_{i=1}^{n} (-1) \, \mathbb{1}_{(-\infty,0)}(x_i - \theta) + \mathbb{1}_{(0,\infty)}(x_i - \theta) = 0.$$

The MLE in a parametric model is the minimizer of the negative log-likelihood function

$$\hat{\theta}_{\text{MLE}} = \arg\min_{\theta} \left(-\sum_{i=1}^{n} \ln L(\theta; x_i) \right)$$

and is (in case that $L(\cdot; \mathbf{x})$ is differentiable) the solution of

$$\sum_{i=1}^{n} \frac{\partial \ln L(\theta; x_i)}{\partial \theta} = 0.$$

In general we can define:

Definition 4.5 (M-estimator) For a sample $X_1, \ldots, X_n$ of i.i.d. r.v.'s the M-estimator $\hat{\gamma}$ with respect to a function $\psi : \mathbb{R} \times \Gamma \to \mathbb{R}$ is defined as the solution of

$$\sum_{i=1}^{n} \psi(X_i, \hat{\gamma}) = 0. \tag{4.9}$$

In typical cases the equation (4.9) corresponds to a minimization problem

$$\hat{\gamma} = \arg\min_{\gamma} \sum_{i=1}^{n} \varrho(X_i, \gamma).$$

If ϱ is differentiable w.r.t. γ, then $\psi(x, \gamma) = c\frac{\partial \varrho(x, \gamma)}{\partial \gamma}$ for some constant c.
Let us consider some examples for the estimation of location parameters. Here we choose functions $\psi(x, \gamma) = \tilde{\psi}(x - \gamma)$ and $\varrho(x, \gamma) = \tilde{\varrho}(x - \gamma)$, respectively. The functions $\tilde{\varrho}$ are shown in Figure 4.6.

Example 4.11 (Least squares estimator) The functions $\tilde{\varrho}(z) = z^2$ and $\tilde{\psi}(z) = z$ lead to the sample mean considered in (4.8).   □

Example 4.12 (Least absolute value estimator) The median is an M-estimate with $\tilde{\varrho}(z) = |z|$ and

$$\tilde{\psi}(z) = \begin{cases} -1 & \text{if} \quad z < 0 \\ 0 & \text{if} \quad z = 0 \\ 1 & \text{if} \quad z > 0 \end{cases}.$$

Consider the function

$$g(\theta) = \sum_{i=1}^{n} \tilde{\varrho}(x_i - \theta) = \sum_{i=1}^{n} |x_i - \theta| = \sum_{i=1}^{n} |x_{[i]} - \theta|.$$

For $\theta \in (x_{[j]}, x_{[j+1]})$ we have

$$\begin{aligned} g(\theta) &= \sum_{i=1}^{j} (\theta - x_{[i]}) + \sum_{i=j+1}^{n} (x_{[i]} - \theta) \\ &= (2j - n)\theta - \sum_{i=1}^{j} x_{[i]} + \sum_{i=j+1}^{n} x_{[i]}. \end{aligned}$$

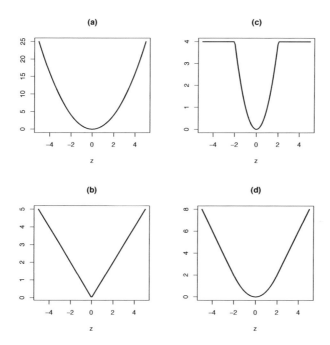

Figure 4.6: Functions defining (a) least squares estimators, (b) least absolute value estimators, (c) a trimmed mean, (d) a Winsorized mean.

Thus, g is piecewise linear on the intervals $(x_{[j]}, x_{[j+1]})$: It is monotone decreasing for $j < \frac{n}{2}$ and monotone increasing for $j > \frac{n}{2}$. For $\theta < x_{[1]}$ g is monotone decreasing, and for $\theta > x_{[n]}$ monotone increasing. Since g is continuous, we have for even n that g is monotone decreasing in $(-\infty, x_{[\frac{n}{2}]})$, constant in $(x_{[\frac{n}{2}]}, x_{[\frac{n}{2}]+1})$ and monotone increasing in $(x_{[\frac{n}{2}]+1}, \infty)$. Thus it has its minimum on each point of the interval $(x_{[\frac{n}{2}]}, x_{[\frac{n}{2}]+1})$. For odd n g is monotone decreasing in $(-\infty, x_{[\frac{n+1}{2}]})$ and monotone increasing in $(x_{[\frac{n+1}{2}]+1}, \infty)$.

Hence for odd n there is a unique minimum at $\hat{\theta} = x_{[\frac{n+1}{2}]}$; for even n every point in the interval $(x_{[\frac{n}{2}]}, x_{[\frac{n}{2}]+1})$ is a minimizer. Thus $\hat{\theta} = x_{\text{med}}$. The function g is shown in Figure 4.7 in Example 4.15.   □

Example 4.13 (A trimmed mean) One of Huber's proposals to eliminate outliers is to use the following function

$$\tilde{\varrho}(z) = \begin{cases} z^2 & \text{if } |z| \le k \\ k^2 & \text{if } |z| > k \end{cases}$$

where k is suitably chosen. The corresponding function $\tilde{\psi}$ is

$$\tilde{\psi}(z) = \begin{cases} z & \text{if} \quad |z| \le k \\ 0 & \text{if} \quad |z| > k \end{cases}.$$

□

Example 4.14 (A Winsorized mean) A further proposal of Huber represents a compromise between the mean and the median. The resulting estimator is a type of Winsorized mean. [1] The following function is used

$$\tilde{\varrho}_{\text{Hub}}(z) = \begin{cases} \frac{1}{2}z^2 & \text{if} \quad |z| \le k \\ k|z| - \frac{1}{2}k^2 & \text{if} \quad |z| > k \end{cases}.$$

The estimate is a solution of $\sum_{i=1}^{n} \tilde{\psi}(x_i - \theta) = 0$:

$$\tilde{\psi}(z) = \begin{cases} -k & \text{if} \quad z \quad < -k \\ z & \text{if} \quad |z| \quad \le k \\ k & \text{if} \quad z \quad > k \end{cases}.$$

This function is proportional to z^2 for $|z| \le k$, but outside this interval it continues at straight lines instead of parabolic arcs. As k becomes larger, $\tilde{\varrho}$ will agree with $z^2/2$ over most of its range, so that the estimator comes closer to the mean. As k becomes smaller, the estimator will come closer to the median. □

Example 4.15 Suppose we have observed the following data:

$$1.4, \ 1.7, \ 2.0, \ 1.5, \ 1.2, \ 4.0, \ 0.9, \ 0.0, \ 1.3.$$

The arithmetic mean of this sample is $\bar{x} = 1.556$. It is influenced by the rather large observation 4.0 which can be considered as an outlier.
The median is $x_{\text{med}} = x_{[5]} = 1.4$. Figure 4.7 shows the observations and the function g defined by (4.12).
The Huber estimate defined in Example 4.13 with $k = 1$ is $\tilde{\theta} = 1.42857$. Note that this is the arithmetic mean of the observations $x_{[i]}$, $i = 2, \ldots, 8$. For all these observations we have $|x_{[i]} - \tilde{\theta}| \le k$. Consider the arithmetic mean of the observations $x_{[i]}$ with $i = 1, \ldots, 8$ and with $i = 2, \ldots, 9$, respectively. Denote them by $\tilde{\theta}_{1.8}$ and $\tilde{\theta}_{2.9}$. We have $\tilde{\theta}_{1.8} = 1.25$ and $|\tilde{\theta}_{1.8} - x_{[1]}| > k$, and with $\tilde{\theta}_{2.9} = 1.75$ we obtain $|\tilde{\theta}_{2.9} - x_{[9]}| > k$.

□

---

[1] Note that there are also other types of trimmed and Winsorized means that belong to the class of L-estimators.

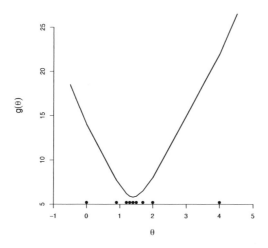

Figure 4.7: Sum of absolute values for the data example 4.15. The points characterize the observations.

This approach can be generalized to the estimation of multidimensional parameters. In this case the functions ψ and ϱ are defined on $\mathbb{R} \times \Gamma$ with $\Gamma \subseteq \mathbb{R}^k$. Let us consider the problem of estimating the parameter in the linear regression model. We will see in Chapter 6 "Linear Model" that the least squares method is the most widely used method. Here one can also define an M-estimator:
Let $Y_1, \ldots, Y_n$ be a sample of independent random variables with

$$\mathsf{E}_\theta Y_i = \sum_{j=1}^{p} x_{ij}\beta_j = x_i^{\mathsf{T}}\beta, \qquad \beta = (\beta_1, \ldots, \beta_k)^{\mathsf{T}}$$

where x_{ij} are certain input variables and the β_j's are unknown parameters. Suppose that $\mathsf{Var}_\theta Y_i = \sigma^2$ for all i (see Chapter 6). The least squares estimator for β is the minimizer of

$$\sum_{i=1}^{n}(Y_i - x_i^{\mathsf{T}}\beta)^2.$$

An estimator which is less sensitive to outliers is defined by

$$\hat{\beta} = \arg\min_{\beta} \sum_{i=1}^{n} \tilde{\varrho}_{\text{Hub}}\left(\frac{Y_i - x_i^{\mathsf{T}}\beta}{\sigma}\right)$$

where $\tilde{\varrho}_{\text{Hub}}$ is defined as in Example 4.14.

4.2   Unbiasedness and Mean Squared Error

In general there is more than one estimator for a parameter of interest. So one needs criteria for the comparison of different estimators. The starting point of the considerations is the fact that one has to compare estimators rather than estimates, i.e., the rules which say how to derive the estimates from the data rather than the estimated values. The difference $T(\mathbf{x}) - \gamma$ is the error when we use $T(\mathbf{x})$ to estimate γ. The error varies from sample to sample because so does $\mathbf{X}$. The distribution of $T(\mathbf{X})$ tells us where and with which probability the estimates lie, when we use the estimator T. The more this distribution is concentrated around γ, the better is the estimator. The most widely used concept to describe this approach is that of the **M**ean **S**quared **E**rror. In the following we assume that the parameter γ is real-valued, i.e., $\Gamma \subseteq \mathbb{R}$.

Definition 4.6 (MSE) Let $\mathcal{P} = \{\mathsf{P}_\theta : \theta \in \Theta\}$ be a statistical model for a random variable $\mathbf{X}$ on $\mathcal{X}$, let $g \colon \Theta \to \Gamma$ be a function and $T \colon \mathcal{X} \to \Gamma$ an estimator for $\gamma = g(\theta)$. The mean squared error (MSE) of T is given by

$$\mathsf{MSE}(T, \theta) = \mathsf{E}_\theta (T - g(\theta))^2.$$

The MSE is the expected squared distance of the estimator from the parameter to be estimated — in other words, estimators with a small MSE lead "in average" to values which are near to γ, where the "average " is taken over all possible samples distributed according to P_θ. The MSE can be decomposed:

$$\mathsf{MSE}(T, \theta) = \mathsf{Bias}^2(T, \theta) + \mathsf{Var}_\theta T, \tag{4.10}$$

where

$$\mathsf{Bias}(T, \theta) = \mathsf{E}_\theta T - g(\theta)$$

is the **bias** of T at θ.

Exercise 4.3 Verify the decomposition (4.10).

Definition 4.7 (Unbiasedness) An estimator T for $\gamma = g(\theta)$ is called **unbiased** if

$$\mathsf{Bias}(T, \theta) = 0 \qquad \text{for all} \quad \theta \in \Theta.$$

A large bias means that the long-run average value of T lies far from $g(\theta)$, and this is undesirable. The choice of an estimator is very often restricted to the class of unbiased estimators. But there are cases where a small bias is accepted, in particular if the bias converges to zero when the sample size tends to infinity. Moreover, there are cases where no unbiased estimator exists.

Example 4.16 (Nonexistence of an unbiased estimator) Let X be a r.v. with distribution $\text{Bin}(n, \theta)$. The parameter $\gamma = 1/\theta$ cannot be estimated without bias. To prove this, note that an unbiased estimator T for γ has to satisfy

$$\mathsf{E}_\theta T = \sum_{k=0}^{n} T(k) \binom{n}{k} \theta^k (1 - \theta)^{n-k} = \frac{1}{\theta}$$

for all $\theta \in (0, 1)$. Consider the l.h.s. of the last equation for $\theta \to 0$:

$$T(0)(1 - \theta)^n + \sum_{k=1}^{n} T(k) \binom{n}{k} \theta^k (1 - \theta)^{n-k} \to T(0),$$

but the r.h.s. converges to ∞. This means that an estimator satisfying the unbiasedness condition does not exist. □

Special case 4.12 (The sample variance) Let $X_1, \ldots, X_n$ be i.i.d. r.v.'s with finite variance. Suppose that the parameter of interest is $\gamma = \sigma^2 = \text{Var}_\theta X_1$. The moment estimator is given by

$$\hat{\sigma}^2_{\text{MME}} = \frac{1}{n} \sum_{i=1}^{n} (X_i - \overline{X})^2.$$

This estimator is not unbiased: Since

$$\mathsf{E}_\theta \overline{X}^2 = \frac{1}{n^2} \sum_{i=1}^{n} \sum_{j=1}^{n} \mathsf{E}_\theta X_i X_j = \frac{1}{n^2} \sum_{i=1}^{n} \mathsf{E}_\theta X_i^2 + \frac{1}{n^2} \sum_{i=1}^{n} \sum_{\substack{j=1 \\ j \neq i}}^{n} \mathsf{E}_\theta X_i \mathsf{E}_\theta X_j$$

$$= \frac{1}{n} m_2 + \frac{1}{n}(n - 1) m_1^2,$$

we obtain

$$\mathsf{E}_\theta \frac{1}{n} \sum_{i=1}^{n} (X_i - \overline{X})^2 = \mathsf{E}_\theta \frac{1}{n} \sum_{i=1}^{n} X_i^2 - \mathsf{E}_\theta \overline{X}^2 = m_2 - \frac{1}{n} m_2 - \frac{1}{n}(n - 1) m_1^2.$$

Thus

$$\mathsf{E}_\theta \hat{\sigma}^2_{\text{MME}} = \frac{n - 1}{n} \sigma^2 \quad \text{and} \quad \text{Bias}(\hat{\sigma}^2, \theta) = -\frac{\sigma^2}{n}.$$

With the correction factor $\frac{n}{n-1}$ we obtain an unbiased estimator, namely the well-known empirical variance (or sample variance)

$$S^2 = \frac{1}{n - 1} \sum_{i=1}^{n} (X_i - \overline{X})^2.$$

Now, let us additionally assume that the X_i's are **normally distributed**. Then we know that

$$V = \sigma^{-2} \sum_{i=1}^{n} (X_i - \overline{X})^2$$

has a χ_{n-1}^2-distribution, and we obtain

$$\mathsf{E}_\theta V = (n-1) \quad \text{and} \quad \mathsf{Var}_\theta V = 2(n-1).$$

Using (4.10) we obtain for the method of moment estimator, which now is also the MLE, that

$$\mathsf{MSE}(\hat{\sigma}_{\mathrm{MLE}}^2, \theta) = \left(-\frac{\sigma^2}{n}\right)^2 + 2\frac{n-1}{n^2}\sigma^4 = \frac{2n-1}{n^2}\sigma^4.$$

The MSE of the sample variance is given by

$$\mathsf{MSE}(S^2, \theta) = 0 + \frac{2}{n-1}\sigma^4 = \frac{2}{n-1}\sigma^4.$$

Hence

$$\mathsf{MSE}(S^2, \theta) > \mathsf{MSE}(\hat{\sigma}_{\mathrm{MLE}}^2, \theta).$$

OBS! The empirical variance S^2 is unbiased, but in the case of normally distributed r.v.'s its MSE is greater than that of $\hat{\sigma}_{\mathrm{MLE}}^2$.   □

Special case 4.13 (Uniform distribution) Let $X_1, \ldots, X_n$ be i.i.d. according to the uniform distribution over the interval $[0, \theta]$. We have seen on page 81 that the MLE for θ is the maximum $\hat{\theta}_{\mathrm{MLE}} = X_{\max}$. Using standard methods one can show

$$\mathsf{E}_\theta X_{\max} = \frac{n}{n+1}\theta \quad \text{and} \quad \mathsf{Var}_\theta X_{\max} = \frac{n}{(n+1)^2(n+2)}\theta^2.$$

Hence

$$\mathsf{MSE}(X_{\max}, \theta) = \frac{2\theta^2}{(n+1)(n+2)}.$$

The MSE of the bias-corrected estimator

$$\widehat{\theta} = \frac{n+1}{n}X_{\max} = X_{\max} + \frac{1}{n}X_{\max}$$

is given by

$$\mathsf{MSE}(\widehat{\theta}, \theta) = \frac{\theta^2}{n(n+2)}$$

and

$$\mathsf{MSE}(\hat{\theta}_{\mathrm{MLE}}, \theta) > \mathsf{MSE}(\widehat{\theta}, \theta).$$

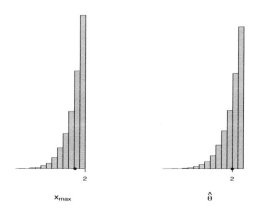

Figure 4.8: Histograms for the simulation study described in Example 4.17; the points indicate the mean of the simulated estimates.

OBS! The MLE for the parameter of the uniform distribution can be improved by the bias corrected estimator!   □

Example 4.17 (Simulation) In a simulation study $M = 10000$ samples of sizes $n = 10$ from $U[0, \theta]$ with $\theta = 2$ were generated. From each sample $x_{\max}$ and the corrected $\hat{\theta}$ were computed. Figure 4.8 shows the histograms of these estimates. The points indicate the mean of the generated estimates. We clearly see that $X_{\max}$ underestimates the true value $\theta = 2$. The mean of the bias-corrected estimates is 1.999 which is almost equal to the true underlying parameter. As estimates for the mean squared errors we obtain

$$\widehat{\text{MSE}}(X_{\max}, 2) = 0.061 \qquad \text{and} \qquad \widehat{\text{MSE}}(\hat{\theta}, 2) = 0.034.$$

                                                                                    □

Example 4.18 (Binomial distribution) Suppose $X \sim \text{Bin}(n, \theta)$. On page 79 it is shown that the MLE for θ is given by

$$\widehat{\theta}_{\text{MLE}} = \frac{X}{n}.$$

This estimator is unbiased and its variance (and MSE) is given by

$$\text{MSE}(\widehat{\theta}_{\text{MLE}}, \theta) = \text{Var}_\theta(\widehat{\theta}_{\text{MLE}}) = n^{-1}\theta(1 - \theta). \qquad (4.11)$$

The variance is maximal at $\theta = 1/2$. Consider now the estimator

$$\widetilde{\theta} = \frac{X+1}{n+2}.$$

The bias of this estimator is

$$\text{Bias}(\widetilde{\theta}, \theta) = \frac{n\theta + 1}{n+2} - \theta = \frac{1 - 2\theta}{n+2}$$

and the MSE is equal to

$$\text{MSE}(\widetilde{\theta}, \theta) = \frac{n\theta(1-\theta) + (1-2\theta)^2}{(n+2)^2}. \qquad (4.12)$$

The comparison of (4.11) and (4.12) shows that the estimator $\widetilde{\theta}$ has the smaller mean squared error for all parameters θ satisfying

$$\frac{(\theta - \frac{1}{2})^2}{\theta(1-\theta)} \leq 1 + \frac{1}{n},$$

and this is fulfilled for $|\theta - \frac{1}{2}| \leq 1/\sqrt{8} \approx 0.35$. (See Figure 4.9.) Thus, if one knows in advance that the unknown parameter θ takes values around $\frac{1}{2}$, it is better to estimate it by $\widetilde{\theta}$. $\qquad\qquad\qquad\qquad\qquad\qquad\qquad\qquad\qquad\qquad\qquad\qquad$ □

4.2.1   Best Unbiased Estimators

When searching for a good estimator with respect to the mean squared error we restrict ourselves to unbiased estimators. Then the best estimator is the one with smallest variance—in other words, the **Best Unbiased Estimator (BUE)**. Sometimes it is also called the **Minimum Variance Unbiased Estimator(MVUE)**.

Remark 4.3 From the decision-theoretic viewpoint the MSE is the risk with respect to the quadratic loss function. A loss function W is defined on $\Theta \times \Gamma$, and $W(\theta, \hat{\gamma}(\mathbf{x}))$ measures the loss from estimating $\gamma = g(\theta)$ by $\hat{\gamma}(\mathbf{x})$. Usually loss functions are convex functions as $W(\theta, \hat{\gamma}(\mathbf{x})) = |\hat{\gamma}(\mathbf{x}) - g(\theta)|^p$ for some $p > 0$. An estimator has to be good for all possible data $\mathbf{x}$. So $\mathbf{x}$ is considered as realization of a r.v. with P_θ , and we measure the usefulness of an estimator by "the mean over all samples," that is, we take the expectation:

$$R(\theta, \hat{\gamma}) = \mathsf{E}_\theta W(\theta, \hat{\gamma}(\mathbf{X})).$$

The function R is called the **risk**. In other words, a MVUE is a best unbiased estimator with respect to the quadratic loss.

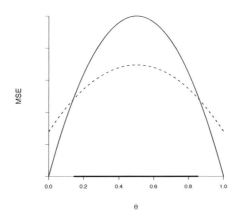

Figure 4.9: MSE as function of θ for estimators considered in Example 4.18; the dashed line is the MSE of the biased estimator, the solid line that of the unbiased estimator. The bold line indicates the region where the biased estimator has a smaller MSE.

Definition 4.8 (Best unbiased estimator) Given a statistical model $\mathcal{P} = \{\mathsf{P}_\theta : \theta \in \Theta\}$, an unbiased estimator T^* for a parameter $\gamma = g(\theta) \in \mathbb{R}$ is called **best unbiased estimator (BUE)**, if for any other unbiased estimator T

$$\mathsf{Var}_\theta T^* \leq \mathsf{Var}_\theta T \qquad \text{for all } \theta \in \Theta.$$

When looking for such best estimators it is natural to ask whether there is a lower bound to the variance of estimators for a parameter γ. An answer to this question is given by the Cramér–Rao inequality. If an estimator attains this bound it is the best estimator. In cases where the bound is not attained or where the assumptions of this approach are not satisfied the application of the Lehmann–Scheffé theorem is helpful.

4.2.2 Cramér–Rao Lower Bound

For simplicity we begin with the case $\Theta \subseteq \mathbb{R}$. To derive a lower bound for the variance we need the regularity conditions introduced in Chapter 3. Let $\mathcal{A} = \{\mathbf{x} : p(\mathbf{x}; \theta) > 0\}$ be the common support of the probability measures P_θ of the underlying family $\mathcal{P}$. Furthermore, we require the estimator T to be a **regular unbiased estimator**, that is

$$\int_{\mathcal{A}} T(\mathbf{x}) \frac{\partial}{\partial \theta} L(\theta; \mathbf{x}) \, d\mathbf{x} = \frac{\partial}{\partial \theta} \int_{\mathcal{A}} T(\mathbf{x}) L(\theta, \mathbf{x}) \, d\mathbf{x}.$$

Now, we can formulate the following theorem:

Theorem 4.2 (Cramér–Rao bound) *Suppose that the regularity conditions* **Reg 3** *and* **Reg 4** *are satisfied and that the Fisher information satisfies* $0 < I_{\mathbf{X}}(\theta) < \infty$. *Set* $\gamma = g(\theta)$, *where* g *is a continuously differentiable real-valued function with derivative* $g' \neq 0$. *If* T *is a regular unbiased estimator for* γ, *then*

$$\text{Var}_\theta T \geq \frac{[g'(\theta)]^2}{I_{\mathbf{X}}(\theta)} \qquad \text{for all} \quad \theta \in \Theta.$$

The equality holds if and only if for $\mathbf{x} \in \mathcal{A}$ *and for all* $\theta \in \Theta$

$$T(\mathbf{x}) - g(\theta) = \frac{g'(\theta)\, V(\theta; \mathbf{x})}{I_{\mathbf{X}}(\theta)} \tag{4.13}$$

where $V(\cdot; \mathbf{x})$ *is the score function.*

PROOF: We obtain

$$\text{Cov}_\theta(T(\mathbf{X}), V(\theta; \mathbf{X})) = \mathsf{E}_\theta(T(\mathbf{X})\, V(\theta; \mathbf{X}))$$

$$= \int_{\mathcal{A}} T(\mathbf{x})\, \frac{\partial \ln p(\mathbf{x}; \theta)}{\partial \theta}\, p(\mathbf{x}; \theta)\, d\mathbf{x} = \int_{\mathcal{A}} T(\mathbf{x}) \frac{\partial p(\mathbf{x}; \theta)}{\partial \theta}\, d\mathbf{x}.$$

Since T is assumed to be a regular estimator we can interchange integration and differentiation. Furthermore, we use the unbiasedness of T and obtain

$$\text{Cov}_\theta(T(\mathbf{X}), V(\theta; \mathbf{X})) = \frac{\partial}{\partial \theta} \int_{\mathcal{A}} T(\mathbf{x}) p(\mathbf{x}, \theta) d\mathbf{x} = \frac{\partial}{\partial \theta} \mathsf{E}_\theta T(\mathbf{X}) = g'(\theta).$$

Set now $c(\theta) = g'(\theta)/I_{\mathbf{X}}(\theta)$. Then we have

$$\begin{aligned}
0 \;\leq\; & \text{Var}_\theta(T - c(\theta)\, V(\theta; \mathbf{X})) \\
=\; & \text{Var}_\theta T + c^2(\theta) \text{Var}_\theta V(\theta; \mathbf{X}) - 2c(\theta) \text{Cov}_\theta(T(\mathbf{X}), V(\theta; \mathbf{X})) \\
=\; & \text{Var}_\theta T + c^2(\theta) I_{\mathbf{X}}(\theta) - 2c(\theta)\, g'(\theta) \\
=\; & \text{Var}_\theta T - [g'(\theta)]^2 / I_{\mathbf{X}}(\theta)
\end{aligned}$$

and the desired statement follows.

The equality holds if the r.v. $T(\mathbf{X}) - c(\theta) V(\theta; \mathbf{X})$ is constant (except on a set having P_θ- measure zero). Of course, this constant is the expectation $g(\theta)$, but since we have assumed that $L(\theta; \mathbf{x})$ is strictly positive for all $\mathbf{x} \in \mathcal{A}$, this means $T(\mathbf{x}) - c(\theta) V(\theta; \mathbf{x}) = g(\theta)$ or, equivalently, $T(\mathbf{x}) - g(\theta) = V(\theta; \mathbf{x}) g'(\theta)/I_{\mathbf{X}}(\theta)$. $\qquad \square$

Exercise 4.4 Repeat the proof of Theorem 4.2 and specialize it for discrete distributions.

Theorem 4.2 shows that the variance of an arbitrary regular unbiased estimator cannot be smaller than a quantity which depends only on the Fisher information of the underlying model. If we can find an unbiased estimator that attains the equality, we need not search any further: no unbiased estimator could perform better. We might also hope that when T has a small bias and its variance is close to the lower bound, it will be difficult to improve it.

Moreover, the larger the value of the Fisher information the smaller the bound for the variance of the estimate. That means that we can estimate more accurately if X contains more information.

For a sample of i.i.d. r.v.'s, i.e., in the model

$$\mathcal{P} = \{P_\theta = P_{1,\theta}^{\otimes n} : \ \theta \in \Theta \subseteq \mathbb{R}\},$$

we get because of the additivity of the Fisher information that $I_X(\theta) = nI_X(\theta)$. Then

$$\text{Var}_\theta T \geq \frac{[g'(\theta)]^2}{nI_X(\theta)} \qquad \text{for all} \quad \theta \in \Theta.$$

This means that the variance of an unbiased regular estimator is at least of order $1/n$.

Definition 4.9 (Efficiency) The **efficiency** of an unbiased estimator T is defined by the ratio of its variance and the Cramér–Rao bound, that is

$$e(T, \theta) = \frac{[g'(\theta)]^2}{I_X(\theta)\text{Var}_\theta T}.$$

An unbiased estimator which attains the Cramér–Rao bound is called an **efficient** estimator.

An efficient estimator is BUE.

Example 4.19 (Binomial distribution) Suppose that $X \sim \text{Bin}(n, \theta)$. The parameter of interest is $\gamma = \theta(1 - \theta)$. We know that the lower bound for the variance of an unbiased estimator for γ is

$$\frac{(1 - 2\theta)^2\theta(1 - \theta)}{n}.$$

An unbiased estimator T is defined by

$$T(x) = \frac{1}{n - 1}x(1 - \frac{x}{n}).$$

Its variance is

$$\frac{\theta}{n} - \frac{\theta^2(5n - 7) - 4\theta^3(2n - 3) + \theta^4(4n - 6)}{n(n - 1)}.$$

Thus this estimator is not efficient. But note that

$$e(T, \theta) = \dfrac{\dfrac{(1 - 2\theta)^2 \theta (1 - \theta)}{n}}{\dfrac{\theta}{n} - \dfrac{\theta^2(5n - 7) - 4\theta^3(2n - 3) + \theta^4(4n - 6)}{n(n - 1)}} \to 1.$$

□

The following theorem is valid for exponential families.

Theorem 4.3 *Suppose that the distribution of* $\mathbf{X} = (X_1, \ldots, X_n)$ *belongs to a one-parameter exponential family in* ζ *and* T. *Then the sufficient statistic* T *is an efficient estimator for the parameter* $\gamma = g(\theta) = \mathsf{E}_\theta T$.

PROOF: The probability function of the sample is given by

$$p(\mathbf{x}; \theta) = A(\theta) \exp(T(\mathbf{x})\zeta(\theta)) \, h(\mathbf{x})$$

and we have for the score function

$$V(\theta; \mathbf{x}) = \frac{\partial}{\partial \theta} \ln p(\mathbf{x}; \theta) = \frac{A'(\theta)}{A(\theta)} + \zeta'(\theta) T(\mathbf{x}).$$

This means that $V(\theta; \mathbf{x})$ and $T(\mathbf{x})$ are linearly related, hence

$$\mathsf{Cor}_\theta(V(\theta; \mathbf{X}), T(\mathbf{X}))^2 = 1. \qquad (4.14)$$

From the proof of Theorem 4.2 we know that for unbiased estimators:

$$\mathsf{Cov}_\theta(V(\theta; \mathbf{X}), T(\mathbf{X})) = g'(\theta),$$

thus by (4.14)

$$\frac{[g'(\theta)]^2}{\mathsf{Var}_\theta T(\mathbf{X}) \, \mathsf{Var}_\theta V(\theta; \mathbf{X})} = 1.$$

Hence because of $\mathsf{Var}_\theta V(\theta; \mathbf{X}) = I_\mathbf{X}(\theta)$

$$\mathsf{Var}_\theta T(\mathbf{X}) = \frac{[g'(\theta)]^2}{I_\mathbf{X}(\theta)}.$$

That means that T attains the Cramér–Rao bound.

□

Special case 4.14 (Normal distribution, known variance)
Suppose that we have a sample of i.i.d. r.v.'s with $X_i \sim$
$N(\mu, \sigma_0^2)$, where the variance σ_0^2 is known. This model is a
one-parameter exponential family with $T(\mathbf{x}) = \sum_{i=1}^n x_i$ and
$\zeta(\mu) = \mu/\sigma_0^2$. The Fisher information is n/σ_0^2. The variance
of the estimator $\hat{\mu} = \frac{1}{n} \sum_{i=1}^n X_i$ is σ_0^2/n. Thus, the bound is
attained and the arithmetic mean is BUE. Note that this model is also an
exponential family in $\tilde{\zeta} = n\zeta$ and $\hat{\mu}$. Thus the efficiency follows also from
Theorem 4.3.

□

Special case 4.15 (Normal distribution, known mean)
Suppose we have a sample of i.i.d. r.v.'s with $X_i \sim N(\mu_0, \sigma^2)$,
where the mean μ_0 is known. This model is a one-parameter
exponential family with $T(\mathbf{x}) = \sum_{i=1}^n (x_i - \mu_0)^2$ and $\zeta(\sigma^2) =$
$-1/(2\sigma^2)$. The Fisher information is $n/(2\sigma^4)$ and the variance
of the estimator $\hat{\sigma}^2 = \frac{1}{n} \sum_{i=1}^n (X_i - \mu_0)^2$ is given by $2\sigma^4/n$.
Thus the lower bound is attained, and the proposed estimator is BUE. Note
that the estimator satisfies

$$\frac{1}{n} \sum_{i=1}^n (X_i - \mu_0)^2 - \sigma^2 = \frac{\frac{\sum_i (X_i - \mu_0)^2}{2\sigma^4} - \frac{1 \cdot n}{2\sigma^2}}{\frac{n}{2\sigma^4}},$$

i.e., (4.13) holds with $g' = 1$ and

$$V(\sigma^2; \mathbf{X}) = \frac{\sum_{i=1}^n (X_i - \mu_0)^2}{2\sigma^4} - \frac{1 \cdot n}{2\sigma^2}.$$

□

Special case 4.16 (Uniform distribution) Let $\mathbf{X} = (X_1, \ldots, X_n)$ be a
sample of i.i.d. r.v.'s which are uniformly distributed over the interval $[0, \theta]$.
An unbiased estimator for θ is given by $T(\mathbf{X}) = \frac{n+1}{n} X_{\max}$. Its variance is
$\theta^2/(n(n+2))$. This variance is of order n^{-2}, but this is not a contradiction
to the Cramér–Rao bound, because the uniform distributions do not form a
regular model.

□

Special case 4.17 (Exponential distribution) Consider a sample of i.i.d. r.v's from $\mathsf{Exp}(\lambda)$ with $\lambda > 0$. We know that exponential distributions are an exponential family. The parameter of interest is the expectation, that is $\gamma(\lambda) = 1/\lambda$. An unbiased estimator for γ is given by the sample mean $\overline{X}$. Its variance is $1/(n\lambda^2)$; since the Fisher information is n/λ^2 we get for the lower bound with $g'(\lambda) = -1/\lambda^2$ the expression

$$\frac{\lambda^{-4}}{n\lambda^{-2}} = \frac{1}{n\lambda^2}.$$

Thus the sample mean is an efficient estimator. $\square$

Remark 4.4 Maximum likelihood estimators are not necessarily unbiased. But considering their behavior for increasing n we will see that the variance of their limiting distribution is just this lower bound. Thus, maximum likelihood estimators are (under regularity conditions) asymptotically efficient (compare also Theorem 4.4).

4.2.3 Best Unbiased Estimators and Cramér–Rao Inequality for Multidimensional Parameters

Suppose now $\Theta \subseteq \mathbb{R}^k$ and that the parameter of interest γ is m-dimensional, that is, $g : \Theta \to \Gamma \subseteq \mathbb{R}^m$. An estimator $T : \mathcal{X} \to \Gamma$ is called **unbiased** for γ, if

$$\mathsf{E}_\theta T_j(\mathbf{X}) = g_j(\theta) \qquad \text{for all} \quad \theta \in \Theta \qquad \text{and all} \quad j = 1, \ldots, m.$$

Definition 4.10 Let T and T^* be two unbiased estimators for γ. We say that T^* has a smaller covariance matrix than T at $\theta \in \Theta$, if the covariance matrices of T and T^* satisfy

$$u^\mathsf{T}(\mathsf{Cov}_\theta T^* - \mathsf{Cov}_\theta T)u \leq 0 \qquad \text{for all} \quad u \in \mathbb{R}^m.$$

We write
$$\mathsf{Cov}_\theta T^* \preceq \mathsf{Cov}_\theta T.$$

Suppose now that the regularity conditions **Reg 1**, **Reg' 2** and **Reg' 4** are satisfied and the Fisher information matrix is not singular. Then the **Cramér–Rao inequality** has the following form:

$$\mathsf{Cov}_\theta T \succeq (D_\theta g)(\theta) I_\mathbf{X}^{-1}(\theta)(D_\theta g)^\mathsf{T}(\theta) \qquad \text{for all} \quad \theta \in \Theta.$$

Here $D_\theta g$ denotes the $m \times k$ matrix of partial derivatives $\frac{\partial g_j(\theta)}{\partial \theta_l}$, $j = 1, \ldots, m$, $l = 1, \ldots, k$, and $I_{\mathbf{X}}^{-1}(\theta)$ is the inverse of the Fisher information matrix

$$I_{\mathbf{X}}(\theta) = \mathsf{Cov}_\theta V(\theta; \mathbf{X}),$$

where $V(\theta; \mathbf{X})$ is the k-dimensional score vector of the partial derivatives $\frac{\partial \ln L(\theta;\mathbf{X})}{\partial \theta_l}$ (see Definition 3.5).

Special case 4.18 (Normal distribution) Suppose that we have a sample of independent $N(\mu, \sigma^2)$-distributed r.v.'s. The parameter of interest is $\theta = (\mu, \sigma^2)$. The Fisher information matrix $I_{\mathbf{X}}(\theta)$ was calculated in (3.9) on page 46:

$$I_{\mathbf{X}}(\theta) = \begin{pmatrix} \frac{n}{\sigma^2} & 0 \\ 0 & \frac{n}{2\sigma^4} \end{pmatrix}.$$

The inverse of $I_{\mathbf{X}}(\theta)$ is the Cramér–Rao bound. The estimators $\overline{X}$ and S^2 are independent and

$$\mathsf{Var}_\theta \overline{X} = \frac{\sigma^2}{n} \quad \text{and} \quad \mathsf{Var}_\theta S^2 = \frac{2\sigma^4}{(n-1)}.$$

Thus the lower bound for the variance is not attained, but as we will see in the next section this estimator is the best estimator in the class of all unbiased estimators. □

We conclude Section 4.2 with the following result on maximum likelihood estimators:

Theorem 4.4 *Under the regularity conditions **Reg 1**, **Reg' 2**–**Reg' 4** a maximum likelihood estimator has the following properties:*

1. *The maximum likelihood estimator depends on the data only via the sufficient statistic.*

2. *If there exists an efficient unbiased estimator $\tilde{\theta}$, then $\tilde{\theta} = \hat{\theta}_{\text{MLE}}$ almost surely.*

PROOF:

1. This is a consequence of Corollary 3.2 on page 55.

2. For simplicity let us prove the second statement for real-valued parameters.

Suppose that $\tilde{\theta}$ is an efficient estimator. Then, from equation (4.13) in Theorem 4.2 it follows that

$$\tilde{\theta}(\mathbf{x}) - \theta = \frac{V(\theta; \mathbf{x})}{I_{\mathbf{x}}(\theta)}$$

for all θ and all $\mathbf{x} \in \mathcal{A}$. That is, also for $\theta = \hat{\theta}_{\mathrm{MLE}}$ we have

$$\tilde{\theta}(\mathbf{x}) - \hat{\theta}_{\mathrm{MLE}}(\mathbf{x}) = \frac{V(\hat{\theta}_{\mathrm{MLE}}(\mathbf{x}); \mathbf{x})}{I_{\mathbf{x}}(\hat{\theta}_{\mathrm{MLE}}(\mathbf{x}))}.$$

Since $\hat{\theta}_{\mathrm{MLE}}(\mathbf{x})$ maximizes $l(\cdot; \mathbf{x})$, we have $V(\hat{\theta}_{\mathrm{MLE}}(\mathbf{x}); \mathbf{x}) = 0$. Thus

$$\tilde{\theta}(\mathbf{x}) - \hat{\theta}_{\mathrm{MLE}}(\mathbf{x}) = 0.$$

$\square$

4.2.4   The Rao–Blackwell and Lehmann–Scheffé Theorems

In the previous section we learned that there is a lower bound for the variance of an unbiased estimator, and we saw examples showing that there are reasonable estimators not attaining this bound. The following question arises: Having an estimator which does not attain the bound, is there another estimator, maybe also not efficient in the sense defined above, but with a smaller variance? Or in other words: When should we stop our search for a better estimator? And moreover, how should we proceed in cases where the regularity assumptions are not fulfilled? The answer is given by the following theorems. The first one says that it is enough to consider estimators based on a sufficient statistic. And it can be also used for the construction of better estimators:

Theorem 4.5 (Rao–Blackwell) *Let T be a sufficient statistic for the statistical model $\mathcal{P}$, and let $\tilde{\gamma}$ be an unbiased estimator for the parameter $\gamma = g(\theta) \in \mathbb{R}^k$. Define*

$$\hat{\gamma}(T) = \mathrm{E}_\theta(\tilde{\gamma}|T).$$

The conditional expectation $\hat{\gamma}$ is independent of θ, i.e., $\hat{\gamma}(T) = \mathrm{E}.(\tilde{\gamma}|T)$. Furthermore, for all $\theta \in \Theta$

$$\mathrm{E}_\theta \, \hat{\gamma} = g(\theta)$$

and

$$\mathrm{Cov}_\theta \, \hat{\gamma} \preceq \mathrm{Cov}_\theta \tilde{\gamma}.$$

If $\mathrm{trace}\,(\mathrm{Cov}_\theta \tilde{\gamma}) < \infty$, *then* $\mathrm{Cov}_\theta \, \hat{\gamma} = \mathrm{Cov}_\theta \tilde{\gamma}$ *iff* $\mathrm{P}_\theta(\hat{\gamma} = \tilde{\gamma}) = 1$.

PROOF: Because of the sufficiency of T (see Definition 3.7) the conditional distribution of the data given T does not depend on θ, and therefore $\hat{\gamma}$ is indeed a statistic. Furthermore, the property of the conditional expectation implies:

$$\mathsf{E}_\theta\,\hat{\gamma} \;=\; \mathsf{E}_\theta\left(\mathsf{E}.(\tilde{\gamma}|T)\right) \;=\; \mathsf{E}_\theta\,\tilde{\gamma} \;=\; g(\theta).$$

Furthermore, we have

$$
\begin{aligned}
\mathsf{Cov}_\theta\tilde{\gamma} \;&=\; \mathsf{E}_\theta(\tilde{\gamma}-\gamma)(\tilde{\gamma}-\gamma)^\mathsf{T} \\
&=\; \mathsf{E}_\theta(\tilde{\gamma}-\hat{\gamma})(\tilde{\gamma}-\hat{\gamma})^\mathsf{T} + \mathsf{E}_\theta(\hat{\gamma}-\gamma)(\hat{\gamma}-\gamma)^\mathsf{T} + 2\mathsf{E}_\theta(\tilde{\gamma}-\hat{\gamma})(\hat{\gamma}-\gamma)^\mathsf{T} \\
&=\; \mathsf{E}_\theta(\tilde{\gamma}-\hat{\gamma})(\tilde{\gamma}-\hat{\gamma})^\mathsf{T} + \mathsf{Cov}_\theta\hat{\gamma} + 2\mathsf{E}_\theta(\tilde{\gamma}-\hat{\gamma})(\hat{\gamma}-\gamma)^\mathsf{T}.
\end{aligned}
$$

Since

$$\mathsf{E}_\theta(\tilde{\gamma}-\hat{\gamma}|T) \;=\; 0,$$

the third summand on the r.h.s. of the last equation is equal to

$$\mathsf{E}_\theta\left((\hat{\gamma}-\gamma)\mathsf{E}_\theta(\tilde{\gamma}-\hat{\gamma}|T)\right) = 0.$$

Thus we have

$$\mathsf{Cov}_\theta\tilde{\gamma} \;=\; \mathsf{E}_\theta(\tilde{\gamma}-\hat{\gamma})(\tilde{\gamma}-\hat{\gamma})^\mathsf{T} + \mathsf{Cov}_\theta\hat{\gamma} \;\succeq\; \mathsf{Cov}_\theta\hat{\gamma}.$$

And the equality holds if $\mathsf{E}_\theta(\tilde{\gamma}-\hat{\gamma})(\tilde{\gamma}-\hat{\gamma})^\mathsf{T} = 0$, but this is true if $\tilde{\gamma} = \hat{\gamma}$ except on sets N with $\mathsf{P}_\theta(N)=0$.

$\square$

Special case 4.19 (Binomial distribution) Suppose that we have a sample of n independent Bernoulli distributed r.v.'s $X_1,\ldots,X_n$ with unknown parameter θ. It is easy to see that X_1 is an unbiased estimator for θ. Furthermore we know that $T(\mathbf{X}) = \sum_{i=1}^n X_i$ is a sufficient statistic (see Special case 3.8). Now let us derive an estimator with smaller variance:

$$
\begin{aligned}
\mathsf{E}_\theta\left(X_1 \Big| \sum_{i=1}^n X_i = t\right) \;&=\; \mathsf{P}_\theta\left(X_1 = 1 \Big| \sum_{i=1}^n X_i = t\right) \\
&=\; \frac{\mathsf{P}_\theta(X_1 = 1, \sum_{i=1}^n X_i = t)}{\mathsf{P}_\theta(\sum_{i=1}^n X_i = t)} \\
&=\; \frac{\mathsf{P}_\theta(X_1 = 1, \sum_{i=2}^n X_i = t-1)}{\binom{n}{t}\theta^t(1-\theta)^{n-t}} \\
&=\; \frac{\mathsf{P}_\theta(X_1 = 1)\mathsf{P}_\theta(\sum_{i=2}^n X_i = t-1)}{\binom{n}{t}\theta^t(1-\theta)^{n-t}} \\
&=\; \frac{\theta \cdot \binom{n-1}{t-1}\theta^{t-1}(1-\theta)^{n-1-(t-1)}}{\binom{n}{t}\theta^t(1-\theta)^{n-t}} \\
&=\; \frac{t}{n}.
\end{aligned}
$$

Thus we get the well-known arithmetic mean $\frac{1}{n}\sum_{i=1}^{n} X_i$ as an estimator for θ.   □

Remark 4.5 For any sample of i.i.d. r.v.'s we have the same result as in the Special case 4.19. Because the sum $\sum_{i=1}^{n} X_i$ is symmetric in $X_1, \ldots, X_n$, we have

$$E_\theta(X_1|\sum_{i=1}^{n} X_i = t) = E_\theta(X_2|\sum_{i=1}^{n} X_i = t) = \cdots = E_\theta(X_n|\sum_{i=1}^{n} X_i = t)$$

and

$$
\begin{aligned}
E_\theta(X_1|\sum_{i=1}^{n} X_i = t) &= \frac{\sum_{j=1}^{n} E_\theta(X_j|\sum_{i=1}^{n} X_i = t)}{n} \\
&= \frac{1}{n} E_\theta(\sum_{j=1}^{n} X_j|\sum_{i=1}^{n} X_i = t) = \frac{t}{n}.
\end{aligned}
$$

Special case 4.20 (Uniform distribution) Consider Special case 3.12 on page 57. It was shown there that the statistic $X_{\max}$ is sufficient for the family of all uniform distributions over $[0, \theta]$ with unknown θ. For an unbiased estimator for θ we have (at least) two candidates, namely $2\overline{X}$ and $\frac{n+1}{n} X_{\max}$. By the Rao–Blackwell theorem we see immediately that the latter is better.   □

Exercise 4.5 (Continuation of Special case 4.20.) Compute the variances of the two estimators $2\overline{X}$ and $\frac{n+1}{n} X_{\max}$. What is the gain when using the estimator based on the sufficient statistic?

The Rao–Blackwell theorem shows that good estimators are based on sufficient statistics. Thus we can restrict our search for optimal estimators to this class. Now the question is: Is it enough for an unbiased estimator to be sufficient to become a best estimator, or should we go on trying to find a better one?

The following theorem states that this is not necessary if in addition the sufficient statistic is complete.

Definition 4.11 (Completeness) A statistical model $\{P_\theta : \theta \in \Theta\}$ is called **complete** if for any function $h : \mathcal{X} \to \mathbb{R}$:

$$\mathsf{E}_\theta h(\mathbf{X}) = 0 \qquad \text{for all} \quad \theta \in \Theta \tag{4.15}$$

implies

$$\mathsf{P}_\theta \left(h(\mathbf{X}) = 0 \right) = 1 \qquad \text{for all} \quad \theta \in \Theta. \tag{4.16}$$

A statistic $T \sim \mathsf{P}_\theta^T$ is called complete if the statistical model $\{\mathsf{P}_\theta^T : \theta \in \Theta\}$ is complete.

The concept of completeness describes the ranges of the parameter space Θ related to the range of the sample space. For "big" sets Θ the condition (4.15) is a strong one and the conclusion (4.16) is not surprising.

Example 4.20 Suppose the statistical model consists only of the following two distributions: $\mathsf{N}(1,2)$ and $\mathsf{N}(0,1)$. This model is not complete. To show that take the function $h(x) = (x-1)^2 - 2$.
If $X \sim \mathsf{N}(1,2)$, then $\mathsf{E}(X-1)^2 - 2 = \mathrm{Var}X - 2 = 0$, and if $X \sim \mathsf{N}(0,1)$, then $\mathsf{E}(X-1)^2 - 2 = \mathsf{E}X^2 - 2\mathsf{E}X + 1 - 2 = 1 - 0 + 1 - 2 = 0$. But $h(x) \neq 0$ for $x > \sqrt{2} + 1$. $\qquad\qquad\square$

Example 4.21 Consider the statistical model $\{P_1, P_2, P_3\}$ consisting of the following three-point distributions on $\{0, 1, 2\}$:

	0	1	2
P_1	0.1	0.1	0.8
P_2	0	1	0
P_3	0.5	0	0.5

Condition (4.15) can be rewritten as a matrix equality

$$\begin{pmatrix} 0.1 & 0.1 & 0.8 \\ 0 & 1 & 0 \\ 0.5 & 0 & 0.5 \end{pmatrix} \begin{pmatrix} h(0) \\ h(1) \\ h(2) \end{pmatrix} = 0. \tag{4.17}$$

Because

$$\det \begin{pmatrix} 0.1 & 0.1 & 0.8 \\ 0 & 1 & 0 \\ 0.5 & 0 & 0.5 \end{pmatrix} = -0.35$$

the only solution of (4.17) is $h(0) = h(1) = h(2) = 0$. Thus $\{P_1, P_2, P_3\}$ is complete. $\qquad\qquad\square$

Example 4.22 The statistical model $\{U[0,\theta],\ \theta \in \mathbb{R}_+\}$ is complete. The condition (4.15) is equivalent to

$$\int_0^\theta \frac{1}{\theta} h(x)\,dx = 0 \qquad \text{for all}\quad \theta \in \mathbb{R}_+.$$

The relation $\int_0^\theta h(x)dx = 0$ for all θ implies $\frac{\partial}{\partial\theta}\int_0^\theta h(x)dx = 0$. But

$$\frac{\partial}{\partial\theta}\int_0^\theta h(x)dx = h(\theta) \quad \text{almost everywhere.}$$

Thus $h(x) = 0$. Furthermore, the statistic $X_{\max}$ is complete. The density of $X_{\max}$ is given by

$$f_\theta(t) = \frac{nt^{n-1}}{\theta^n} \mathbb{1}_{(0,\theta)}(t).$$

For an arbitrary function h the equality

$$E_\theta h(X_{\max}) = \int_{-\infty}^\infty h(t) f_\theta(t)dt = \frac{n}{\theta^n}\int_0^\theta h(t)t^{n-1}dt = 0$$

is satisfied for all θ if

$$\int_0^\theta h^-(t)t^{n-1}dt = \int_0^\theta h^+(t)t^{n-1}dt,$$

where h^+ and h^- are the positive and negative parts of h, respectively. But this implies $h^+(t) = h^-(t)$ and, therefore, $h(t) = 0$. $\qquad\qquad\square$

Exercise 4.6 Let $\mathbf{X}$ be a sample of i.i.d. r.v.'s from $U[\theta-\frac{1}{2},\theta+\frac{1}{2}]$ with $\theta \in \mathbb{R}$. Show that $T = (T_1,T_2) = (X_{\min},X_{\max})$ is sufficient but not complete. Hint: Choose the function h given by $h(t_1,t_2) = t_2 - t_1 - \frac{n-1}{n+1}$.

Theorem 4.6 *Assume that $\mathcal{P}$ is a k-parameter exponential family with natural parameter $\zeta = (\zeta_1,\ldots,\zeta_k)$ and the natural parameter space $\mathcal{Z}^*$ contains a nonempty k-dimensional interval. Then the statistic $T(\mathbf{X})$ is sufficient and complete.*

The proof can be found in Witting (1978). The following corollary is useful for applications. Theorem 2.3 on page 21 and the theorem above imply:

Corollary 4.1 *Let us assume that P_θ belongs to a strictly k-parameter exponential family, then the statistic $T(\mathbf{X})$ is sufficient and complete.*

Example 4.23 (Flowers) The numbers of white and violet flowers together are sufficient and complete. □

Theorem 4.7 (Lehmann–Scheffé) *Let T be a sufficient and complete statistic for the statistical model $\mathcal{P}$, and let $\tilde{\gamma}_1$ be an unbiased estimator for the parameter $\gamma = g(\theta) \in \mathbb{R}^k$. Then the estimator*

$$\hat{\gamma} = \hat{\gamma}(T) = \mathsf{E}.\,(\tilde{\gamma}_1|T)$$

has the smallest covariance matrix among all unbiased estimators for the parameter $\gamma = g(\theta)$. That is, for all estimators $\tilde{\gamma}$ with $\mathsf{E}_\theta\tilde{\gamma} = g(\theta)$ we have

$$\mathsf{Cov}_\theta\hat{\gamma} \preceq \mathsf{Cov}_\theta\tilde{\gamma} \qquad \text{for all } \theta \in \Theta.$$

PROOF: Let us carry out an indirect proof. Suppose that there exists an estimator $\tilde{\gamma}_2$ with

$$\mathsf{Cov}_{\theta_0}\tilde{\gamma}_2 \prec \mathsf{Cov}_{\theta_0}\hat{\gamma}$$

for some θ_0. The Rao–Blackwell theorem implies for

$$\hat{\gamma}_2 = \hat{\gamma}_2(T) = \mathsf{E}_\theta(\tilde{\gamma}_2|T)$$

that

$$\mathsf{Cov}_{\theta_0}\hat{\gamma}_2 \preceq \mathsf{Cov}_{\theta_0}\tilde{\gamma}_2 \prec \mathsf{Cov}_{\theta_0}\hat{\gamma}. \tag{4.18}$$

On the other hand, $\hat{\gamma}_2$ as well as $\hat{\gamma}$ are unbiased estimators, and the completeness of the statistic T implies that $\mathsf{P}_\theta(\hat{\gamma}_2 = \hat{\gamma}) = 1$.
But then $\mathsf{Cov}_{\theta_0}\hat{\gamma}_2 = \mathsf{Cov}_{\theta_0}\hat{\gamma}$, which is a contradiction to (4.18) and therefore to the assumption that $\hat{\gamma}$ is not the best estimator.

□

Special case 4.21 (Normal distribution) Let $\mathbf{X}$ be a sample of independent $\mathsf{N}(\mu, \sigma^2)$-distributed r.v.'s with parameter of interest $\theta = (\mu, \sigma^2)$. We know that the arithmetic mean and the sample variance

$$\overline{X} = \frac{1}{n}\sum_{i=1}^{n} X_i \qquad \text{and} \qquad S^2 = \frac{1}{n-1}\sum_{i=1}^{n}(X_i - \overline{X})^2$$

are unbiased estimators. From Special case 2.1 and Corollary 4.1 follows that the statistic $(\sum_{i=1}^{n} X_i, \sum_{i=1}^{n} X_i^2)$ is sufficient and complete for the class of normal distributions. Since the estimators depend only on this statistic they are the best estimators. For the estimation of the mean this result follows from the Cramér–Rao inequality, because this estimator is efficient. But the optimality of S^2 is a consequence of the Lehmann–Scheffé theorem.   □

Special case 4.22 (Uniform distribution) We continue Special case 4.20. Since the statistic $X_{\max}$ is not only sufficient, but also complete, the unbiased estimator $\frac{n+1}{n} X_{\max}$ is not only better than the estimator constructed from the sample mean, but it has minimal variance in the class of all unbiased estimators. Note that this is an example where the Cramér–Rao inequality does not apply.   □

Special case 4.23 (Poisson distribution) We return to Special case 4.6 on page 78. There we constructed two estimators for the parameter λ of a Poisson distribution by the method of moments:

$$\hat{\lambda} = \frac{1}{n} \sum_{i=1}^{n} x_i \quad \text{and} \quad \tilde{\lambda} = \frac{1}{n} \sum_{i=1}^{n} (x_i - \bar{x})^2.$$

Since the family of Poisson distributions forms a strictly 1-parametric exponential family with $A(\lambda) = \exp(-\lambda)$, $h(x) = 1/x!$, $T(x) = x$ and $\zeta(\lambda) = \log \lambda$, we get by Corollary 4.1 that the arithmetic mean is sufficient and complete. Thus, by the Lehmann–Scheffé theorem we get that the arithmetic mean $\hat{\lambda}$ is the best unbiased estimator. Let us compute the Cramér–Rao bound: For a single observation we have

$$V_x(\lambda) = \frac{x}{\lambda} - 1 \quad \text{and} \quad I_x(\lambda) = \lambda^{-1}.$$

Thus the lower bound is $\frac{\lambda}{n}$. But this is also the variance of the arithmetic mean. Hence, the arithmetic mean is efficient. Note that the efficiency can also be concluded from Theorem 4.3.   □

Let us use Special case 4.23 as an illustration: Namely, let us apply the Rao–Blackwell theorem and compute the conditional expectation of the bias-corrected $\tilde{\lambda}$ given the sufficient statistic $\sum_{i=1}^{n} X_i$. Note that the bias-corrected $\tilde{\lambda}$ is the sample variance S^2.

Special case 4.24 (Poisson distribution, Rao–Blackwellization) For this purpose we derive the conditional distribution of X_i given $\sum_{j=1}^{n} X_j = k$. Since the sum of r independent Poisson-distributed r.v.'s with parameter λ is Poisson-distributed with parameter $r\lambda$, we get for $m \leq k$:

$$
\begin{aligned}
\mathsf{P}_\lambda(X_i = m \mid \sum_{j=1}^{n} X_j = k) &= \frac{\mathsf{P}_\lambda(X_i = m, \sum_{j=1}^{n} X_j = k)}{\mathsf{P}_\lambda(\sum_{j=1}^{n} X_j = k)} \\
&= \frac{\mathsf{P}_\lambda(X_i = m, \sum_{\substack{j=1 \\ j\neq i}}^{n} X_j = k - m)}{\mathsf{P}_\lambda(\sum_{j=1}^{n} X_j = k)} \\
&= \frac{\lambda^m \left((n-1)\lambda\right)^{k-m} k! \exp(-\lambda(1 + (n-1)))}{(n\lambda)^k (k-m)! m! \exp(-n\lambda)} \\
&= \binom{k}{m} \left(1 - \frac{1}{n}\right)^{k-m} \left(\frac{1}{n}\right)^m .
\end{aligned}
$$

Thus the conditional distribution is the binomial distribution with parameters k and $\frac{1}{n}$:

$$
X_i \Big| \sum_{j=1}^{n} X_j = k \sim \mathrm{Bin}(k, \frac{1}{n}).
$$

Hence we get for $S^2 = \frac{n}{n-1}\tilde{\lambda}$

$$
\begin{aligned}
\mathsf{E}_\lambda\left(S^2 \Big| \sum_{j=1}^{n} X_j = k\right) &= \frac{n}{n-1}\left(\mathsf{E}_\lambda(X_1^2 \Big| \sum_{j=1}^{n} X_j = k) - \frac{k^2}{n^2}\right) \\
&= \frac{n}{n-1}\left(\frac{k}{n}(1 - \frac{1}{n}) + \frac{k^2}{n^2} - \frac{k^2}{n^2}\right) \\
&= \frac{k}{n}.
\end{aligned}
$$

Thus, starting from the sample variance as an unbiased estimator for λ we arrive by using the Rao–Blackwell theorem at the arithmetic mean. □

Special case 4.25 (Negative binomial distribution) There exists no unbiased estimator for the parameter $\gamma = 1/\theta$ in the binomial model considered in the Example 4.16. Taking a negative binomial sampling scheme we can find a BUE for γ. Here sampling is continued until a specific number of successes, say k, has been obtained. Let $X + k$ denote the required number of trials. Then X has a negative binomial distribution, i.e.,

$$
\mathsf{P}_\theta(X = x) = \binom{k + x - 1}{k - 1} \theta^k (1 - \theta)^x .
$$

The expectation and the variance are

$$\mathsf{E}_\theta X = \frac{k}{\theta} - k \qquad \mathsf{Var}_\theta X = \frac{k}{\theta^2} - \frac{k}{\theta}.$$

Thus, an unbiased estimator for γ is

$$\hat{\gamma}_{\mathrm{MME}} = \frac{X}{k} + 1.$$

Since X is sufficient and complete (compare Exercise 4.7) this estimator is a BUE.   □

Exercise 4.7 Suppose X has a negative binomial distribution with (unknown) parameter θ and known parameter k. Verify that X is sufficient and complete.

4.3   Asymptotic Properties of Estimators

Intuition tells us that the bigger our sample the more we can trust our inferences, because that way the sample contains more information about the underlying distribution. To characterize the properties of an estimator for a large sample size one can apply the different notions of convergence defined in probability theory:

Definition 4.12 (Consistency) We say that a sequence $\{T_n\}$ of estimators for a parameter $\gamma = g(\theta)$ is **weakly consistent**, if T_n converges in probability to γ, that is: If for any $\varepsilon > 0$ and for all $\theta \in \Theta$

$$\lim_{n \to \infty} \mathsf{P}_\theta \left(|T_n - g(\theta)| > \varepsilon \right) = 0.$$

If $\{T_n\}$ converges with probability one or almost surely (a.s.) to γ, that is, for all $\theta \in \Theta$

$$\mathsf{P}_\theta \left(\lim_{n \to \infty} T_n = g(\theta) \right) = 1,$$

then it is **strongly consistent**.

Strong consistency implies weak consistency. We restrict our considerations to weak consistency and write $T_n \xrightarrow{\mathrm{P}} \gamma$. Consistency is very often a consequence of the laws of large numbers. The simplest case is the convergence of the sample mean. By the continuous mapping theorem we obtain the consistency of other estimators.

Theorem 4.8 (Continuous mapping theorem) *Let S_n be a sequence of r.v.'s, S_0 be a r.v. and h be a continuous function.*

$$\text{If} \quad S_n \xrightarrow{\text{P}} S_0, \qquad \text{then} \quad h(S_n) \xrightarrow{\text{P}} h(S_0),$$

$$\text{if} \quad S_n \xrightarrow{\text{a.s.}} S_0, \qquad \text{then} \quad h(S_n) \xrightarrow{\text{a.s.}} h(S_0).$$

The proof can be found in Gut (2005).

An important tool for the verification of consistency is the Chebyshev inequality (proved for example in Gut (2005)): For a r.v. Z with finite mean we have for all constants τ

$$P(|Z - EZ| > \tau) \le \frac{\text{Var}\, Z}{\tau^2}.$$

Let $X_1, \ldots, X_n$ be i.i.d. copies of a r.v. with distribution function F. Then we have:

1. The arithmetic mean converges to $E_F X = \mu$: $\overline{X}_n \xrightarrow{\text{P}} \mu$.

2. The empirical variance and the standard deviation converge to $\text{Var}_F X = \sigma^2$ and σ, respectively:

$$S_n^2 \xrightarrow{\text{P}} \sigma^2 \qquad S_n \xrightarrow{\text{P}} \sigma.$$

3. The relative frequency of an event A converges to its probability. Define $\nu_n(A) = \frac{1}{n} \sum_{i=1}^{n} 1_i(A)$, where $1_i(A) = 1$ if A occurs in the i-th trial, and zero otherwise. Then $\nu_n(A) \xrightarrow{\text{P}} P(A)$.

4. From the last result it follows immediately that the empirical distribution function is consistent at each point t. This result can be extended, namely the convergence is not only pointwise, but uniform. We have the following theorem:

Theorem 4.9 (Glivenko) *Let $\hat{F}_n$ be the empirical distribution function of the i.i.d. $X_1, \ldots, X_n$. Then*

$$\sup_{-\infty < t < \infty} |\hat{F}_n(t) - F(t)| \longrightarrow 0 \qquad a.s.$$

For the proof see again Gut (2005).

5. If the j-th moment $m_j = E_F X^j$ exists, then the empirical moment $\hat{m}_{jn} = \frac{1}{n} \sum_{i=1}^{n} X_i^j$ is a consistent estimator for m_j. Hence, the method of moments estimator for $\gamma = h(m_1, \ldots, m_r)$ is a consistent estimator if the function h is continuous. Examples are the empirical versions of the skewness, the kurtosis and the correlation.

OBS! Bigger samples do not always give better estimates. Consider the following case:

Special case 4.26 (Cauchy distribution) The Cauchy density

$$f(x;\theta) = \frac{1}{\pi[1 + (x - \theta)^2]}, \qquad -\infty < x < \infty, \quad -\infty < \theta < \infty$$

is symmetric with mode θ. None of the moments exists, and the average $\overline{X}$ of independent Cauchy-distributed r.v.'s $X_1, \ldots, X_n$ has the same distribution as a single observation. So it would be useless to estimate θ by $\overline{X}$. We might as well use X_1.

To demonstrate this we simulated $M = 500$ samples of size $n = 200$. From each sample $\overline{x}_n$ was computed. This arithmetic mean and the observation x_1 are shown in Figure 4.10. The horizontal lines show the true $\theta = 1$. We see the arithmetic mean does not concentrate around θ. It behaves like x_1. On the second row of this figure you see the same simulation for the normal distribution. □

Another concept of describing asymptotic behavior is that of convergence in distribution. In the context of estimation we consider the asymptotic normality of a sequence $\{T_n\}$ of estimators for γ:

Definition 4.13 (Asymptotic normality) A sequence of estimators $\{T_n\}$ for a m-dimensional parameter $\gamma = g(\theta)$ is **asymptotically normal** if for all $\theta \in \Theta$ the distribution of $\sqrt{n}(T_n - g(\theta))$ converges to a m-dimensional normal distribution with mean zero and a (positive definite) covariance matrix $\Sigma(\theta)$. We write shortly

$$\sqrt{n}\,(T_n - g(\theta)) \xrightarrow{\mathcal{D}} \mathsf{N}_m(0, \Sigma(\theta)). \tag{4.19}$$

Remark 4.6 Be careful, the usual short "$\xrightarrow{\mathcal{D}}$"-notation forgets that the distribution of the statistic on the l.h.s. of (4.19) depends on the underlying probability measure P_θ.

Asymptotic normality of an estimator is important in the two following aspects:

First, it is useful for **approximation of the distribution** of the estimator. Consider for simplicity the case $m = 1$. Then an equivalent form of (4.19) is

$$\mathsf{P}_\theta\left(\frac{\sqrt{n}\,(T_n - g(\theta))}{\sigma(\theta)} \leq t\right) \longrightarrow \Phi(t) \qquad \text{for } n \to \infty$$

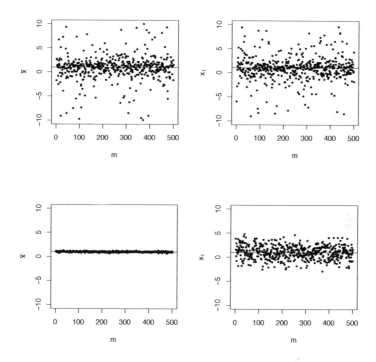

Figure 4.10: The left figure on the first row shows the arithmetic mean of 500 simulated samples from a Cauchy distribution with location parameter $\mu = 1$; the right figure shows the first observation of these samples. On the second row the same is shown for samples from a $N(1, 2)$-distribution. The sample size is $n = 200$.

for all t and for all θ, where $\sigma^2(\theta) > 0$. Now using a consistent estimator for $\sigma^2(\theta)$, say $\hat{\sigma}_n^2$, we get by the Slutsky theorem

$$P_\theta \left(\frac{\sqrt{n}\,(T_n - g(\theta))}{\hat{\sigma}_n} \leq t \right) \longrightarrow \Phi(t) \qquad \text{for } n \to \infty$$

for all t and for all θ. Thus we can use the asymptotic normality to approximate the distribution of the standardized estimator by the standard normal distribution. This will lead to asymptotic α-tests for $\gamma = g(\theta)$ and asymptotic confidence intervals, respectively. Often the quantity $\hat{\sigma}_n/\sqrt{n}$ is considered as the standard error of T_n, denoted by s.e.(T_n).

Secondly, asymptotic normality is a useful tool in stating **asymptotic efficiency**. In Section 4.2.1 we have defined an estimator to be efficient if it attains the Cramér–Rao bound (compare Definition 4.9). We call an estimator asymptotically efficient, if it is asymptotically normal with

$$\Sigma(\theta) = (D_\theta g)(\theta) I_{\mathbf{X}}^{-1}(\theta)(D_\theta g)^{\mathsf{T}}(\theta) \quad \text{for all } \theta,$$

where $I_{\mathbf{X}}(\theta)$ is the information matrix of the underlying distribution.

OBS! Note, here it is not supposed that $\mathrm{Var}_\theta T_n$ exists. The only assumption is the asymptotic normality of the estimator.

Remark 4.7 Under the regularity conditions formulated in Chapter 3 and some further regularity conditions one can show that the MLE is consistent. Then a Taylor expansion of the log-likelihood function together with the application of the central limit theorem to the score function and the law of large numbers to the observed information imply the asymptotic normality of the MLE. The variance of the limiting distribution is the inverse of the Fisher information. Thus the MLE is asymptotically efficient.

> Under regularity conditions maximum likelihood estimators are consistent, asymptotically normal and efficient. Thus, the maximum likelihood estimator is the best we can get.

Before giving an example, let us make the following remark: The proof for asymptotic normality of estimators which are sums of i.i.d. r. v.'s is based on the central limit theorem (CLT). Estimators of more complex structure can be handled by the following theorem which we formulate here for the case $m = 1$. The formulation for $k > 1$ can be found in Rao (1973).

Theorem 4.10 (Delta method) *Suppose T_n is an estimator of the form $T_n = h(S_n)$ where the sequence $\{S_n\}$ is asymptotically normal, i.e.,*

$$\sqrt{n}\,(S_n - \mu) \xrightarrow{\ D\ } \mathsf{N}(0, \tau^2)$$

for some constants μ and $\tau^2 > 0$.
If h has a continuous nonzero derivative h' at μ, then

$$\sqrt{n}\,(T_n - h(\mu)) \xrightarrow{\ D\ } \mathsf{N}\left(0, [h'(\mu)]^2 \tau^2\right).$$

Example 4.24 Let $X_1, \ldots, X_n$ be i.i.d. r.v.'s with $\mathsf{E}X_i = \mu \neq 0$ and $\mathrm{Var}X_i = \sigma^2 < \infty$. The parameter $\gamma = \log \mu$ is estimated by $\hat{\gamma}_n = \log \overline{X}_n$. This estimator is consistent and asymptotically normal: With $h(s) = \log s$ and $h'(s) = 1/s$ we obtain from $\sqrt{n}\,(\overline{X} - \mu) \xrightarrow{\ D\ } \mathsf{N}(0, \sigma^2)$, that

$$\sqrt{n}\,(\log \overline{X}_n - \log \mu) \xrightarrow{\ D\ } \mathsf{N}\left(0, \frac{\sigma^2}{\mu^2}\right).$$

From this statement on the limit we conclude: If the underlying parameter μ is small the estimator $\hat{\gamma}_n$ spreads more around γ. $\square$

Example 4.25 (Poisson distribution) Let $X_1, \ldots, X_n$ be i.i.d. according to $\mathsf{Poi}(\theta)$. The Fisher information is $I_X(\theta) = \theta^{-1}$. The Cramér–Rao bound for estimating the parameter $\gamma = g(\theta) = \theta^2$ is given by $\frac{[g'(\theta)]^2 \theta}{n} = \frac{4\theta^3}{n}$. Furthermore, we know that the sample mean $\overline{X}$ is a sufficient and complete statistic. Thus, the Lehmann–Scheffé theorem implies that the unbiased estimator

$$T_n = \overline{X}_n^2 - \frac{\overline{X}_n}{n}$$

has minimal variance in the class of all unbiased estimators. But since

$$\mathsf{Var}_\theta T_n = \frac{4\theta^3}{n} + \frac{2\theta^2}{n^2} > \frac{4\theta^3}{n},$$

the Cramér–Rao bound is not attained. From the CLT (applied to $\overline{X}_n$ with asymptotic variance θ), the Slutsky theorem and Theorem 4.10 for $h(s) = s^2$ we obtain

$$\sqrt{n}\,(T_n - \theta^2) \xrightarrow{D} \mathsf{N}(0, 4\theta^3).$$

Thus the estimator T_n is asymptotically efficient.   □

4.4   List of Problems

1. Given a sample of i.i.d. r.v.'s with distribution function with $F(t) = t^\alpha$ for $t \in (0,1)$, $F(t) = 0$ for $t \leq 0$ and $F(t) = 1$ for $t \geq 1$. Find an estimator for $\alpha > 0$ a) using the method of moments, b) using the maximum likelihood method.

2. Consider the genetic model in Examples 2.8 and 2.14. Formulate the statistical model for observations $(x_1, \ldots, x_n)$ and derive the MLE for the parameter θ.

3. Consider independent r.v.'s with:

$$X_i \sim \mathsf{N}(\mu_1, \sigma^2), \quad i = 1, \ldots, n_1 \qquad \text{and} \qquad Y_j \sim \mathsf{N}(\mu_2, \sigma^2), \quad j = 1, \ldots, n_2.$$

Formulate the statistical model and derive the MLE for the parameter $(\mu_1, \mu_2, \sigma^2) \in \mathbb{R} \times \mathbb{R} \times \mathbb{R}_+$.

4. Let $X_1, \ldots, X_n$ be i.i.d. according to $\mathsf{N}(\mu, \sigma^2)$. Show that the estimators

$$\overline{X} = \frac{1}{n} \sum_{i=1}^n X_i \qquad \text{and} \qquad S^2 = \frac{1}{n-1} \sum_{i=1}^n (X_i - \overline{X})^2$$

are uncorrelated.

5. Let T_{1n} and T_{2n} be unbiased estimators for θ.

a) Show that for each $\alpha \in [0,1]$ $T_n = \alpha T_{1n} + (1-\alpha) T_{2n}$ is an unbiased estimator for θ.

b) Suppose that T_{1n} and T_{2n} are uncorrelated and that $\mathsf{Var}\, T_{1n} = \sigma_{1n}^2$ and $\mathsf{Var}\, T_{2n} = \sigma_{2n}^2$. Calculate $\mathsf{Var}\, T_n$.

c) Find the value of $\alpha \in [0,1]$ such that $\mathsf{Var} T_n$ is minimal.

d) Assume $\sigma_{1n}^2 \to 0$ and $\sigma_{2n}^2 \to 0$ as $n \to \infty$. Show that T_n is consistent.

6. Let $X_1..., X_n$ be independent r.v.'s distributed according to $\mathsf{U}[-\theta, \theta]$. Find the MLE for θ.

7. Consider independent r.v.'s X_i and Y_j, $i,j = 1, \ldots, n$ with $X_i \sim \mathsf{N}(\mu, \sigma^2)$ and $Y_i \sim \mathsf{N}(\mu, \lambda \sigma^2)$, where μ is known.

a) Suppose $\sigma^2 > 0$ is known. Derive the MLE for $\lambda > 0$.

b) Assume both σ^2 and λ are unknown. Derive the MLE for $\theta = (\sigma^2, \lambda)$.

8. Consider independent r.v.'s X_i and Y_j, $i,j = 1, \ldots, n$ with $X_i \sim \mathsf{N}(\mu, \sigma_1^2)$ and $Y_i \sim \mathsf{N}(\mu, \sigma_2^2)$, where σ_1^2 and σ_2^2 are known. Find the MLE for μ and derive its variance.

9. Let $X_1, \ldots, X_n$ be i.i.d. r.v.'s with density $f(x; \theta) = \frac{\theta}{x^2} \mathbb{1}_{[\theta, \infty)}(x)$. Find a MME for θ.

10. Consider the Kungsängslilja example, where p_1 and p_2 are unknown parameters.

a) Calculate the correlation between the numbers of white and pink flowers.

b) Derive the 2×2 Fisher information matrix.

c) The maximum likelihood estimators are unbiased. Are they also efficient?

11. Let $X_1, \ldots, X_n$ be i.i.d. according to $\mathsf{U}[0, \theta]$.
a) Compute the MSE of the method of moment estimator (based on $\overline{X}$) for θ. b) Compute the MSE of the bias-corrected MLE for θ. c) Compare both mean squared errors, also for $n \to \infty$. d) Which estimator is optimal? Explain!

12. Let $X_1, \ldots, X_n$ be i.i. Raleigh-distributed, i.e., their density is given by

$$f(x; a) = \frac{2}{a} x \exp(-\frac{x^2}{a}) \, \mathbb{1}_{(0,\infty)}(x), \qquad a > 0.$$

a) Calculate the Fisher information.

b) Derive the maximum likelihood estimator for a.

c) Derive method of moment estimators (based on the mean, variance, second moments).

d) Is the maximum likelihood estimator unbiased?

e) Which of the moment estimators are unbiased?

f) Calculate the MSE of the maximum likelihood estimator.

g) Calculate the MSE of one of the method of moment estimators.

h) Compare the MSE's in f) and g).

i) Is the BUE unique?

Hint: $\mathsf{E}_a X_1 = \frac{\sqrt{\pi a}}{2}$, $\mathsf{Var}_a X_1 = a(1 - \pi/4)$, $\mathsf{E}_a X_1^4 = 2a^2$.

4.5 Further Reading

In this chapter the maximum likelihood method for estimating a parameter is presented for parametric models. It can be extended in several directions to handle semiparametric and fully nonparametric models. Let us mention the following: The partial likelihood was introduced by Cox (1972) to analyze the semiparametric proportional hazard models for right-censored failure data. Penalized likelihood estimation methods are applied in semiparametric regression models by Davison (2003) and for the estimation of the density and the regression function in completely nonparametric models. Here a penalty term which measures the smoothness of the function to be estimated is incorporated in the likelihood, see Eggermont and LaRiccia (2001),Eggermont and LaRiccia (2009). A lot of examples of maximum likelihood estimators also in complex data structures can be found in Pawitan (2001).

M-estimators are introduced in Section 4.1.3. For a deeper understanding of these estimators and their role for the construction of robust statistical procedures we recommend the book of Jurečková and Sen (1996). In the book of Hampel et al. (1986) an approach to robustness which is based on the influence function is presented.

We have not included confidence estimators. One can show that there is a duality between testing hypotheses and constructing confidence regions. We refer to the book of Bickel and Doksum (2007).

The mean squared error (MSE) considered here as criterion to describe the quality of an estimator is a special case of a risk investigated in a decision-theoretic approach to statistical inference. As in the case of the MSE, where we restricted ourselves to the unbiased estimators, also in general it is possible to find an estimator which minimizes the risk at every value of θ only if one restricts the class of estimators to be considered. Another possibility is to modify the notion of optimality: One can define a weighted risk for some suitable nonnegative weight function over the parameter space Θ, or one looks for an estimator that minimizes the maximum risk, where the maximum is taken w.r.t. the parameter. The first approach leads to Bayes estimators, the second to minimax estimators. Both approaches are presented in detail and discussed in the book of Lehmann and Casella (1998).

The asymptotic behavior of estimators is only shortly covered in Section 4.3. For a more detailed presentation of asymptotic methods, of results concerning consistency and limiting distributions we refer to the book of Lehmann (1998). Here also very interesting examples are given, and the limiting statements are applied to the construction of asymptotic confidence regions and asymptotic test procedures. In van der Vaart (1998) asymptotic methods from a more theoretic point of view are presented. Starting from the idea of approximation by limit experiments the author covers maximum likelihood estimators, M-estimators, asymptotic efficiency, but also semiparametric models and empirical processes. Ibragimov and Has'minski (1981) present a decision-theoretic based approach to the asymptotic estimation theory.

Chapter 5

Testing Hypotheses

A **hypothesis** is a statement about an unknown state of nature. We describe the state of nature with the help of a **statistical model**. In our context a hypothesis is a specification of the supposed model. In other words: First we believe in a model and then we want to find out the truth about some more details within it. A **test** of a hypothesis is a procedure based on data to find out whether this hypothesis is true or not. We want that a test gives us the unique answer, like "the fixpoint in the space from which we can move the earth." That, of course, is an absolute unrealistic dream. The data are random, so the test decision is random too. Nevertheless tests are useful. They give us random decisions, which imply the right answer with high probability. But it is very important to learn the theory behind testing procedures, to know the properties of a test in order to be able to **interpret the test results right and carefully**. In this chapter we take up two classical approaches for testing hypotheses:

- providing evidence by the data for or against a hypothesis

- testing as a two-action decision problem.

5.1 Test Problems

Suppose a statistical model for a sample $\mathbf{X} = (X_1, ..., X_n)$:

$$\mathcal{P} = \{\mathsf{P}_\theta : \theta \in \Theta\}.$$

Then hypotheses are subsets of $\mathcal{P}$ formulated by dividing the parameter space into two parts:

$$\Theta = \Theta_0 \cup \Theta_1 \text{ with } \Theta_0 \cap \Theta_1 = \varnothing.$$

The **null hypothesis** is usually denoted by H_0. Under H_0 we state that the sample $\mathbf{X}$ is distributed according to P_θ, where the parameter θ belongs to Θ_0. Thus the specified model under H_0 is

$$\mathcal{P}_0 = \{\mathsf{P}_\theta : \theta \in \Theta_0\}. \tag{5.1}$$

We write

$$H_0 : \theta \in \Theta_0.$$

Sometimes we abbreviate $H_0 : \Theta_0$. We say that the hypothesis H_0 **is true**, iff the sample is distributed according to $P_\theta \in \mathcal{P}_0$. The name "null" hypothesis comes from the fact that H_0 traditionally is the statement of "no effect", often formulated by $\theta = 0$. The alternative hypothesis describes the other case and is usually denoted by H_1 or H_A. Under H_1 we state that the sample $\mathbf{X}$ is distributed according to P_θ and the parameter θ belongs to Θ_1. We write

$$H_1 : \theta \in \Theta_1$$

or $H_1 : \Theta_1$.

A hypothesis is called **simple** if it completely specifies the distribution; otherwise it is called **composite**.

For parameter spaces with an order relation we can consider one-sided test problems. For $\Theta \subseteq \mathbb{R}$ we say: H_0 and H_1 are **one-sided** iff

$$H_0 : \theta \geq \theta_0 \text{ versus } H_1 : \theta < \theta_0 \quad \text{ or } \quad H_0 : \theta \leq \theta_0 \text{ versus } H_1 : \theta > \theta_0.$$

We call H_1 **two-sided** iff

$$H_0 : \theta = \theta_0 \text{ versus } H_1 : \theta \neq \theta_0. \tag{5.2}$$

We also say H_1 is one-sided for $\Theta = [\theta_0, \infty)$ and

$$H_0 : \theta = \theta_0 \text{ versus } H_1 : \theta > \theta_0 \tag{5.3}$$

and for $\Theta = (-\infty, \theta_0]$ and

$$H_0 : \theta = \theta_0 \text{ versus } H_1 : \theta < \theta_0. \tag{5.4}$$

Let us start with some examples.

Example 5.1 (Ballpoint pens) Continuation of Example 2.1 and Example 2.9 on page 9. We are interested in quality control. The pens should have a guaranty of 2%. We wish to guarantee that in average from 100 pens not more than 2 pens are defective. Suppose the binomial model (2.2):

$$\mathcal{P} = \{\text{Bin}(400, \theta)^{\otimes 7} : \theta \in (0, 1)\}.$$

Let us consider different test problems: First we test our claim

$$H_0 : \theta = 0.02 \text{ versus } H_1 : \theta \neq 0.02. \tag{5.5}$$

Here $\Theta = (0, 1)$ and $\Theta_0 = \{0.02\}, \Theta_1 = (0, 0.02) \cup (0.02, 1)$. The null hypothesis is simple, since under H_0 the sample distribution is $\text{Bin}(400, 0.02)^{\otimes 7}$. Otherwise, it is nice if less than 2 pens of 100 are defective; thus we are interested in

$$H_0 : \theta \geq 0.02 \text{ versus } H_1 : \theta < 0.02. \tag{5.6}$$

Now we have another decomposition of the parameter space in

$$\Theta = (0,1) = (0,0.02) \cup [0.02,1) \tag{5.7}$$

with $\Theta_0 = [0.02,1)$ and $\Theta_1 = (0,0.02)$. Both hypotheses in (5.7) are composite. Maybe it is better to formulate the test problem in the other direction:

$$H_0 : \theta \le 0.02 \text{ versus } H_1 : \theta > 0.02. \tag{5.8}$$

Now $\Theta_0 = (0,0.02]$ and $\Theta_1 = (0.02,1)$. Later on page 139 we will explain the rule in which direction the test problem should be formulated. $\quad\square$

Example 5.2 (Flowers) Continuation of Example 2.2 on page 5, Example 2.10 on page 10 and Example 2.17 on page 14. We have the multinomial distribution $\mathsf{Mult}(n,p_1,p_2,p_3)$

$$P_\theta(\mathbf{n}) = p(\mathbf{n};\theta) = \frac{n!}{n_1! \, n_2! \, n_3!} p_1^{n_1} p_2^{n_2} p_3^{n_3}, \ \theta = (p_1,p_2), \ \mathbf{n} = (n_1,n_2,n_3),$$

with $n = n_1 + n_2 + n_3$ and with $\Theta = \{(p_1,p_2) : 0 < p_1 + p_2 < 1, 0 < p_1 < 1, 0 < p_2 < 1\}$. The theory has doubts that the part of white flowers is $\frac{1}{3}$. We formulate the test problem as

$$H_0 : p_1 = \frac{1}{3} \text{ versus } H_1 : p_1 \ne \frac{1}{3}. \tag{5.9}$$

Here the parameter space Θ is the open triangle in the square $[0,1]^2$ shown in Figure 5.1. Θ_0 is a line and H_0 is not simple. The model under H_0 is

$$\mathcal{P}_0 = \left\{ \mathsf{Mult}(n, \frac{1}{3}, \vartheta, 1 - \vartheta - \frac{1}{3}) : \vartheta \in (0, \frac{2}{3}) \right\}.$$

Theoretically, it should be impossible that all colors have the same chance. Testing this opinion we would consider

$$H_0 : p_1 = \frac{1}{3}, \ p_2 = \frac{1}{3} \text{ versus } H_1 : p_1 \ne \frac{1}{3} \text{ or } p_2 \ne \frac{1}{3}. \tag{5.10}$$

Now we have a simple null hypothesis. Under H_0 the sample distribution is $\mathsf{Mult}(n, \frac{1}{3}, \frac{1}{3}, \frac{1}{3})$. $\quad\square$

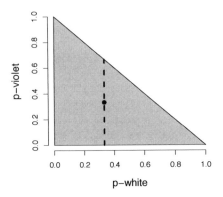

Figure 5.1: The parameter space and the hypotheses in Example 5.2.

 Example 5.3 (Pendulum) Continuation of Example 2.3 and Example 2.11 on page 11 with the model:

$$\mathcal{P} = \{\mathsf{N}(\mu, \sigma^2)^{\otimes 10} : \theta = (\mu, \sigma^2) \in \mathbb{R} \times \mathbb{R}_+\}. \tag{5.11}$$

We are interested in the exact knowledge about the expected value μ. The theory says it should be different from μ_0. Thus we have

$$H_0 : \mu = \mu_0 \text{ versus } H_1 : \mu \neq \mu_0. \tag{5.12}$$

The parameter space in model (5.11) is $\Theta = \mathbb{R} \times \mathbb{R}_+$. H_0 describes a line, not a single point! The model under H_0 is

$$\mathcal{P}_0 = \{\mathsf{N}(\mu_0, \sigma^2)^{\otimes 10} : \sigma^2 \in \mathbb{R}_+\}.$$

Assume now that the variance is σ_0^2 and is known. Then instead of model (5.11) we have the following model

$$\mathcal{P} = \{\mathsf{N}(\mu, \sigma_0^2)^{\otimes 10} : \mu \in \mathbb{R}\} \tag{5.13}$$

and the test problem is written in the same way

$$H_0 : \mu = \mu_0 \text{ versus } H_1 : \mu \neq \mu_0. \tag{5.14}$$

But under (5.13) the parameter space is the real line, $\Theta = \mathbb{R}$ and $\Theta_0 = \{\mu_0\}$ is a single point. Under the null hypothesis in (5.14) the distribution of $\mathbf{X}$ is $\mathsf{N}(\mu_0, \sigma_0^2)^{\otimes 10}$. □

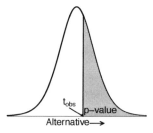 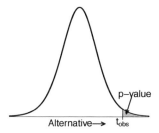

Figure 5.2: In the left picture the observed value of the statistic lies in the center of the null distribution. There is no contradiction to the null hypothesis. This is measured by a high p-value. In the right picture the observed value lies in the tail of the null distribution in the direction of the alternative. That provides an evidence against the null hypothesis. The p-value is small.

OBS! In Example 5.3 we have seen that the hypotheses depend on the underlying model; they can be different also in cases where the formulation is the same!

5.2   Tests: Assessing Evidence

This approach is more or less intuitive. We will describe it shortly and present a collection of examples and special cases.

The basic idea underlying this approach is that testing hypotheses involves comparing samples results $\mathbf{x} = (x_1, ..., x_n)$ with the model $\mathcal{P}_0$ under the null hypothesis. The comparison is based on a statistic $T(\mathbf{x})$ which measures the discrepancy or inconsistency between data $\mathbf{x}$ and model $\mathcal{P}_0$. $T(\mathbf{x})$ is called a **test statistic**. Two main questions arise:

1. How can we find an appropriate measure $T(\mathbf{x})$?
2. What conclusions can we draw from $T(\mathbf{x})$?

Sometimes the intuition immediately suggests a useful measure. Often we take $T(\mathbf{x}) = \widehat{\theta} - \theta_0$, where $\widehat{\theta}$ is a reasonable estimate of θ. Especially we need $T(\mathbf{x})$ to have a **unique and completely known** distribution P_0^T under $\mathcal{P}_0$. This distribution is called a **null distribution.** In the typical situation of a suitable test statistic we expect that the null distribution is unimodal, tailing off in one or both directions away from the mode. In assessing the credibility of H_0 we draw conclusions how far the observed value $T(\mathbf{x}) = t_{\text{obs}}$ is from what is typical when H_0 is true, measured by the p-value; compare Figure 5.2.

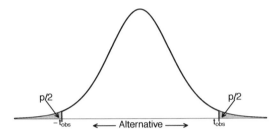

Figure 5.3: Under symmetric null distribution the p-value of a two-sided test is the sum of the p-values of the respective two one-sided tests.

Definition 5.1 The **p-value** corresponding to an observed $T(\mathbf{x}) = t_{\text{obs}}$ is the probability of $T(\mathbf{X})$ lying at and beyond t_{obs} in the "direction of the more extreme values" of the alternative, computed from the null distribution.

Let us specify the "direction of the more extreme values" when the test statistic $T(\mathbf{x})$ is chosen in such a way that we expect large values of t_{obs} when the underlying parameter θ is large.

Consider the test problem (5.3) with the one-sided alternative $H_1 : \theta > \theta_0$, then

$$\text{p-value} = \mathsf{P}_0^T \left(T \geq t_{\text{obs}} \right) = \mathsf{P}_{\theta_0} \left(T(\mathbf{X}) \geq t_{\text{obs}} \right).$$

Otherwise, for the test problem (5.4) with $H_1 : \theta < \theta_0$ we have

$$\text{p-value} = \mathsf{P}_0^T \left(T \leq t_{\text{obs}} \right) = \mathsf{P}_{\theta_0} \left(T(\mathbf{X}) \leq t_{\text{obs}} \right). \tag{5.15}$$

The definition of a p-value for a two-sided alternative is quite complicated for discrete null distributions. Let us assume that the null distribution is continuous. Then the p-value is given by

$$\text{p-value} = 2 \min \left\{ \mathsf{P}_0^T \left(T \leq t_{\text{obs}} \right), \mathsf{P}_0^T \left(T \geq t_{\text{obs}} \right) \right\}. \tag{5.16}$$

This definition of the p-value for the two-sided alternative in (5.2) is illustrated in Figure 5.4. For a continuous and symmetric null distribution around θ_0 we get

$$\text{p-value} = \mathsf{P}_0^T \left(|T| > |t_{\text{obs}}| \right),$$

illustrated in Figure 5.3.

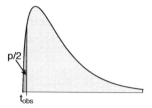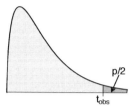

Figure 5.4: The alternative is two-sided. The p-value is twice of the smaller tail probability; compare (5.16).

Remark 5.1 The p-value is a conditional probability. Let z be the realization of a r.v. Z with distribution P_θ^T, $\theta \in \Theta$. Further let Z_0 be a random variable independent of Z distributed according to the null distribution P_0^T, with distribution function F_0. Then the p-value defined in (5.15) is

$$\text{p-value} = \mathsf{P}(Z_0 \le z \mid Z = z) = \mathsf{P}_0^T(Z_0 \le z) = F_0(z).$$

Until now the p-value was defined for given data $\mathbf{x}$ and fixed t_{obs}. For different experiments we get different p-values. Thus the p-value is a random variable and its distribution is relevant. We consider continuous distributions only. Assume that the test statistic $T(\mathbf{X})$ has a distribution function F_θ with density $f(.; \theta)$. Under H_0 the distribution of $T(\mathbf{X})$ is the null distribution with distribution function F_0 and with density f_0. Then for the test problem (5.4) the p-value is the random variable given by

$$P = F_0(T(\mathbf{X})).$$

For $q \in (0, 1)$ we have

$$\mathsf{P}_\theta(P \le q) = \mathsf{P}_\theta\left(F_0(T(\mathbf{X})) \le q\right) = \mathsf{P}_\theta(T(\mathbf{X}) \le F_0^{-1}(q)) = F_\theta\left(F_0^{-1}(q)\right).$$

Under H_0 we get $\mathsf{P}_0(P \le q) = q$ thus

$$\text{under } H_0: \ P \sim \mathsf{U}[0, 1].$$

In general, the density g_θ of the p-value is

$$g_\theta(q) = \frac{d}{dq} F_\theta\left(F_0^{-1}(q)\right) = \frac{f_\theta(F_0^{-1}(q))}{f_0(F_0^{-1}(q))}, \ q \in (0, 1).$$

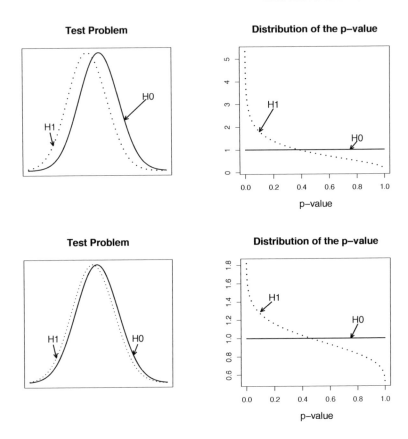

Figure 5.5: In the first row the distance between the distributions under the null hypothesis and under the alternative is relatively large. The p-value density under the alternative is high for small values. With high probability we can find strong evidence against the null hypothesis. In the second row it is more complicated to differ between the hypotheses.

In Figure 5.5 we see the densities of a p-value based on a normally distributed test statistic with $F_0(z) = \Phi(z)$ and $F_\theta(z) = \Phi(z - \theta)$. In this case we have $g_\theta(q) = \exp(\theta \, \Phi^{-1}(q) - \frac{\theta^2}{2})$.

Exercise 5.1 Suppose a continuous null distribution. Show that under the null hypothesis the p-value defined in (5.16) is uniformly distributed.

How to draw the conclusion. Small p-values mean that the observed test statistic does not lie in the central region of the null distribution; compare again Figure 5.2. We do not believe in the hypothesis H_0. For instance, for a p-value of 0.0004 we formulate: "The data provide very strong evidence against the null hypothesis." In other words:

> For small p-values we have a strong evidence against H_0.

A more quantified interpretation can be given with the help of the following table:

p-value > 0.12	no evidence against H_0
$0.05 <$ p-value < 0.12	weak evidence against H_0
p-value ≈ 0.05	some evidence against H_0
$0.01 <$ p-value < 0.05	evidence against H_0
p-value < 0.010	strong evidence against H_0
p-value < 0.001	very strong evidence against H_0

Table 5.1: Interpretation of the p-value

OBS! The method which consists in repeating an experiment until you get a p-value which delivers evidence against the hypothesis is a kind of modern cheating. Assume the experiments are repeated mutually independent under the same conditions. Then we obtain i.i.d. p-values $P_1, \ldots, P_j, \ldots$. Introduce $Z_j = \mathbb{1}_{[0,\alpha]}(P_j)$, then under H_0 it holds that $Z_j \sim \text{Ber}(\alpha)$ i.i.d. Thus the number of trials up to the first success is geometrical distributed $\text{Geo}(\alpha)$ with expectation $\frac{1}{\alpha}$. That means for $\alpha = 0.05$ the expected value for the number of needed experiments is 20!

Let us show several examples to demonstrate the use of the p-value. We start with a historical one.

Example 5.4 (Arbuthnott's Data) Arbuthnott (1667-1735) was a physician in London. He collected the yearly number of male and female Christenings in London from 1629-1710 and observed that every year the number of male Christenings was higher. Arbuthnott (1712) calculated under the assumption that the chance for males and females is the same:

$$P_0(\text{"more males than females in all 82 years"}) = \left(\frac{1}{2}\right)^{82} = 2.068 * 10^{-25}.$$

He continued: "From where it follows it is ART not CHANCE, that governs." Let us reformulate his study step by step. Every year, whenever the number of males was higher than the number of females he set a $+$. Let x be the number

Figure 5.6: The yearly number of male and female Christenings in London from 1629–1710. The dashed line presents the number of female Christenings. The connected line presents the number of male Christenings.

of +. (By the way that was the invention of the sign test!) He considered a binomial model for X

$$\mathcal{P} = \left\{ \mathrm{Bin}\,(n, p) : p \in \left[\frac{1}{2}, 1 \right) \right\}.$$

The null hypothesis is that the chance for girls and boys is the same. As alternative we formulate the suspicion that more boys are born:

$$H_0 : p = \frac{1}{2} \text{ versus } H_1 : p > \frac{1}{2}.$$

As a measure for discrepancy between sample and hypothesis we take the difference between the estimator and the hypothetical value $T(x) = \frac{x}{n} - \frac{1}{2}$. The null distribution is unique, because

$$\mathcal{P}_0 = \left\{ \mathrm{Bin}\left(82, \frac{1}{2} \right) \right\}.$$

The observed value of the test statistic is $T(82) = \frac{82}{82} - \frac{1}{2} = \frac{1}{2}$. Then, the p-value is

$$\text{p-value} = \mathsf{P}_0^T (T \geq \frac{1}{2}) = \mathsf{P}_{\frac{1}{2}}\,(X \geq 82) = \left(\frac{1}{2} \right)^{82}.$$

Thus Arbuthnott merely calculated the p-value and concluded that he had a very strong evidence against the hypothesis.   □

Exercise 5.2 Carry out the tests for assessing evidence in Example 5.1 for the test problems (5.5) and (5.8).

Example 5.5 (Pendulum) Continuation of Example 5.3 with model (5.13), where $\sigma_0^2 = (0.012)^2$. We will carry out a test for the two-sided test problem (5.14) with $\mu_0 = 2.1$. As a measure for discrepancy we choose

$$T(\mathbf{x}) = \widehat{\mu}_{\mathrm{MLE}} - \mu_0. \qquad (5.17)$$

The maximum likelihood estimate $\widehat{\mu}_{\mathrm{MLE}}$ is the sample mean $\overline{x}$. The null distribution of $T(\mathbf{X})$ is

$$\mathsf{P}_0^T = \mathsf{N}\left(0, \frac{1}{n}\sigma_0^2\right). \qquad (5.18)$$

From Example 4.4, page 76, we know $\widehat{\mu}_{\mathrm{MLE}} = 1.998$. Thus $T(\mathbf{x}) = t_{\mathrm{obs}} = -0.102$ and

$$
\begin{aligned}
\text{p-value} &= 1 - \mathsf{P}_0^T\left(|T| < |t_0|\right) = 1 - \mathsf{P}_0^T\left(\left|\frac{\sqrt{n}T}{\sigma_0}\right| < \frac{\sqrt{n}}{\sigma_0}|t_0|\right) \\
&= 2\Phi\left(-\frac{\sqrt{10}\,(0.102)}{0.012}\right) = 2\Phi\left(-26.879\right) \approx 0.
\end{aligned}
$$

There is a very strong evidence against the hypothesis $H_0 : \mu = 2.1$.   □

Example 5.6 (Pendulum) Consider Example 5.3 with the model (5.11), where σ^2 is unknown too. We will carry out a test for the two-sided test problem (5.12) with $\mu_0 = 2.0$. The test statistic (5.17) is not useful, because the null distribution of (5.17) still depends on the unknown variance. The null hypothesis is not a simple one. Instead of (5.17) we take the studentized statistic:

$$T(\mathbf{x}) = \frac{\widehat{\mu}_{\mathrm{MLE}} - \mu_0}{\text{s.e.}\left(\widehat{\mu}_{\mathrm{MLE}}\right)},$$

where s.e. stands for standard error, here

$$\text{s.e.}\left(\widehat{\mu}_{\mathrm{MLE}}\right) = \frac{1}{\sqrt{n}}s,$$

where s is the sample deviation. Under the null hypothesis we have that $T(\mathbf{X})$ is t-distributed with 9 degrees of freedom. **OBS!** We get a unique null

distribution of the test statistic, also in case of a composite null hypothesis!
Using $s = 0.013$, the observed value of the test statistic is

$$T(\mathbf{x}) = \frac{\sqrt{10}\,(1.998 - 2)}{0.013} = -0.4865.$$

The p-value is $1 - P_0^T(|T| < 0.4865) = 0.642$. There is no evidence against
$H_0 : \mu = 2.0$.   □

Exercise 5.3 Compare the test results of Example 5.5 and Example 5.6. Is
there a contradiction?

5.2.1 Simulation Tests

Let us very shortly explain the main idea of simulation tests with the help of
a constructed example. Assume that the null distribution is continuous and
has a density f_0. Let the p-value be given by

$$p = \int \mathbb{1}_{A_{\mathrm{obs}}}(t)\, f_0(t)\, dt,$$

where A_{obs} describes the region in the direction of the extreme values. $\mathbb{1}_A(t)$
is the indicator function.
If $f_0(t)$ is completely known, then it remains to calculate an integral. When
the integral is complicated then the p-value can be approximated by Monte
Carlo methods. These tests are called **Monte Carlo tests**. For instance, by
using independent Monte Carlo we carry out the following steps:
Algorithm for independent Monte Carlo:

1. Draw K random variables $t_j, j = 1, \ldots, K$ from P_0^T.
2. Approximate the p-value by

$$p_{\mathrm{MC}} = \frac{\#\{j : t_j \in A_{\mathrm{obs}}\}}{K}.$$

In case of a simple hypothesis $H_0 : P_0$ we can draw K samples of sample size
n from P_0: $\mathbf{x}_j = 1, \ldots, K$ and calculate $t_j = T(\mathbf{x}_j)$ for each sample in step 1.

Example 5.7 (Monte Carlo test) Consider data coming from two independent samples:

$$
\begin{aligned}
\mathbf{x} &= (0.969, 2.047, 2.349, 1.366, 1.399, -0.001, 0.371, 1.817, 0.902, 1.114) \\
\mathbf{y} &= (0.967, 1.214, -0.673, 1.275, 3.668, -0.478, 5.265, 0.579, 0.556, 5.989).
\end{aligned}
$$

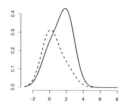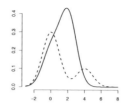

Figure 5.7: The left picture shows the density of both samples under the null hypothesis. The right picture shows the underlying distribution of the data.

The first sample consists of i.i.d. r.v.'s from a mixture of normal distributions:

$$(X_1, ..., X_n) \text{ i.i.d. with } X_i = U_{1i}Z_{1i} + (1 - U_{1i})Z_{2i},$$

where $U_{1i} \sim \text{Ber}(\frac{1}{2})$ i.i.d. and $Z_{1i} \sim N(0, 1)$ i.i.d., $Z_{2i} \sim N(\mu, 1)$ i.i.d.. The second sample comes from another mixture of normal distributions

$$(Y_1, ..., Y_n) \text{ i.i.d. with } Y_i = U_{2i}V_{1i} + (1 - U_{2i})V_{2i},$$

where $U_{2i} \sim \text{Ber}(\frac{3}{4})$ i.i.d. and $V_{1i} \sim N(0, 1)$ i.i.d., $V_{2i} \sim N(\mu + \Delta, 1)$ i.i.d. All Bernoulli variables and all normal variables are mutually independent. For the Monte Carlo test we assume that $\mu = 2$. Thus the model is

$$\mathcal{P} = \{P^{\text{sample1}} \otimes P_\Delta^{\text{sample2}} : \Delta \in \mathbb{R}\}.$$

The test problem is

$$H_0 : \Delta = 0 \text{ versus } H_1 : \Delta > 0.$$

The null hypothesis $H_0 : P_0$ is simple. An appropriate test statistic is $T = 4\overline{Y} - 2\overline{X}$, because

$$
\begin{aligned}
ET &= 4E\overline{Y} - 2E\overline{X} \\
&= 4EU_2EV_1 + 4(1 - EU_2)EV_2 - 2EU_1EZ_1 - 2(1 - EU_1)EZ_2 \\
&= 4(1 - \frac{3}{4})(\mu + \Delta) - 2\frac{1}{2}\mu = \Delta.
\end{aligned}
$$

We have $t_{\text{obs}} = 4 * 1.8362 - 2 * 1.2333 = 4.8782$. The p-value is defined as $P_0 (T > t_{\text{obs}})$. Now carry out a Monte Carlo approximation: We generate 1000 samples under P_0 and calculate the respective t_{obs}. Only two samples of 1000 have a value higher than the observed 4.8782. Hence, the approximated p-value is 0.002. We conclude that there is strong evidence against the null hypothesis $H_0 : \Delta = 0$. That is the right conclusion, because the data were simulated with $\Delta = 3$; compare Figure 5.7.   □

Exercise 5.4 In Example 5.7 the test statistic $T = 4\overline{Y} - 2\overline{X}$ is chosen for a better comparison of the different simulation tests; compare also the following Example 5.8. Propose a more appropriate test statistic for Example 5.7 using the knowledge: $\mu = 2$. Why is it better?

The bootstrap ideas are applied in cases where $f_0(t)$ is not completely known. Assume that the distribution of the sample $\mathsf{P}_{(\theta, \lambda)}$ depends on the parameter of θ to be tested, and on a parameter λ. The null distribution of the test statistic $T(\mathbf{X})$ depends also on λ: $T \sim \mathsf{P}^T_{(\theta_0, \lambda)}$. Davison and Hinkley (2003) proposed as a natural approach to take the following p-value

$$\text{p-value} = \mathsf{E}_{(\theta_0, \widehat{\lambda})} \, \mathbb{1}_{A_{\text{obs}}}(T).$$

Algorithm for a parametric bootstrap test:

1. Calculate an estimate $\widehat{\lambda} = \widehat{\lambda}(\mathbf{x})$.

2. Draw B random samples of size n from $\mathsf{P}^{\mathbf{X}}_{(\theta_0, \widehat{\lambda})}$:

$$X^*_{i,j} = x^*_{i,j}, \ \ \mathbf{X}^*_j \sim \mathsf{P}^{\mathbf{X}}_{(\theta_0, \widehat{\lambda})}, \ \ i = 1, \ldots, n, \ \ j = 1, \ldots, B.$$

   OBS! The samples are drawn under the hypothesis!

3. For each bootstrap sample calculate the test statistic

$$t^*_j = T(\mathbf{x}^*_j), \ j = 1, \ldots, B.$$

4. Approximate the p-value by

$$p_{boot} = \frac{\#\{j : t^*_j \in A_{\text{obs}}\}}{B}.$$

Example 5.8 (Bootstrap Test)
Continuation of Example 5.7. Unlike in the example above, the expected value μ is not known. The model is now

$$\mathcal{P} = \{\mathsf{P}^{\text{sample1}}_{\mu} \otimes \mathsf{P}^{\text{sample2}}_{(\mu, \Delta)} : (\mu, \Delta) \in \mathbb{R}^2\}.$$

We estimate μ by $\widehat{\mu} = 2\overline{X} = 2.4666$ and generate 1000 samples under $\mathsf{P}^{\text{sample1}}_{2.46} \otimes \mathsf{P}^{\text{sample2}}_{(2.46, 0)}$. This time 11 samples of 1000 have a value of the respective t_{obs} higher than 4.8782. Hence the bootstrap p-value p_{boot} is 0.011. We conclude that we have evidence against the null hypothesis $H_0 : \Delta = 0$. $\qquad\qquad\square$

5.3   Tests: Decision Rules

In what follows we consider a test as a two action decision rule between the null hypothesis H_0 and the alternative H_1. We begin with nonrandomized tests.

Definition 5.2 A (nonrandomized) **test** φ is a statistic from the sample space $\mathcal{X}$ to $\{0,1\}$:

$$\varphi(\mathbf{x}) = \begin{cases} 1 & \text{if } \mathbf{x} \in C_1 \quad (\text{reject } H_0) \\ 0 & \text{if } \mathbf{x} \in C_0 \quad (\text{do not reject } H_0) \end{cases},$$

where $\mathcal{X} = C_1 \cup C_0$, with $C_1 \cap C_0 = \emptyset$. C_1 is called the **critical region**.

Whenever we observe $\mathbf{x} \in C_1$ we reject the null hypothesis H_0. Each partition of the sample space $\mathcal{X}$ defines a test. Each subset of $\mathcal{X}$ can be a critical region. For $C_1 = \mathcal{X}$ we will always reject H_0; for $C_1 = \emptyset$ we never reject H_0. Just for illustration we come back to Example 3.1 on page 25.

Example 5.9 (Lion's appetite) Let us suppose a lion has only two stages, lethargic θ_0 (former θ_3) and hungry θ_1. We consider the test problem:

$$H_0 : \theta = \theta_0 \text{ versus } H_1 : \theta = \theta_1. \tag{5.19}$$

The sample space is $\mathcal{X} = \{0, 1, 2, 3, 4\}$. The test φ_1 with critical region $C_{1,1} = \{0, 1, 2, 3, 4\}$ always rejects H_0. The test φ_2 with critical region $C_{1,2} = \emptyset$ never rejects H_0. The test φ_3 with critical region $C_{1,3} = \{0\}$ rejects that the lion is lethargic iff he eats no person, which seems not very reasonable. The test φ_4 with the critical region $C_{1,4} = \{4\}$:

$$\varphi_4(x) = \begin{cases} 1 & \text{if } \quad x \in \{4\} \quad (\text{reject } H_0) \\ 0 & \text{if } \quad x \in \{0, 1, 2, 3\} \quad (\text{do not reject } H_0) \end{cases}$$

or the test φ_5 with the critical region $C_{1,5} = \{3, 4, \}$,

$$\varphi_5(x) = \begin{cases} 1 & \text{if } \quad x \in \{3, 4\} \quad (\text{reject } H_0) \\ 0 & \text{if } \quad x \in \{0, 1, 2\} \quad (\text{do not reject } H_0) \end{cases}$$

seems to be more useful. The test φ_6 with $C_{1,6} = \{0, 4\}$ looks useless in this context:

$$\varphi_6(x) = \begin{cases} 1 & \text{if } \quad x \in \{0, 4\} \quad (\text{reject } H_0) \\ 0 & \text{if } \quad x \in \{1, 2, 3\} \quad (\text{do not reject } H_0) \end{cases}.$$

Altogether we have in this example $32 = 2^5$ different nonrandomized tests.□

Example 5.10 (Pendulum) Continued Example 5.3 and 5.5. Consider the model $\mathcal{P} = \{N(\mu, \sigma_0^2)^{\otimes 10} : \mu \in \mathbb{R}\}$ and the test problem $H_0 : \mu = \mu_0$ versus $H_1 : \mu \neq \mu_0$. Then, for each c we can define a critical region with the help of the test statistic $T(\mathbf{x}) = \widehat{\mu}_{\text{MLE}} - \mu_0$ by

$$C_1 = \{\mathbf{x} : |\widehat{\mu}_{\text{MLE}} - \mu_0| > c\}. \tag{5.20}$$

□

The goal of this section is to define useful properties which we will require from a test. We will construct tests, which possess these properties. For this approach it makes sense to introduce the more general notation of a randomized test, especially for statistical models which are subsets of discrete distributions. In the following we always consider a randomized test when we speak about a test.

Definition 5.3 (Test) A randomized **test** φ is a step function on $C_1, C_=$ and C_0 to $[0, 1]$, where $\mathcal{X} = C_1 \cup C_= \cup C_0$ and $C_1, C_=, C_0$ are mutually disjunct:

$$\varphi(\mathbf{x}) = \begin{cases} 1 & \text{if } \mathbf{x} \in C_1 & (\text{reject } H_0) \\ \gamma & \text{if } \mathbf{x} \in C_= & (\text{reject } H_0 \text{ with probability } \gamma) \\ 0 & \text{if } \mathbf{x} \in C_0 & (\text{do not reject } H_0) \end{cases}.$$

To carry out the test obtain data $\mathbf{x}$ and then sample u from $\text{U}[0, 1]$ if $u < \varphi(\mathbf{x})$ reject H_0; otherwise do not reject H_0.

A test decision done by a randomized test depends on two independent random variables: the sample $\mathbf{X} \sim P_\theta$ and $U \sim \text{U}[0, 1]$. Thus the rejection probability

Figure 5.8: Randomization in real life.

is

$$
\begin{aligned}
\mathsf{P}_\theta(\text{``reject } H_0\text{''}) \;&=\; \mathsf{P}_\theta^{(\mathbf{X},U)}(U < \varphi(\mathbf{X})) = \mathsf{E}_\theta\,\mathbb{1}_{(0,\varphi(\mathbf{X}))}(U) \\
&=\; \mathsf{E}_\theta\mathsf{E}(\,\mathbb{1}_{(0,\varphi(\mathbf{X}))}(U)\mid \mathbf{X}) \\
&=\; \mathsf{E}_\theta\varphi(\mathbf{X}) = 1\,\mathsf{P}_\theta(C_1) + \gamma\,\mathsf{P}_\theta(C_=) + 0\,\mathsf{P}_\theta(C_0) \\
&=\; \mathsf{P}_\theta(C_1) + \gamma\mathsf{P}_\theta(C_=).
\end{aligned}
$$

Note, the nonrandomized test of Definition 5.2 is a special case of the Definition 5.3 for $\gamma = 0$ or $\gamma = 1$. Maybe the application of a randomized decision feels uncommon, but we randomize also in real life; compare Figure 5.8!

Example 5.11 (Lion's appetite) Continuation of Example 5.9. An example for a randomized test for H_0 : "The lion was lethargic" versus H_1 : "The lion was hungry" could be

$$
\varphi_{\text{rand}}(x) = \left\{
\begin{array}{lll}
1 & \text{if } x \in \{3,4\} & (\text{``The lion was hungry.''}) \\
\frac{1}{2} & \text{if } x \in \{2,1\} & (\text{reject } H_0 \text{ with probability } \frac{1}{2}) \\
0 & \text{if } x \in \{0\} & (\text{``The lion was lethargic.''}).
\end{array}
\right.
\qquad (5.21)
$$

For $x \in \{2,1\}$ generate u from $\mathsf{U}[0,1]$. If $u < \frac{1}{2}$, then reject H_0 and decide that the lion was hungry. $\qquad\square$

There are two possibilities for a wrong decision.
First, H_0 is true, but we reject H_0. That is the **error of type I**. Second, H_0 is not true, but we do not reject H_0. That is the **error of type II**.

	H_0 is true	H_0 is wrong
reject H_0	first type	no error
do not reject H_0	no error	second type

We define:

Definition 5.4 For the test problem H_0: Θ_0 versus H_1: Θ_1 we define

1. **error of type I:** reject H_0, but $\theta \in \Theta_0$.

2. **error of type II:** do not reject H_0, but $\theta \in \Theta_1$.

3. The function defined on Θ by

$$\pi(\theta) := \mathsf{E}_\theta \varphi(\mathbf{X}) = \mathsf{P}_\theta(C_1) + \gamma \mathsf{P}_\theta(C_=)$$

is called the **power function** of the test φ.

The test result depends on the sample. That means making an error is a random event. What we would like to try is to minimize the probabilities for the errors of first and second type. For $\theta \in \Theta_0$ we have that $\pi(\theta)$ is the probability of making the error of first type. For $\theta \in \Theta_1$, $1 - \pi(\theta)$ is the probability of making the error of second type. Thus we want tests with

$$\pi(\theta) \text{ is minimal for } \theta \in \Theta_0 \text{ and } \pi(\theta) \text{ is maximal for } \theta \in \Theta_1. \tag{5.22}$$

Let us illustrate this by our Lion example.

Example 5.12 (Lion's appetite) Consider the statistical model in Example 3.1 on page 25 and the test problem (5.19). Let us calculate the power functions for the tests φ_4 and φ_6 in Example 5.9. We have for the test φ_4: $\pi(\theta_0) = \mathsf{E}_{\theta_0} \varphi_4(\mathbf{X}) = \mathsf{P}_{\theta_0}(4) = 0$ and $\pi(\theta_1) = \mathsf{E}_{\theta_1} \varphi_4(\mathbf{X}) = \mathsf{P}_{\theta_1}(4) = 0.1$. For the test φ_6 we obtain $\pi(\theta_0) = 0.9$ and $\pi(\theta_1) = 0.1$. If we compare φ_4 and φ_6, then we see that φ_6 is really a bad test, with a high chance for both type of errors. Consider now test φ_{rand} in Example 5.11. We obtain

$$
\begin{aligned}
\pi(\theta_0) &= \mathsf{P}_{\theta_0}(C_1) + \gamma \mathsf{P}_{\theta_0}(C_=) = \\
&= \mathsf{P}_{\theta_0}(4) + \mathsf{P}_{\theta_0}(3) + \frac{1}{2}(\mathsf{P}_{\theta_0}(2) + \mathsf{P}_{\theta_0}(1)) = 0.05
\end{aligned}
$$

and $\pi(\theta_1) = 0.95$. This result looks much better. $\square$

Example 5.13 (Pendulum) Continuation of Example 5.10. Consider the nonrandomized test with the critical region (5.20). Using (5.18) we obtain

$$\pi(\theta_0) = \mathsf{P}_{\theta_0}(C_1) = 1 - \mathsf{P}_{\theta_0}(|\widehat{\theta} - \theta_0| \le c) = 2\Phi\left(\frac{-\sqrt{n}}{\sigma_0}c\right) \tag{5.23}$$

and

$$\pi(\theta_1) = 1 - P_{\theta_1}(|\widehat{\theta} - \theta_0| \leq c) = 1 - \Phi\left(\sqrt{n}\frac{-c + \theta_0 - \theta_1}{\sigma_0}\right) + \Phi\left(\sqrt{n}\frac{-c + \theta_0 - \theta_1}{\sigma_0}\right).$$
(5.24)

If we want to construct a test with a critical region of type (5.20), then we have the free choice of the critical value c. The best would be to find a c^* which minimizes the probability of both types of errors. **OBS!** We cannot have (5.22)! Consider the power function as a function of c. We see $\pi(\theta_0)$ is decreasing in c and for all θ_1 $\pi(\theta_1)$ is decreasing in c too. We have to find a way out!   □

Exercise 5.5 Plot the probabilities for the error of first type and for the error of second type as functions of the value c in Example 5.13. Discuss the monotony properties of the functions.

In Example 5.13 we saw that it is not possible to optimize the probability of both errors simultaneously. The way out is to treat the errors asymmetrically. We will see that in test theory the error of first type will get the greater importance! The rejection of the null hypotheses becomes the safe decision. Therefore the test problem has to be formulated, such that the goal is to reject the null hypothesis. That means a test decision rule is like an indirect proof. When the aim of an experiment is to establish an assertion, then the negation of the assertion should be taken as null hypothesis. The assertion becomes the alternative.

> H_1 is the claim we want to establish.
> H_0 is the negation of it.

Exercise 5.6 Compare the possible test problems in Example 5.1 on page 122. Firstly consider the consumer point of view: You want to show that there are more defective pens than guaranteed by the producer. Secondly take the position of the producer, who will be sure that the production fulfills the guarantee conditions. Which test problem is related to the consumer point of view? Which test problem is important for the producer?

5.3.1   The Neyman–Pearson Test

We study the case where the statistical model consists of two probability measures only: $\mathcal{P} = \{P_0, P_1\}$. Consider the simple hypotheses

$$H_0 : P_0 \text{ versus } H_1 : P_1.$$
(5.25)

In order to define and to find an "optimal" test for this simple test problem we will study the probabilities of wrong decisions. We define:

> **Definition 5.5** For testing simple H_0 versus H_1, the **size** of the type I error is $\alpha = P_0\,(\text{"reject } H_0\text{"})$. The **size** of the type II error is $\beta = P_1\,(\text{"do not reject } H_0\text{"})$.

It holds

$$\alpha = P_0\,(\text{"reject } H_0\text{"}) = P_0(C_1) + \gamma P_0(C_=)$$

and

$$\beta = P_1\,(\text{"do not reject } H_0\text{"}) = P_1(C_0) + (1-\gamma)P_1(C_=).$$

We can write it as

$$\alpha = E_0\varphi(\mathbf{X}) \text{ and } \beta = 1 - E_1\varphi(\mathbf{X}). \tag{5.26}$$

Example 5.14 (Normal distribution) Consider the test problem of two normal distributions

$$H_0 : N(0,4)^{\otimes 25} \text{ versus } H_1 : N(1,4)^{\otimes 25}$$

and the class of tests based on the test statistic $T(\mathbf{x}) = \overline{x}$

$$\varphi(\mathbf{x}) = \left\{ \begin{array}{lll} 1 & \text{if} & T(\mathbf{x}) > c \\ \gamma & \text{if} & T(\mathbf{x}) = c \\ 0 & \text{if} & T(\mathbf{x}) < c \end{array} \right. .$$

Under $H_0 : T(\mathbf{X}) \sim N(0, \frac{4}{25})$ and under $H_1 : T(\mathbf{X}) \sim N(1, \frac{4}{25})$. **OBS!** The randomization is meaningless, because: $P_0(C_=) = P_1(C_=) = 0!$ Thus we have a test

$$\varphi(\mathbf{x}) = \left\{ \begin{array}{lll} 1 & \text{if} & \overline{x} > c \\ 0 & \text{if} & \overline{x} \le c \end{array} \right. \tag{5.27}$$

with

$$\alpha = P_0(C_1) = P_0(\overline{X} > c) = 1 - \Phi\left(\frac{5}{2}c\right) \tag{5.28}$$

and

$$\beta = P_1(C_0) = P_1(\overline{X} < c) = \Phi\left(\frac{5}{2}(c-1)\right). \tag{5.29}$$

Again we see $\alpha = \alpha(c) \searrow$ and $\beta = \beta(c) \nearrow$. Compare Figure 5.9. $\square$

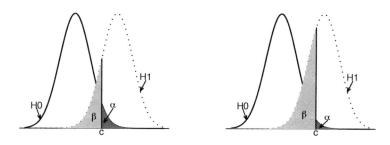

Figure 5.9: Illustration to Example 5.14 contradictory relation between α and β when changing the critical region by choosing a larger c.

Example 5.15 (Binomial distribution) Consider the problem of testing two binomial distributions

$$H_0 : \text{Bin}(10, \frac{1}{2}) \text{ versus } H_1 : \text{Bin}(10, \frac{1}{4})$$

and tests based on the test statistic $T(\mathbf{x}) = x$

$$\varphi(x) = \begin{cases} 1 & \text{if } x < c \\ \gamma & \text{if } x = c \\ 0 & \text{if } x > c \end{cases}. \qquad (5.30)$$

OBS! The randomization makes a difference, because $P_0(C_=) = \text{bin}(c, 10, \frac{1}{2}) = \binom{10}{c}\left(\frac{1}{2}\right)^{10} \neq 0$ and $P_1(C_=) = \text{bin}(c, 10, \frac{1}{4}) = \binom{10}{c}\left(\frac{1}{4}\right)^c\left(\frac{3}{4}\right)^{10-c} \neq 0$! We get

$$\alpha = P_0(C_1) + \gamma P_0(C_=) = \sum_{k=0}^{c-1} \text{bin}(k, 10, \frac{1}{2}) + \gamma \text{bin}(c, 10, \frac{1}{2})$$

and

$$\beta = \sum_{k=c+1}^{10} \text{bin}(k, 10, \frac{1}{4}) + (1 - \gamma)\text{bin}(c, 10, \frac{1}{4}). \qquad (5.31)$$

Again we see $\alpha = \alpha(c) \searrow$ and $\beta = \beta(c) \nearrow$.   □

The size of error of type I and the size of error of type II behave contradictorily. Let us study for a given test problem (5.25) all possible randomized tests φ and the relating sizes $(\alpha(\varphi), \beta(\varphi))$. The area $\{(\alpha(\varphi), \beta(\varphi)) : \varphi$ is a test$\}$ is called the set of $\alpha\beta$-**representations**. See Figure 5.10.

Theorem 5.1 *The set of $\alpha\beta$-representations is convex, is included in the closed unit square and includes the points $(0, 1)$ and $(1, 0)$.*

PROOF: A set is convex iff it contains all line segments connecting two points. Take two arbitrary points $(\alpha_1, \beta_1) = (\alpha(\varphi_1), \beta(\varphi_1))$ and $(\alpha_2, \beta_2) = (\alpha(\varphi_2), \beta(\varphi_2))$. Define tests

$$\varphi(\mathbf{x}) = \lambda\varphi_1(\mathbf{x}) + (1 - \lambda)\varphi_2(\mathbf{x}), \ 0 \le \lambda \le 1.$$

Using (5.26) we get

$$\alpha(\lambda) = \mathsf{E}_0\varphi(\mathbf{X}) = \mathsf{E}_0\left(\lambda\varphi_1(\mathbf{X}) + (1 - \lambda)\varphi_2(\mathbf{X})\right) = \lambda\alpha_1 + (1 - \lambda)\alpha_2$$

$$\begin{aligned}
\beta(\lambda) &= \mathsf{E}_1\left(1 - \varphi(\mathbf{X})\right) \\
&= \mathsf{E}_1\left(\lambda(1 - \varphi_1(\mathbf{X})) + (1 - \lambda)(1 - \varphi_2(\mathbf{X}))\right) \\
&= \lambda\beta_1 + (1 - \lambda)\beta_2.
\end{aligned}$$

Hence the segment $(\alpha(\lambda), \beta(\lambda)), \ 0 \le \lambda \le 1$ belongs to the set of $\alpha\beta$-representations.

The point $(0, 1)$ is related to the test $\varphi(\mathbf{x}) \equiv 0$ because $\alpha(\varphi) = \mathsf{E}_0\varphi(\mathbf{X}) = 0$ and $\beta(\varphi) = \mathsf{E}_1\left(1 - \varphi(\mathbf{X})\right) = 1$. The point $(1, 0)$ is related to the test $\varphi(\mathbf{x}) \equiv 1$.

□

The points in a plane are not totally ordered. We cannot say that a pair (α, β) corresponds to a better test. However, tests with the same α_0 are ordered and we should prefer the test with the smallest β; see Figure 5.11. Furthermore, we can see in the picture that by monotony of the lower boundary line it is the smallest β for all tests with an α less than α_0. The preassigned bound α_0 is called **significance level**.

Definition 5.6 (MP test) A test φ^* is called **most powerful** (MP test) of size α iff $\mathsf{E}_0\varphi^*(\mathbf{X}) = \alpha$ and $\mathsf{E}_1\varphi^*(\mathbf{X}) \ge \mathsf{E}_1\varphi(\mathbf{X})$ for all tests φ with $\mathsf{E}_0\varphi(\mathbf{X}) \le \alpha$.

Recall $\pi(\theta_1) = \mathsf{E}_1\varphi(\mathbf{X})$ is the power of the test φ under the alternative $\mathsf{P}_1 = \mathsf{P}_{\theta_1}$. We have

$$\beta = 1 - \pi(\theta_1);$$

thus a most powerful test minimizes the chance for an error of second type. What we are doing once more, now in other words.

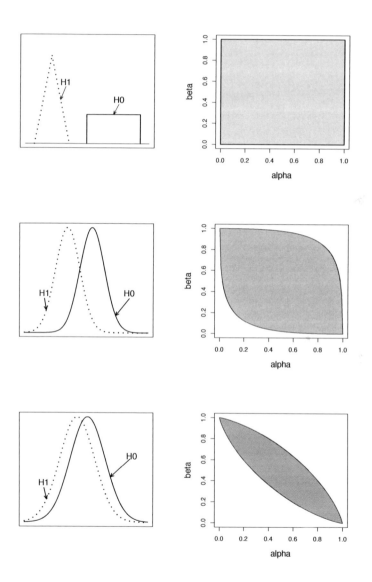

Figure 5.10: In the first row the alternative distributions have disjunct support. Here it is possible to have an error free decision! In the second row the distance between the distribution is still large. It is possible to find good tests with small α and small β. The test problem in the last row is worse. Tests with small α have a large β.

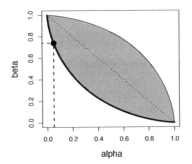

Figure 5.11: For a fixed $\alpha_0 = 0.05$ the smallest β is marked with the thick dot. The corresponding test is most powerful; compare Definition 5.6.

First we set a bound for the probability of the error of first type, then we search the test with the smallest probability of the error of second type.

This means we control only the error of first type. Only the probability of rejecting the null hypotheses is under control!

The "**safe decision**" is the rejection of H_0.

Before we go further in looking for an MP test let us go back to the notion "p-value," in Definition 5.1. When we interpret the formulation: "We have evidence against H_0" as the decision to reject H_0, then the size of the test is bounded by $\alpha_0 = 0.05$ with respect to the table on page 129.
Consider a test φ with continuous distributed test statistic $T(\mathbf{X})$ and with significance level α_0 of the form

$$\varphi(\mathbf{x}) = \begin{cases} 1 & \text{if} \quad T(\mathbf{x}) \geq c(\alpha_0) \\ 0 & \text{if} \quad T(\mathbf{x}) < c(\alpha_0) \end{cases}, \tag{5.32}$$

with

$$P_0(T(\mathbf{X}) \geq c(\alpha_0)) = \alpha_0.$$

For $T(\mathbf{x}) = t_{\text{obs}}$ the p-value is given by p-value $= P_0(T(\mathbf{X}) \geq t_{\text{obs}})$. Then

$$\text{p-value} \leq \alpha_0 \Longleftrightarrow T(\mathbf{x}) \geq c(\alpha_0);$$

compare Figure 5.12. The following correspondence holds:

The decision of φ with $E_0\varphi(\mathbf{X}) = \alpha_0$ to reject H_0
is equivalent to the decision to reject H_0 iff p-value $\leq \alpha_0$.

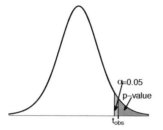 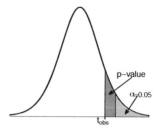

Figure 5.12: For a fixed $\alpha_0 = 0.05$ the comparison of the areas and the comparison of the observed value T_{obs} with the 0.95-quantile are equivalent. In the left picture we reject H_0 but in the right picture we cannot reject H_0.

Now let us return to the search of a best test: Which test is most powerful? The answer will be given by the next theorem. We begin by defining good candidates. Let $p_0(.)$ be the probability function related to P_0 and $p_1(.)$ be the one related to P_1. This notation includes discrete distributions as well as continuous distributions.

Definition 5.7 For $k \geq 0$ and $k = \infty$ and $\gamma \in [0, 1]$ tests of the form

$$\varphi(\mathbf{x}) = \begin{cases} 1 & \text{if} \quad p_0(\mathbf{x}) < k\, p_1(\mathbf{x}) \\ \gamma & \text{if} \quad p_0(\mathbf{x}) = k\, p_1(\mathbf{x}) \\ 0 & \text{if} \quad p_0(\mathbf{x}) > k\, p_1(\mathbf{x}) \end{cases} \tag{5.33}$$

are called **Neyman–Pearson tests**.

The test statistic in (5.33) is the likelihood quotient

$$\Lambda^*(\mathbf{x}) = \frac{p_0(\mathbf{x})}{p_1(\mathbf{x})};$$

thus a Neyman–Pearson test is a **likelihood ratio test**. For $p_1(\mathbf{x}) = 0$ we set $\frac{p_0(\mathbf{x})}{p_1(\mathbf{x})} = \infty$. The data $\mathbf{x}$ with $p_0(\mathbf{x}) = 0$ and $p_1(\mathbf{x}) = 0$ have probability zero and are excluded from $\mathcal{X}$. Note that for $k = \infty$ we obtain the test $\varphi(\mathbf{x}) \equiv 1$. The following theorem gives the answer: The Neyman–Pearson tests are MP tests. This result has also its own importance outside test theory. By historical reasons it is called a "lemma": the Neyman–Pearson lemma.

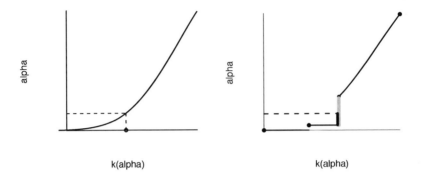

Figure 5.13: The left picture shows the choice of $k(\alpha)$ in (5.34), when randomization is not needed. In the right picture the distribution function of P_0 has a jump at $k(\alpha)$. $\gamma(\alpha)$ is the ratio of the step height up to α to the whole step height at $k(\alpha)$.

Theorem 5.2 (Neyman–Pearson lemma) *For all $\alpha \in [0,1]$ exist $\gamma(\alpha)$ and $k(\alpha)$ such that the Neyman–Pearson test with $\gamma(\alpha), k(\alpha)$ has size α and is most powerful of size α.*
Especially for $\alpha > 0$:

1. *If there exists a $k = k(\alpha)$ such that*

$$P_0(p_0(\mathbf{X}) < k\, p_1(\mathbf{X})) = \alpha, \qquad (5.34)$$

   then $\gamma = \gamma(\alpha) = 0$.
2. *Otherwise take $k = k(\alpha)$ with*

$$P_0(p_0(\mathbf{X}) < k\, p_1(\mathbf{X})) < \alpha < P_0(p_0(\mathbf{X}) \le k\, p_1(\mathbf{X})) \qquad (5.35)$$

   and $\gamma = \gamma(\alpha)$ with

$$\gamma = \frac{\alpha - P_0(p_0(\mathbf{X}) < k\, p_1(\mathbf{X}))}{P_0(p_0(\mathbf{X}) = k\, p_1(\mathbf{X}))}. \qquad (5.36)$$

PROOF: First we show that (5.33) is an α-test, that is $E_0\varphi(\mathbf{X}) = \alpha$. Take $\alpha = 0$, then the Neyman–Pearson test is $\varphi(\mathbf{x}) \equiv 0$, where $k = 0$, $\gamma = 0$. For $\alpha > 0$ we have two cases; compare Figure 5.13. If (5.34) holds, then the Neyman–Pearson test is not randomized:

$$E_0\varphi(\mathbf{X}) = P_0(C_1) = P_0(p_0(\mathbf{X}) < k\, p_1(\mathbf{X})) = \alpha.$$

Otherwise, we have a randomized test with (5.35) and (5.36):

$$\mathsf{E}_0\varphi(\mathbf{X}) = \mathsf{P}_0(C_1) + \gamma \mathsf{P}_0(C_=) =$$

$$\mathsf{P}_0(C_1) + \frac{\alpha - \mathsf{P}_0(C_1)}{\mathsf{P}_0(C_=)}\mathsf{P}_0(C_=) = \alpha.$$

Second, we show that for all tests ψ with $\mathsf{E}_0\psi(\mathbf{X}) \leq \alpha$ it holds

$$\mathsf{E}_1\varphi(\mathbf{X}) \geq \mathsf{E}_1\psi(\mathbf{X}).$$

We calculate

$$\mathsf{E}_1(\varphi(\mathbf{X}) - \psi(\mathbf{X})) \quad = \quad \int (\varphi(\mathbf{x}) - \psi(\mathbf{x}))\, p_1(\mathbf{x})d\mathbf{x} = I_1 + I_= + I_0,$$

where

$$I_1 = \int_{C_1} (\varphi(\mathbf{x}) - \psi(\mathbf{x}))\, p_1(\mathbf{x})d\mathbf{x}$$

and

$$I_= = \int_{C_=} (\varphi(\mathbf{x}) - \psi(\mathbf{x}))\, \frac{p_1(\mathbf{x})}{p_0(\mathbf{x})}p_0(\mathbf{x})d\mathbf{x}$$

and

$$I_0 = \int_{C_0} (\varphi(\mathbf{x}) - \psi(\mathbf{x}))\, \frac{p_1(\mathbf{x})}{p_0(\mathbf{x})}p_0(\mathbf{x})d\mathbf{x}$$

with

$$C_1 = \{\mathbf{x} : p_0(\mathbf{x}) < k\, p_1(\mathbf{x})\}$$

and

$$C_= = \left\{\mathbf{x} : \frac{1}{k}p_0(\mathbf{x}) = p_1(\mathbf{x})\right\} \text{ and } C_0 = \left\{\mathbf{x} : \frac{1}{k}p_0(\mathbf{x}) > p_1(\mathbf{x})\right\}.$$

For $\mathbf{x} \in C_1$ the Neyman–Pearson test is $\varphi(\mathbf{x}) = 1$ and

$$I_1 \geq \frac{1}{k}\int_{C_1} (\varphi(\mathbf{x}) - \psi(\mathbf{x}))\, p_0(\mathbf{x})d\mathbf{x}.$$

Furthermore,

$$I_= = \frac{1}{k}\int_{C_1} (\varphi(\mathbf{x}) - \psi(\mathbf{x}))\, p_0(\mathbf{x})d\mathbf{x}.$$

For $\mathbf{x} \in C_0$ the Neyman–Pearson test is $\varphi(\mathbf{x}) = 0$ and

$$I_0 \geq \frac{1}{k}\int_{C_0} (\varphi(\mathbf{x}) - \psi(\mathbf{x}))\, p_0(\mathbf{x})d\mathbf{x}.$$

Summarizing

$$\mathsf{E}_1\left(\varphi(\mathbf{X}) - \psi(\mathbf{X})\right) \geq \frac{1}{k}\int (\varphi(\mathbf{x}) - \psi(\mathbf{x}))\, p_0(\mathbf{x})d\mathbf{x} = \frac{1}{k}\mathsf{E}_0\left(\varphi(\mathbf{X}) - \psi(\mathbf{X})\right).$$

Since $E_0\varphi(\mathbf{X}) = \alpha$ and $E_0\psi(\mathbf{X}) \le \alpha$ it holds

$$E_0\left(\varphi(\mathbf{X}) - \psi(\mathbf{X})\right) = E_0\varphi(\mathbf{X}) - E_0\psi(\mathbf{X}) \ge \alpha - \alpha = 0.$$

Thus $E_1\left(\varphi(\mathbf{X}) - \psi(\mathbf{X})\right) \ge 0$.

$\square$

Corollary 5.1 *Let* $\pi(\theta_1)$ *be the power of the most powerful test with size* $\alpha \in (0,1)$ *for testing* $H_0 : P_{\theta_0}$ *versus* $H_1 : P_{\theta_1}$. *Then*

$$\alpha < \pi(\theta_1) \text{ unless } P_{\theta_0} = P_{\theta_1}.$$

PROOF: Since the test $\psi(\mathbf{x}) \equiv \alpha$ has power α, we get for the most powerful test φ that $\alpha \le \pi(\theta_1)$. If $\alpha = \pi(\theta_1) < 1$ then $\psi(\mathbf{x}) \equiv \alpha$ is most powerful and a Neyman–Pearson test. That means, there exists a constant k and $p_0(x) \equiv kp_1(x)$. That works only for $k = 1$ and $P_{\theta_0} = P_{\theta_1}$. See also Figure 5.11.

$\square$

Let us see which of the tests in Example 5.9 on page 135 are most powerful.

Example 5.16 (Lion's appetite) Continuation of Example 5.9 on page 135. The distributions are given in Example 3.1 on page 25. The simple test problem is

$$H_0 : p_0 \text{ (lethargic) versus } H_1 : p_1 \text{ (hungry)}.$$

Recall

x	0	1	2	3	4
$p_0(x)$	0.9	0.05	0.05	0	0
$p_1(x)$	0	0.05	0.05	0.8	0.1
$\frac{p_0(x)}{p_1(x)}$	∞	1	1	0	0

The critical region of a Neyman–Pearson test is $C_1 = \{x : p_0(x) < k\, p_1(x)\}$. We get for $k = 0$: $C_1 = \{x : p_0(x) < 0\} = \emptyset$ for $0 < k \le 1$: $C_1 = \left\{x : \frac{p_0(x)}{p_1(x)} = 0\right\} = \{3,4\}$, for $1 < k < \infty$: $C_1 = \left\{x : \frac{p_0(x)}{p_1(x)} = 0, \frac{p_0(x)}{p_1(x)} = 1\right\} = \{1,2,3,4\}$, for $k = \infty$: $C_1 = \{x : p_0(x) < \infty\} = \{0,1,2,3,4\} = \mathcal{X}$. Only 4 of the 32 possible critical regions are related to a Neyman–Pearson test. The tests φ_4 and φ_6 in Example 5.9 are no Neyman–Pearson tests. The test $\varphi_{\text{rand}}(x)$ in (5.21) is a Neyman–Pearson test with $k = 1$ and $\gamma = \frac{1}{2}$ and so the most powerful test of size $\alpha = 0.05$.

$\square$

Example 5.17 (Binomial distribution) Continuation of Example 5.15 on page 141. Under H_0 the probability function is

$$p_0(x) = \binom{10}{x} \exp(-n \ln 2)$$

and under H_1 the probability function is

$$p_1(x) = \binom{10}{x} \exp\left(-x \ln 4 + (10 - x) \ln \left(\frac{3}{4}\right)\right).$$

We calculate the region C_1 of the Neyman–Pearson test by

$$\{x : p_0(x) < k \, p_1(x)\}$$
$$= \{x : \exp(-10 \ln 2) < k \, \exp(-x \ln 4 + (10 - x) \ln \left(\frac{3}{4}\right))\}$$
$$= \{x : x < c\},$$

where $c = \frac{10 \ln 2 + 10 \ln\left(\frac{3}{4}\right) + \ln k}{\ln 3}$. Thus, the test (5.30) in Example 5.15 is a Neyman–Pearson test. Let us specify this test (5.30) for $\alpha = 0.05$. We have:

c	0	1	2	3	4	5	6	7	8	9	10	11
$P_0 (x < c)$	0	$\frac{1}{1024}$	$\frac{11}{1024}$	$\frac{7}{128}$	$\frac{11}{64}$	$\frac{193}{512}$	$\frac{319}{512}$	$\frac{53}{64}$	$\frac{121}{128}$	$\frac{1013}{1024}$	$\frac{1023}{1024}$	1.

Note that $\frac{11}{1024} = 0.0107$ and $\frac{7}{128} = 0.0547$. Thus we take $c = 2$ with

$$P_0 (x = c) = \binom{10}{2} \left(\frac{1}{2}\right)^{10} = \frac{45}{1024}$$

and calculate

$$\gamma = \frac{\alpha - P_0 (x < c)}{P_0 (x = c)} = \frac{0.05 - \frac{11}{1024}}{\frac{45}{1024}} = 0.893.$$

Thus the most powerful test of size $\alpha_0 = 0.05$ is

$$\varphi(x) = \begin{cases} 1 & \text{if } x < 2 \\ 0.893 & \text{if } x = 2 \\ 0 & \text{if } x > 2 \end{cases} . \tag{5.37}$$

We can also calculate the size of the error of second type by (5.31):

$$\beta = \sum_{k=3}^{10} \binom{10}{k} \left(\frac{1}{4}\right)^k \left(\frac{3}{4}\right)^{10-k} + (1 - 0.893) \binom{10}{2} \left(\frac{1}{4}\right)^2 \left(\frac{3}{4}\right)^{10-2}$$
$$= 0.504.$$

OBS! This size is really big! The chance is more than $\frac{1}{2}$ to make a wrong decision if the alternative is true. Nevertheless, it is the best what we can do when we require that $\alpha = 0.05$. □

Exercise 5.7 Show that a test with $C_1 = \{\mathbf{x} : \overline{x} > c\}$ is a Neyman–Pearson test for the test problem $H_0 : \mathrm{Ber}\left(\frac{1}{4}\right)^{\otimes 10}$ versus $H_1 : \mathrm{Ber}\left(\frac{1}{2}\right)^{\otimes 10}$. Specify the test for the level $\alpha = 0.05$.

Example 5.18 (Normal distribution) Continuation of Example 5.14 on page 140. Under H_0 the density is

$$f_0(\mathbf{x}) = \left(\frac{1}{2\sqrt{2\pi}}\right)^{25} \exp\left(-\frac{1}{8}\sum_{i=1}^{25} x_i^2\right)$$

and under H_1 we have the density

$$f_1(\mathbf{x}) = \left(\frac{1}{2\sqrt{2\pi}}\right)^{25} \exp\left(-\frac{1}{8}\sum_{i=1}^{25} (x_i - 1)^2\right).$$

Then the critical region $\{\mathbf{x} : f_0(\mathbf{x}) < k\, f_1(\mathbf{x})\}$ of the Neyman–Pearson test is

$$= \left\{\mathbf{x} : \exp\left(-\frac{1}{8}\sum_{i=1}^{25} x_i^2\right) < k \exp\left(-\frac{1}{8}\sum_{i=1}^{25}(x_i - 1)^2\right)\right\}$$

$$= \{\mathbf{x} : \overline{x} > c\},$$

with $c = \frac{1}{2} - \frac{4\ln k}{25}$. The test (5.27) in Example 5.14 on page 140 is a Neyman–Pearson test. Let us specify the test (5.27) for $\alpha = 0.05$. The condition on c is (5.28): $\mathrm{P}_0\left(\overline{X} > c\right) = 1 - \Phi(\frac{5}{2}c) = 0.05$. Thus $c = 0.658$. The Neyman–Pearson test is

$$\varphi(\mathbf{x}) = \begin{cases} 1 & \text{if } \overline{x} > 0.658 \\ 0 & \text{if } \overline{x} \leq 0.658 \end{cases}, \tag{5.38}$$

with $\alpha = 0.05$ and with β calculated by (5.29): $\beta = \Phi(\frac{5}{2}(c - 1)) = 0.196$. $\square$

Exercise 5.8 Show that the test with the critical region $\{\mathbf{x} : \overline{x} < c\}$ is a Neyman–Pearson test for the test problem $H_0 : \mathsf{N}(1,1)^{\otimes n}$ versus $H_1 : \mathsf{N}(-1,1)^{\otimes n}$.

Example 5.19 (Uniform distribution) We are interested in the test problem

$$H_0 : \mathsf{U}[-0.5, 0.5] \text{ versus } H_1 : \mathsf{U}[0,1].$$

The sample space is given by $\mathcal{X} = [-0.5, 1]$, the only interval where at least one of the probability functions is not zero. The likelihood ratio is

$$\frac{f_0(x)}{f_1(x)} = \begin{cases} \infty & \text{if } x \in [-0.5, 0] \\ 1 & \text{if } x \in (0, 0.5] \\ 0 & \text{if } x \in (0.5, 1] \end{cases}.$$

There are Neyman–Pearson tests for $k = \infty, 1, 0$. For $k = \infty$ the test is $\varphi \equiv 1$. Consider the case $k = 1$:

$$\varphi(x) = \begin{cases} 1 & \text{if} \quad x \in (0.5, 1] \\ \gamma & \text{if} \quad x \in (0, 0.5] \\ 0 & \text{if} \quad x \in [-0.5, 0] \end{cases}.$$

The $\gamma(\alpha)$ is determined by

$$\mathsf{E}_0 \varphi(x) = \mathsf{P}_0((0.5, 1]) + \gamma \mathsf{P}_0((0, 0.5]) = 0 + \gamma 0.5 = \alpha.$$

Thus, the most powerful test is

$$\varphi(x) = \begin{cases} 1 & \text{if} \quad x \in (0.5, 1] \\ 2\alpha & \text{if} \quad x \in (0, 0.5] \\ 0 & \text{if} \quad x \in [-0.5, 0] \end{cases}. \tag{5.39}$$

The size of the error of second type is

$$\beta = \mathsf{P}_1([-0.5, 0]) + (1 - \gamma)\,\mathsf{P}_1((0, 0.5]) = 0 + (1 - 2\alpha)\,0.5 = 0.5 - \alpha.$$

Consider now $k = 0$. Then

$$\varphi(x) = \begin{cases} \gamma & \text{if} \quad x \in [0.5, 1] \\ 0 & \text{if} \quad x \in [-0.5, 0.5) \end{cases}.$$

We get $\alpha = \mathsf{E}_0 \varphi(x) = \gamma \mathsf{P}_0([0.5, 1]) = 0$ and $\mathsf{E}_1 \varphi(x) = \gamma \mathsf{P}_1([0.5, 1]) = \frac{\gamma}{2}$. Thus $\beta = 1 - \frac{\gamma}{2}$; β is minimal for $\gamma = 1$. We obtain the test (5.39) with $\alpha = 0$. $\square$

5.3.2   Uniformly Most Powerful Tests

Consider a statistical model for a sample $\mathbf{X} = (X_1, \ldots, X_n)$: $\mathcal{P} = \{\mathsf{P}_\theta : \theta \in \Theta\}$ with $\Theta = \Theta_0 \cup \Theta_1$ with $\Theta_0 \cap \Theta_1 = \varnothing$. We will study now a test problem which also includes composite hypotheses

$$H_0 : \theta \in \Theta_0 \text{ versus } H_1 : \theta \in \Theta_1. \tag{5.40}$$

The Definition (5.6) is valid only for simple hypotheses. We will generalize it to the case (5.40). The chance for an error of first type should be bounded for all $\theta \in \Theta_0$. We define:

Definition 5.8 A test φ is called **test of size** α iff

$$\alpha = \sup_{\theta \in \Theta_0} \mathsf{E}_\theta \varphi(\mathbf{X}). \tag{5.41}$$

For a given $\alpha \in (0,1)$ a test φ is called a **level α-test** iff $\alpha = \sup_{\theta \in \Theta_0} E_\theta \varphi(\mathbf{X})$. Especially we say in this case that a test φ attains the level α; compare Liese and Miescke (2008).

In the following two examples the power functions are monotone and it is possible to calculate the size α of the tests.

Example 5.20 (Binomial distribution) Compare also Example 5.15 on page 141. Consider the statistical model $\mathcal{P} = \{\text{Bin}(10, \theta) : \theta \in (0,1)\}$ and the composite hypotheses

$$H_0 : \theta \geq 0.5 \text{ versus } H_1 : \theta < 0.5. \tag{5.42}$$

Take the test (5.37)

$$\varphi(x) = \begin{cases} 1 & \text{if} \quad x < 2 \\ 0.893 & \text{if} \quad x = 2 \\ 0 & \text{if} \quad x > 2 \end{cases}. \tag{5.43}$$

Then the condition (5.41) is

$$\alpha = \sup_{\theta \in [0.5,1)} [P_\theta(X < 2) + \gamma P_\theta(X = 2)]$$

$$= \sup_{\theta \in [0.5,1)} \left[\sum_{k=0}^{1} \text{bin}(k, 10, \theta) + 0.893 \, \text{bin}(2, 10, \theta) \right],$$

with $\text{bin}(k, 10, \theta) = \binom{10}{k} \theta^k (1 - \theta)^{10-k}$ and

$$\frac{\partial}{\partial \theta} \ln(\text{bin}(k, 10, \theta)) = \frac{\partial}{\partial \theta} (k \ln(\theta) + (10 - k) \ln(1 - \theta))$$

$$= k \frac{1}{\theta} - (10 - k) \frac{-1}{10}$$

$$= \frac{k - 10\theta}{\theta(1 - \theta)} < 0 \text{ for } k = 0, 1, 2 \text{ and } \theta \in [0.5, 1).$$

Hence $\text{bin}(k, 10, \theta)$ is monotone decreasing for $k = 0, 1, 2$ and $\theta \in [0.5, 1)$. We get

$$\alpha = \left[\sum_{k=0}^{1} \text{bin}(k, 10, 0.5) + 0.893 \, \text{bin}(2, 10, 0.5) \right] = 0.05.$$

The maximum of the power function lies on the border to the alternative. $\square$

Example 5.21 (Normal distribution) Compare also Example 5.18 on page 150. Consider the statistical model $\mathcal{P} = \left\{ \mathsf{N}(\theta, 4)^{\otimes 25} : \theta \in \mathbb{R} \right\}$ and the composite hypotheses

$$H_0 : \theta \le 0.5 \text{ versus } H_1 : \theta > 0.5. \qquad (5.44)$$

Take the test (5.38) on page 150:

$$\varphi(\mathbf{x}) = \left\{ \begin{array}{ll} 1 & \text{if } \overline{x} > 0.658 \\ 0 & \text{if } \overline{x} \le 0.658 \end{array} \right. . \qquad (5.45)$$

Note, $\overline{X} \sim \mathsf{N}(\theta, \frac{1}{25}4)$. We have

$$\sup_{\theta \le 0.5} \mathsf{P}_\theta(C_1) = \sup_{\theta \le 0.5} \mathsf{P}_\theta(\overline{x} > 0.658) = \sup_{\theta \le 0.5} \left(1 - \Phi\left(2.5\left(0.658 - \theta\right)\right)\right).$$

It holds that $\Phi(z) \nearrow$. Thus $\Phi\left(\frac{5}{2}\left(\theta - 0.658\right)\right) \nearrow$ for $\theta \le 0.5$ and the maximum of the power function lies on the border to the alternative. We obtain the size $\alpha = 0.35$ which is a high chance for the error of first type. Thus we would not recommend to apply the test (5.45) for the test problem (5.44). □

Exercise 5.9 Apply the test (5.45) for testing $H_0 : \theta \le 0$ versus $H_1 : \theta > 0$. Calculate the size α. Explain the result!

The other problem is how to generalize the requirement in Definition 5.6 of minimizing the probability β of the error of second type? Also the probability of the error of second type depends on the concrete parameter $\theta \in \Theta_1$. The way chosen in test theory is to require that for every fixed single alternative the most powerful test is taken. We give the definition.

Definition 5.9 (UMP test) A test φ^* is called **uniformly most powerful (UMP) of size α** iff

$$\sup_{\theta \in \Theta_0} \mathsf{E}_\theta \varphi^*(\mathbf{X}) = \alpha$$

and

$$\mathsf{E}_\theta \varphi^*(\mathbf{X}) \ge \mathsf{E}_\theta \varphi(\mathbf{X}) \text{ for all } \theta \in \Theta_1 \text{ and for all tests } \varphi \text{ of size at most } \alpha.$$
$$(5.46)$$

OBS! The condition (5.46) is a very strong one.
Let us go back to the lion example:

Example 5.22 (Lion's appetite) Review Example 3.1 on page 25. The statistical model consists of three distributions. Consider now the test problem

$$H_0 : \theta_2 \text{ versus } H_1 : \{\theta_1, \theta_3\}.$$

Recall the distributions in Example 3.1 on page 25

x	0	1	2	3	4
θ_1(hungry)	0	0.05	0.05	0.8	0.1
θ_2(moderate)	0.05	0.05	0.8	0.1	0
θ_3(lethargic)	0.9	0.05	0.05	0	0

Let $\alpha = 0.05$ and let us compare the best tests for each single alternative. The Neyman–Pearson test for $H_0 : \theta_2$ versus $H_1 : \theta_1$ is

$$\varphi(x) = \begin{cases} 1 & \text{if} & x \in \{4\} \\ \frac{1}{2} & \text{if} & x \in \{3\} \\ 0 & \text{if} & x \in \{0, 1, 2\} \end{cases}, \qquad (5.47)$$

with the power function $\pi(\theta_1) = 0.5$, $\pi(\theta_2) = 0.05$, $\pi(\theta_3) = 0$. The Neyman–Pearson test for $H_0 : \theta_2$ versus $H_1 : \theta_3$ is

$$\varphi(x) = \begin{cases} 1 & \text{if} & x \in \{0\} \\ 0 & \text{if} & x \in \{1, 2, 3, 4\} \end{cases}, \qquad (5.48)$$

with the power function $\pi(\theta_1) = 0$, $\pi(\theta_2) = 0.05$, $\pi(\theta_3) = 0.9$. Comparing the power functions

θ	θ_1	θ_2	θ_3
test (5.47)	0.5	0.05	0
test (5.48)	0	0.05	0.9

we see there exists no UMP test of size 0.05.   □

The situation is different when we consider another test problem.

Exercise 5.10 Continue Example 5.22. Consider now the test problem $H_0 : \theta_1$ versus $H_1 : \{\theta_2, \theta_3\}$ and derive the UMP test for $\alpha = 0.05$.

The following notation of a monotone likelihood ratio is useful for parameter sets with a relation of order. For instance, we cannot apply this approach for the flower example because there is no order between the colors.

Definition 5.10 (MLR) A model $\{P_\theta, \theta \in \Theta\}$ with $\Theta \subseteq \mathbb{R}$ is said to have a **monotone likelihood ratio** (MLR) in the statistic T iff for all θ, θ' with $\theta > \theta'$ there exists a nondecreasing function $F_{\theta,\theta'}$ such that

$$\frac{L(\theta; \mathbf{x})}{L(\theta'; \mathbf{x})} = F(T(\mathbf{x})).$$

When we already know that the distributions belong to a one-parameter exponential family then we can apply the following result.

Theorem 5.3 Let $\mathbf{X} = (X_1, \ldots, X_n)$ be an i.i.d. sample from P_θ. If the distribution P_θ belongs to a one-parameter exponential family with

$$p(x; \theta) = A(\theta)h(x)\exp(\zeta(\theta)R(x)),$$

where $\zeta(.)$ is monotone nondecreasing then the model of the sample $\{P_\theta^{\otimes n}, \theta \in \Theta\}$ has an **MLR** in $T(\mathbf{x}) = \sum_{i=1}^{n} R(x_i)$.

PROOF:

The likelihood function of $\mathbf{X}$ is

$$L(\theta; \mathbf{x}) = A(\theta)^n \exp\left(\zeta(\theta)\sum_{i=1}^{n} R(x_i)\right) h(\mathbf{x}) = A(\theta)^n \exp\left(\zeta(\theta)T(\mathbf{x})\right) h(\mathbf{x}).$$

Then the likelihood ratio is

$$\frac{L(\theta; \mathbf{x})}{L(\theta'; \mathbf{x})} = \frac{A(\theta)^n}{A(\theta')^n} \exp\left((\zeta(\theta) - \zeta(\theta'))T(\mathbf{x})\right).$$

As a function of $T(\mathbf{x}) = t$ the likelihood ratio is nondecreasing in t if and only if $\zeta(\theta) - \zeta(\theta') > 0$. That holds because of $\theta > \theta'$ and the monotony of ζ.

$\square$

The following theorem contains the main result of this subsection:

Theorem 5.4 (Blackwell) *Suppose an MLR family in T.*
(a) For the test problem $H_0 : \theta \geq \theta_0$ versus $H_1 : \theta < \theta_0$ the test

$$\varphi(T(\mathbf{x})) = \begin{cases} 1 & if \quad T(\mathbf{x}) < k \\ \gamma & if \quad T(\mathbf{x}) = k \quad , \\ 0 & if \quad T(\mathbf{x}) > k \end{cases} \qquad (5.49)$$

with γ, k such that $\mathsf{E}_{\theta_0}\varphi(\mathbf{X}) = \alpha$ is UMP test of size α.
(b) For the test problem $H_0 : \theta \leq \theta_0$ versus $H_1 : \theta > \theta_0$ the test

$$\varphi(T(\mathbf{x})) = \begin{cases} 1 & if \quad T(\mathbf{x}) > k \\ \gamma & if \quad T(\mathbf{x}) = k \quad , \\ 0 & if \quad T(\mathbf{x}) < k \end{cases} \qquad (5.50)$$

with γ, k such that $\mathsf{E}_{\theta_0}\varphi(\mathbf{X}) = \alpha$ is UMP test of size α.
(c) For any θ of the respective alternative the test (5.49) and the test (5.50) have minimal size of the error of second type among all tests ψ with $\mathsf{E}_{\theta_0}\psi(\mathbf{X}) = \alpha$.

PROOF: * Let us show (a). Consider the simple hypotheses $H_0 : \theta = \theta_0$ versus $H_1 : \theta = \theta_1$, where $\theta_1 < \theta_0$. Applying the Neyman–Pearson lemma (Theorem 5.2), one finds the most powerful test with

$$C_1 = \left\{ \mathbf{x} : \frac{p(\mathbf{x}; \theta_0)}{p(\mathbf{x}; \theta_1)} < C \right\}.$$

Because we have an MLR in T, this region is equivalent to $\{\mathbf{x} : T(\mathbf{x}) < k\}$; thus the test φ in (5.49) is a Neyman–Pearson test. It follows from Theorem 5.2 that there exist γ, k such that $\mathsf{E}_{\theta_0}\varphi(\mathbf{X}) = \alpha$ and $\mathsf{E}_{\theta_1}\varphi(\mathbf{X}) \geq \mathsf{E}_{\theta_1}\psi(\mathbf{X})$ for all α-tests ψ. Note the test in (5.49) is independent of θ_1! Thus the test (5.49) is a Neyman–Pearson test for arbitrary alternatives $H_1 : \theta = \theta_1$, where $\theta_1 < \theta_0$ and $\mathsf{E}_{\theta_1}\varphi(\mathbf{X}) \geq \mathsf{E}_{\theta_1}\psi(\mathbf{X})$ for all $\theta \in \Theta_1$ and for all tests ψ with $\mathsf{E}_{\theta_0}\psi(\mathbf{X}) \leq \alpha$. It remains to show that the test (5.49) fulfills: $\alpha = \sup_{\theta \in \Theta_0} \mathsf{E}_{\theta}\varphi(\mathbf{X})$. Consider now the simple hypotheses $H_0 : \theta'$ versus $H_1 : \theta_0$, where $\theta' > \theta_0$. The test (5.49) is a Neyman–Pearson test for this test problem with

$$\alpha(\theta') = \mathsf{E}_{\theta'}\varphi(\mathbf{X}) \text{ and } \pi(\theta_0) = \mathsf{E}_{\theta_0}\varphi(\mathbf{X}) = \alpha.$$

From Corollary 5.1 follows that $\alpha(\theta') \leq \alpha$. Because θ', $\theta' > \theta_0$ was arbitrary chosen we get $\alpha = \sup_{\theta \in \Theta_0} \mathsf{E}_{\theta}\varphi(\mathbf{X})$.
Consider (b). By interchanging throughout all inequalities in the proof of (a) we obtain (b).
Let us show (c). (c) follows from the fact that the tests φ in (5.49) and in

(5.50) are Neyman–Pearson tests for the test problem $H_0 : \theta_0$ versus $H_1 : \theta_1$. Thus $\mathsf{E}_{\theta_1}\varphi(\mathbf{X}) \geq \mathsf{E}_{\theta_1}\psi(\mathbf{X})$ for all tests ψ with $\mathsf{E}_{\theta_0}\psi(\mathbf{X}) = \alpha$.

□

Exercise 5.11 Write down the proof for the test (5.50).

The exponential families with MLR fulfill the condition of Theorem 5.4. Let us formulate this important case as corollary.

Corollary 5.2 Let $\mathbf{X} = (X_1, ..., X_n)$ be an i.i.d. sample from

$$p(x; \theta) = B(\theta)h(x)\exp(\zeta(\theta)R(x))$$

with monotone $\zeta \uparrow$. Set $T(\mathbf{x}) = \sum_{i=1}^{n} R(x_i)$. Then for

$$H_0 : \theta \leq \theta_0 \text{ versus } H_1 : \theta > \theta_0$$

the UMP test of size α is

$$\varphi(T(\mathbf{x})) = \begin{cases} 1 & \text{if } T(\mathbf{x}) > k \\ \gamma & \text{if } T(\mathbf{x}) = k \\ 0 & \text{if } T(\mathbf{x}) < k \end{cases} \quad , \tag{5.51}$$

with γ, k such that $\mathsf{E}_{\theta_0}\varphi(\mathbf{X}) = \alpha$. For the test problem

$$H_0 : \theta \geq \theta_0 \text{ versus } H_1 : \theta < \theta_0$$

the UMP test of size α is

$$\varphi(T(\mathbf{x})) = \begin{cases} 1 & \text{if } T(\mathbf{x}) < k \\ \gamma & \text{if } T(\mathbf{x}) = k \\ 0 & \text{if } T(\mathbf{x}) > k \end{cases} \quad , \tag{5.52}$$

with γ, k such that $\mathsf{E}_{\theta_0}\varphi(\mathbf{X}) = \alpha$. For any θ of the respective alternative the test (5.49) and the test (5.50) have minimal size of the error of second type among all tests ψ with $\mathsf{E}_{\theta_0}\psi(\mathbf{X}) = \alpha$.

Example 5.23 (Normal distribution) Continued Example 5.14 on page 140: Assume the statistical model $\mathcal{P} = \left\{ \mathsf{N}(\theta, 4)^{\otimes 25} : \theta \in \mathbb{R} \right\}$. The distribution $\mathsf{N}(\theta, 4)$ belongs to a one-parameter exponential family with $R(x) = x$ and $\zeta(\theta) = \frac{\theta}{4}$; compare Special case 2.1. Consider now the composite hypotheses

$$H_0 : \theta \leq 0 \text{ versus } H_1 : \theta > 0.$$

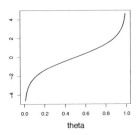

Figure 5.14: The natural parameter $\zeta(\theta)$ of $\text{Bin}(n,\theta)$ is calculated by the logit function, which is monotone.

Then the test

$$\varphi(\mathbf{x}) = \begin{cases} 1 & \text{if } \overline{x} > k/n \\ 0 & \text{if } \overline{x} \le k/n \end{cases} \tag{5.53}$$

is UMP. For $k/n = 0.658$ we have

$$\mathsf{E}_0\varphi(\mathbf{X}) = \mathsf{P}_0\left(\sqrt{n}\frac{\overline{X}}{\sigma_0} > \frac{k}{\sigma_0\sqrt{n}}\right) = 1 - \Phi(\frac{k}{10}) = 0.05.$$

$\square$

Exercise 5.12 Consider Example 5.21 on page 153. The test (5.45) is UMP test of size $\alpha = 0.35$ for the test problem (5.44). Derive the UMP test of size $\alpha_0 = 0.05$ for the test problem (5.44).

Special case 5.1 (One-sided Z-test) Consider the normal model with known variance

$$\mathcal{P} = \left\{ \mathsf{N}(\theta, \sigma^2)^{\otimes n} : \theta \in \mathbb{R} \right\} \tag{5.54}$$

and the test problem

$$H_0 : \theta \le \theta_0 \text{ versus } H_1 : \theta > \theta_0.$$

From Special case 2.1 on page 15 we know that $\{\mathsf{N}(\theta,\sigma^2), \theta \in \mathbb{R}\}$ belongs to a one-parameter exponential family with $\zeta(\theta) = \theta\frac{1}{\sigma^2}$ and $T(x) = x$. From Theorem 5.3 it follows that $\mathcal{P}$ in (5.54) has an MLR in $\overline{x}$. The distribution of $\overline{X}$ under $\theta = \theta_0$ is $\mathsf{N}(\theta_0, \frac{1}{n}\sigma^2)$ and $\mathsf{P}_{\theta_0}(\overline{X} = k) = 0$. Thus from Blackwell theorem 5.4 follows, that the following test $\varphi(\mathbf{x})$ is UMP test of size α

$$\varphi(\mathbf{x}) = \begin{cases} 1 & \text{if } \overline{x} > k \\ 0 & \text{if } \overline{x} \le k \end{cases},$$

where k is chosen such that $E_{\theta_0}\varphi(\mathbf{X}) = \alpha$. We have

$$P_{\theta_0}\left(\overline{X} > k\right) \;=\; 1 - \Phi\left(\frac{\sqrt{n}\,(k - \theta_0)}{\sigma}\right) = \alpha$$

and $k = \frac{\sigma}{\sqrt{n}}z_{1-\alpha} + \theta_0$, where $\Phi(z_{1-\alpha}) = 1 - \alpha$. Summarizing, the UMP test of size α is the well-known Z-test

$$\varphi(\mathbf{x}) = \begin{cases} 1 & \text{if} \quad Z > z_{1-\alpha} \\ 0 & \text{if} \quad Z \le z_{1-\alpha} \end{cases}, \text{ with } Z = \frac{\sqrt{n}(\overline{x} - \theta_0)}{\sigma}.$$

□

Special case 5.2 (One-sided variance test) Consider an i.i.d. sample from a normal distribution with known expectation and unknown variances,

$$\mathcal{P} = \left\{N(0, \theta)^{\otimes n}, \theta \in \mathbb{R}_+\right\} \tag{5.55}$$

and the test problem

$$H_0 : \theta \le \theta_0 \text{ versus } H_1 : \theta > \theta_0.$$

From Special case 2.1 we know that $\{N(0, \theta), \theta \in \mathbb{R}_+\}$ belongs to a one-parameter exponential family with $\zeta(\theta) = -\frac{1}{2}\theta^{-1}$ and $T(x) = x^2$. $\zeta(\theta)$ is monotone increasing in θ. From Theorem 5.3 follows that $\mathcal{P}$ in (5.55) has an MLR in $\sum_{i=1}^{n} x_i^2$. The distribution of $\sum_{i=1}^{n} X_i^2$ is continuous. From Blackwell theorem 5.4 it follows that the following test $\varphi(\mathbf{x})$ is UMP test of size α:

$$\varphi(\mathbf{x}) = \begin{cases} 1 & \text{if} \quad \sum_{i=1}^{n} x_i^2 > k \\ 0 & \text{if} \quad \sum_{i=1}^{n} x_i^2 \le k \end{cases},$$

with k such that $E_{\theta_0}\varphi(\mathbf{X}) = \alpha$. Under $\theta = \theta_0$ the statistic $\frac{1}{\theta_0}\sum_{i=1}^{n} X_i^2$ has a χ^2-distribution with n degrees of freedom. Thus

$$P_{\theta_0}\left(\sum_{i=1}^{n} X_i^2 > k\right) \;=\; 1 - F_{\chi_n^2}\left(\frac{k}{\theta_0}\right) = \alpha$$

and $k = \theta_0\chi_{n;1-\alpha}^2$, where $F_{\chi_n^2}\left(\chi_{n,1-\alpha}^2\right) = 1 - \alpha$. Summarizing, the UMP test of size α is the well-known χ^2-test for variance in a normal distribution

$$\varphi(\mathbf{x}) = \begin{cases} 1 & \text{if} \quad \sum_{i=1}^{n} x_i^2 > \theta_0\chi_{n;1-\alpha}^2 \\ 0 & \text{if} \quad \sum_{i=1}^{n} x_i^2 \le \theta_0\chi_{n;1-\alpha}^2 \end{cases}.$$

□

Special case 5.3 (Binomial distribution) Consider the statistical model

$$\mathcal{P} = \{\text{Bin}(n, \theta), \theta \in (0, 1)\} \tag{5.56}$$

and the test problem $H_0 : \theta \geq \theta_0$ versus $H_1 : \theta < \theta_0$. From Special case 2.3 we know that $\{\text{Bin}(n, \theta), \theta \in (0, 1)\}$ belongs to a one-parameter exponential family with $\zeta(\theta) = \text{logit}(\theta) = \ln(\frac{\theta}{1-\theta}), (\nearrow)$ and $T(x) = x$. The logit function is monotone nondecreasing; see Figure 5.14. From Theorem 5.3 it follows that $\mathcal{P}$ in (5.56) has an MLR in x. Thus from Blackwell Theorem 5.4 we get the following UMP test of size α

$$\varphi(x) = \begin{cases} 1 & \text{if } x < k \\ \gamma & \text{if } x = k \\ 0 & \text{if } x > k \end{cases},$$

where γ, k such that $\mathsf{E}_{\theta_0}\varphi(X) = \alpha$. Here the distribution of X under $\theta = \theta_0$ is $\text{Bin}(n, \theta_0)$. Thus the condition on γ, k is:

$$
\begin{aligned}
\mathsf{E}_{\theta_0}\varphi(\mathbf{X}) &= \mathsf{P}_{\theta_0}(X < k) + \gamma \mathsf{P}_{\theta_0}(X = k) \\
&= \sum_{x=0}^{k-1} \text{bin}(n, \theta_0, x) + \gamma \, \text{bin}(n, \theta_0, k) = \alpha.
\end{aligned}
$$

$\square$

Exercise 5.13 Continue Example 5.20 on page 152. Show that the test (5.43) is UMP of size $\alpha = 0.05$ for the test problem (5.42).

Example 5.24 (Dolphins) Consider the model,

$$\mathcal{P} = \{\text{N}(\theta, \sigma_1^2)^{\otimes n} \otimes \text{N}(a\theta, \sigma_2^2)^{\otimes m}, \theta \in \mathbb{R}\}, \tag{5.57}$$

where $a, \sigma_1^2, \sigma_2^2$ are known. We are interested in testing

$$H_0 : \theta \leq \theta_0 \text{ versus } H_1 : \theta > \theta_0.$$

Because we have not an i.i.d sample we cannot apply Theorem 5.3 directly. But the distribution of the whole sample belongs to a one-parameter exponential family. Denote the joint sample by $\mathbf{Z} = (\mathbf{X}, \mathbf{Y})$ then $f(\mathbf{z}; \theta)$ is

$$
\begin{aligned}
&= \left(\frac{1}{2\pi\sigma_1^2}\right)^{\frac{m}{2}} \left(\frac{1}{2\pi\sigma_2^2}\right)^{\frac{n}{2}} \exp\left(-\frac{1}{2\sigma_1^2}\sum_{i=1}^{m}(x_i - \theta)^2 - \frac{1}{2\sigma_2^2}\sum_{i=1}^{n}(y_i - a\theta)^2\right) \\
&= A(\theta) \exp\left(\zeta(\theta)T(\mathbf{z})\right) h(\mathbf{z})
\end{aligned}
$$

with

$$A(\theta) = \left(\frac{1}{2\pi\sigma_1^2}\right)^{\frac{m}{2}} \left(\frac{1}{2\pi\sigma_2^2}\right)^{\frac{n}{2}} \exp\left(-\frac{m}{2\sigma_1^2}\theta^2 - \frac{n}{2\sigma_2^2}a^2\theta^2\right)$$

and

$$h(\mathbf{z}) = \exp\left(-\frac{1}{2\sigma_1^2}\sum_{i=1}^{m}x_i^2 - \frac{1}{2\sigma_2^2}\sum_{i=1}^{n}y_i^2\right)$$

and $\zeta(\theta) = \theta$ and $T(\mathbf{z}) = \frac{m}{\sigma_1^2}\bar{x} + \frac{an}{\sigma_2^2}\bar{y}$. Thus $\mathcal{P}$ in (5.57) has an MLR in $T(\mathbf{z})$ and we can apply the Blackwell theorem 5.4. The UMP test of size α is

$$\varphi(\mathbf{z}) = \begin{cases} 1 & \text{if} \quad \frac{m}{\sigma_1^2}\bar{x} + \frac{an}{\sigma_2^2}\bar{y} > k \\ 0 & \text{if} \quad \frac{m}{\sigma_1^2}\bar{x} + \frac{an}{\sigma_2^2}\bar{y} \leq k \end{cases},$$

with $k = k(\alpha)$ such that $\mathsf{E}_{\theta_0}\varphi(\mathbf{Z}) = \alpha$. Under $\theta = \theta_0$ the statistic $T(\mathbf{Z})$ is normally distributed with

$$\mathsf{E}_{\theta_0}\left(\frac{m}{\sigma_1^2}\bar{X} + \frac{an}{\sigma_2^2}\bar{Y}\right) = c(m,n)\theta_0$$

and

$$\mathsf{Var}_{\theta_0}\left(\frac{m}{\sigma_1^2}\bar{X} + \frac{an}{\sigma_2^2}\bar{Y}\right) = \left(\frac{m}{\sigma_1^2}\right)^2 \mathsf{Var}_{\theta_0}(\bar{X}) + \left(\frac{an}{\sigma_2^2}\right)^2 \mathsf{Var}_{\theta_0}(\bar{Y})$$
$$= c(m,n),$$

where $c(m,n) = \left(\frac{m}{\sigma_1^2} + \frac{a^2 n}{\sigma_2^2}\right)$. Thus $k = k(\alpha) = \sqrt{c(m,n)}u_{1-\alpha} + c(m,n)\theta_0$. $\square$

Example 5.25 (Genotypes) Consider Example 2.14 on page 12. The Hardy–Weinberg model states that the genotypes AA, Aa and aa occur with the following probabilities:

$$p_\theta(\text{aa}) = \theta^2, \quad p_\theta(\text{Aa}) = 2\theta(1-\theta), \quad p_\theta(\text{AA}) = (1-\theta)^2, \quad (5.58)$$

where θ is an unknown parameter in $\Theta = (0,1)$. Suppose we have an i.i.d. sample $\mathbf{X} = (X_1, ..., X_n)$ and each observation has the distribution (5.58). On page 20, Example 2.21 we have shown that the distribution of $\mathbf{X}$ belongs to a one-parameter exponential family with the sufficient statistic $T(\mathbf{x}) = 2n_1 + n_2$, where n_1 is the number of genotype aa in the sample $\mathbf{x}$ and n_2 is the number of genotype Aa. Suppose we are interested in testing $p_\theta(\text{aa}) < p_\theta(\text{AA})$. Because $\theta^2 < (1-\theta)^2$, iff $\theta < \frac{1}{2}$ we have the following test problem:

$$H_0 : \theta = \frac{1}{2} \text{ versus } H_1 : \theta < \frac{1}{2}.$$

Consider now two different tests: An empirical test based on the test statistic $T_{emp}(\mathbf{x}) = n_1$ and the UMP test, derived from Theorem 5.4. As toy example let $n = 4$. Set $\alpha = 0.05$. The null distribution of N_1 is $\text{Bin}(4, \frac{1}{4})$, that is

0	1	2	3	4
$\frac{81}{256}$	$\frac{27}{64}$	$\frac{27}{128}$	$\frac{3}{64}$	$\frac{1}{256}$

with $\frac{81}{256} = 0.3164$. Then the empirical test is

$$\varphi_{emp}(\mathbf{x}) = \begin{cases} 0.158 & \text{if} \quad n_1 = 0 \\ 0 & \text{if} \quad n_1 > 0 \end{cases}.$$

From Theorem 5.4 follows that the optimal test uses the sufficient statistic as the test statistic. The null distribution of $2N_1 + N_2$ is calculated from the multinomial distribution given by

$$p((n_1, n_2, n_3); \theta) = \frac{n!}{n_1! \, n_2! \, n_3!} p_1^{n_1} p_2^{n_2} p_3^{n_3}, \; p_1 = \theta^2, p_2 = 2\theta(1-\theta), p_3 = (1-\theta)^2.$$

Under H_0 it holds $p_1 = \frac{1}{4}, p_2 = \frac{1}{2}, p_3 = \frac{1}{4}$. For $n = 4$ the null distribution of $T(\mathbf{X}) = 2N_1 + N_2$ is $p^T(k; \frac{1}{2}), k = 0, \ldots, 8$:

k	0	1	2	3	4
$p^T(k; \frac{1}{2})$	0.00390625	0.03125	0.109375	0.21875	0.2734375

where $p^T(8 - k; \frac{1}{2}) = p^T(k; \frac{1}{2})$ for $k = 0, \ldots, 3$. We obtain the UMP test of size 0.05:

$$\varphi_{\text{UMP}}(\mathbf{x}) = \begin{cases} 1 & \text{if} \quad 2n_1 + n_2 < 2 \\ 0.135 & \text{if} \quad 2n_1 + n_2 = 2 \\ 0 & \text{if} \quad 2n_1 + n_2 > 2 \end{cases}.$$

In Figure 5.15 the power functions of both tests are plotted. We can see that the UMP test is importantly better!   □

Exercise 5.14 Consider Example 5.25. Compare both tests, when $n = 10$ and $\alpha = 0.05$. Discuss the main difference!

The following case about the noncentral F-distribution is important for the optimality properties of the F-test in linear models (see Chapter 6).

Special case 5.4 (F-test) Consider the probability family of noncentral F-distributions, with noncentrality parameter λ, $\lambda \geq 0$ and with n_1 and n_2 degrees of freedom. The density is given by

$$f(x; \lambda) = \exp(-\frac{1}{2}\lambda^2) \sum_{k=0}^{\infty} c_k \frac{(\frac{1}{2}\lambda^2)^k}{k!} \frac{x^{\frac{n_1-1}{2}+k}}{(1+x)^{\frac{1}{2}(n_1+n_2+1)+k}}, \quad \text{for } x \geq 0,$$

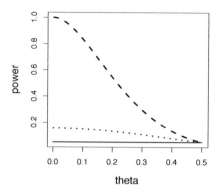

Figure 5.15: The dotted line is the power function of the empirical test, based on an insufficient test statistic in Example 5.25. The broken line is the power function of the UMP test. Under the small sample size of $n = 4$ (unrealistically small!) the improvement is important.

where

$$c_k = \frac{\Gamma\left(\frac{1}{2}(n_1 + n_2 + 1) + k\right)}{\Gamma\left(\frac{n_1+1}{2} + k\right)\Gamma\left(\frac{n_2}{2}\right)}.$$

The density for the central F-distribution ($\lambda = 0$) is

$$f(x; 0) = c_0 \frac{x^{\frac{n_1-1}{2}}}{(1+x)^{\frac{1}{2}(n_1+n_2-1)}}.$$

Then the likelihood ratio is

$$\frac{f(x; \lambda)}{f(x; 0)} = \frac{\exp(-\frac{1}{2}\lambda^2) \sum_{k=0}^{\infty} c_k \frac{\left(\frac{1}{2}\lambda^2\right)^k}{k!} \frac{x^{\frac{n_1-1}{2}+k}}{(1+x)^{\frac{1}{2}(n_1+n_2-1)+k}}}{c_0 \frac{x^{\frac{n_1-1}{2}}}{(1+x)^{\frac{1}{2}(n_1+n_2-1)}}}$$

$$= \exp(-\frac{1}{2}\lambda^2) \sum_{k=0}^{\infty} \frac{c_k}{c_0} \frac{\left(\frac{1}{2}\lambda^2\right)^k}{k!} \frac{x^k}{(1+x)^k}.$$

For all $k > 0$, $\frac{x^k}{(1+x)^k}$ is an increasing function in x, so $\frac{f(x,\lambda)}{f(x;0)}$ is increasing in x. Consider the test problem

$$H_0 : \lambda = 0 \text{ versus } H_1 : \lambda > 0 \qquad (5.59)$$

and one observation x. We cannot apply Theorem 5.4 directly, because we have not shown the monotony of the likelihood ratio for arbitrary pairs λ, λ'. But we can apply the argumentation in proof of Theorem 5.4 for testing $H_0 : \lambda = 0$ versus $H_1 : \lambda > 0$. We obtain the UMP test of size α:

$$\varphi(x) = \begin{cases} 1 & \text{if} \quad x > F_{n_1,n_2;1-\alpha} \\ 0 & \text{if} \quad x \leq F_{n_1,n_2;1-\alpha} \end{cases},$$

where $F_{n_1,n_2;1-\alpha}$ is a quantile of the central F-distribution with n_1 and n_2 degrees of freedom, $P_0(X < F_{n_1,n_2;1-\alpha}) = 1 - \alpha$. $\qquad\qquad\square$

5.3.3 Unbiased Tests

Recall Example 5.22 on page 154. There we calculated that there is no uniform most powerful test for a lion's appetite for testing the moderate mode against the two extremes, hungry and lethargic. The same situation occurs whenever we want to test a simple null hypothesis against a two-sided alternative. Let us consider one more example:

Example 5.26 (Normal distribution) Continuation of Example 5.23 on page 157. The test φ_1 given in (5.53) is UMP of size $\alpha = 0.05$ for the one-sided test problem $H_0 : \theta \leq 0$ versus $H_1 : \theta > 0$. The following test

$$\varphi_2(\mathbf{x}) = \begin{cases} 1 & \text{if} \quad \bar{x} < -0.658 \\ 0 & \text{if} \quad \bar{x} \geq -0.658 \end{cases} \qquad (5.60)$$

is UMP test of size $\alpha = 0.05$ for the one-sided test problem $H_0 : \theta \geq 0$ versus $H_1 : \theta < 0$. Consider now the test problem

$$H_0 : \theta = 0 \text{ versus } H_1 : \theta \neq 0. \qquad (5.61)$$

A UMP φ^* test of size $\alpha = 0.05$ for (5.61) has to fulfill

$$E_{\theta_0}\varphi^*(\mathbf{X}) = \alpha = 0.05$$

and

$$E_\theta\varphi^*(\mathbf{X}) \geq E_\theta\varphi(\mathbf{X}) \text{ for all tests } \varphi \text{ of size at most } \alpha \text{ and all } \theta \neq 0.$$

Because φ_1 and φ_2 are of size α it implies

$$E_\theta\varphi^*(\mathbf{X}) \geq E_\theta\varphi_1(\mathbf{X}) \text{ for all } \theta \neq 0$$

and

$$E_\theta\varphi^*(\mathbf{X}) \geq E_\theta\varphi_2(\mathbf{X}) \text{ for all } \theta \neq 0.$$

Otherwise, φ_1 and φ_2 being UMP imply they have highest power function on the related alternative regions. Thus,

$$E_\theta \varphi^*(\mathbf{X}) \le E_\theta \varphi_1(\mathbf{X}) \text{ for all } \theta < 0$$

$$E_\theta \varphi^*(\mathbf{X}) \le E_\theta \varphi_2(\mathbf{X}) \text{ for all } \theta > 0.$$

Summarizing, we obtain

$$E_\theta \varphi^*(\mathbf{X}) = E_\theta \varphi_1(\mathbf{X}) \text{ for all } \theta \le 0$$

and

$$E_\theta \varphi^*(\mathbf{X}) = E_\theta \varphi_2(\mathbf{X}) \text{ for all } \theta \ge 0.$$

The family of normal distributions is complete, hence

$$P_\theta(\varphi^*(\mathbf{X}) = \varphi_1(\mathbf{X})) = 1 \text{ for all } \theta \le 0$$

and

$$P_\theta(\varphi^*(\mathbf{X}) = \varphi_2(\mathbf{X})) = 1 \text{ for all } \theta \ge 0.$$

But the tests $\varphi_1(\mathbf{x})$ and $\varphi_2(\mathbf{x})$ in (5.53) and in (5.60) are different with positive probability P_0. That gives the contradiction. There exists no UMP test for the test problem (5.61). □

The way out is to constrain the class of competing tests. The one-sided UMP tests should be excluded. That is done with the help of the following notation.

Definition 5.11 (unbiased α-test) A test φ is called an **unbiased α-test** iff $\alpha = \sup_{\theta \in \Theta_0} E_\theta \varphi(\mathbf{X})$ and $\inf_{\theta \in \Theta_1} E_\theta \varphi(\mathbf{X}) \ge \alpha$.

Note, it is reasonable to require that the power function is not smaller than α in all points of the alternative; compare Figure 5.16.

Definition 5.12 (UMPU α-test) A test φ^* is called a **uniform most powerful unbiased α-test** iff $\sup_{\theta \in \Theta_0} E_\theta \varphi^*(\mathbf{X}) = \alpha$ and $E_\theta \varphi^*(\mathbf{X}) \ge E_\theta \varphi(\mathbf{X})$ for all $\theta \in \Theta_1$ and for all unbiased α-tests φ.

The following theorem gives us the **uniform most powerful unbiased α-test** for a two-sided test problem if the underlying model belongs to a one-parameter exponential family. Note that Example 5.26 belongs to this case.

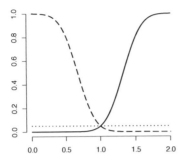 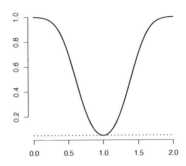

Figure 5.16: In the left picture the power functions of the two optimal one-sided tests are plotted. The right picture shows the power function of the optimal unbiased two-sided test.

Theorem 5.5 *Let* $\mathbf{X} = (X_1, \ldots, X_n)$ *be a sample from distribution* P_θ *belonging to a one-parameter exponential family with*

$$p(\mathbf{x}; \theta)(\mathbf{x}) = C(\theta) \exp(\theta T(\mathbf{x})) h(\mathbf{x}), \qquad \theta \in \mathbb{R}. \qquad (5.62)$$

For the test problem

$$H_0 : \theta = \theta_0 \ versus \ H_1 : \theta \neq \theta_0 \qquad (5.63)$$

the UMPU α-*test is given by*

$$\varphi(T(\mathbf{x})) = \begin{cases} 1 & if \ \ T(\mathbf{x}) > k_{\text{upper}}; T(\mathbf{x}) < k_{\text{lower}} \\ \gamma_1 & if \qquad T(\mathbf{x}) = k_{\text{upper}} \\ \gamma_2 & \qquad \ \ T(\mathbf{x}) = k_{\text{lower}} \\ 0 & if \quad k_{\text{lower}} < T(\mathbf{x}) < k_{\text{upper}} \end{cases}, \qquad (5.64)$$

with $\gamma_1, \gamma_2, k_{\text{upper}}, k_{\text{lower}}$, *such that*

$$E_{\theta_0} \varphi(\mathbf{X}) = \alpha \qquad (5.65)$$

and

$$E_{\theta_0} \varphi(\mathbf{X}) T(\mathbf{X}) = E_{\theta_0} T(\mathbf{X}) \alpha. \qquad (5.66)$$

Remark 5.2 The condition (5.65) is the condition to be an α-test. The condition (5.66) is related to the unbiasedness. The power function $\pi(\theta)$ of an unbiased test is minimal at θ_0. The condition $\pi'(\theta_0) = 0$ implies (5.66); compare the Figure 5.16. For more details see the proof of the Theorem 5.5.

The following lemma is used for proving the Theorem 5.5. We will quote it without proof and for once by using a measure theoretical formulation.

Theorem 5.6 (Generalized Fundamental lemma) * *Let $f_1, .., f_{m+1}$ be real-valued functions defined on a Euclidean space and integrable with respect to a measure μ, and suppose that for given constants $c_1, ..., c_m$ there exists a function ϕ satisfying*

$$\int f_i \phi d\mu = c_i, \quad i = 1, ..., m. \tag{5.67}$$

Let C be the class of functions ϕ with (5.67). Then it holds:

1. *There is one member of C that maximizes $\int f_{m+1} \phi d\mu$.*

2. *A sufficient condition for a member of C to maximize $\int f_{m+1} \phi d\mu$ is the existence of constants $k_1, ..., k_m$ such that*

$$\phi(x) = \begin{cases} 1 & if \quad f_{m+1}(x) > \sum_{i=1}^{m} k_i f_i(x) \\ 0 & if \quad f_{m+1}(x) < \sum_{i=1}^{m} k_i f_i(x) \end{cases}. \tag{5.68}$$

3. *If a member of C satisfies (5.68) with $k_1, ..., k_m \geq 0$, then it maximizes $\int f_{m+1} \phi d\mu$ among all functions satisfying*

$$\int f_i \phi d\mu \leq c_i, \quad i = 1, ..., m. \tag{5.69}$$

4. *The set $M = \{(\int f_1 \phi d\mu, ..., \int f_m \phi d\mu) : \phi\}$ is convex and closed. If $(c_1, ..., c_m)$ is an inner point of M, then there exist constants $k_1, ..., k_m$ and a test ϕ satisfying (5.67) and (5.68), and a necessary condition for a member of C to maximize $\int f_{m+1} \phi d\mu$ is that (5.68) holds a.s..*

PROOF: The proof of the Generalized Fundamental lemma is, for instance, given in Lehmann and Romano (2005).

□

Now let us return to Theorem 5.5 and its proof.

PROOF OF THEOREM 5.5:* Consider the test problem $H_0 : \theta_0$ versus $H_1 : \theta_1$ where θ_1 is arbitrary fixed such that $\theta_1 \neq \theta_0$. Because of (5.62), the distribution of $T = T(\mathbf{X})$ is of the form

$$p^T(t; \theta) = C(\theta) \exp(\theta t).$$

The unbiasedness of a test $\varphi(T(\mathbf{x})) = \varphi(t)$ implies that the power function

$$\pi(\theta) = \int \varphi(t)C(\theta)\exp(\theta t)\,dt$$

must have a minimum at $\theta = \theta_0$. Thus $\pi'(\theta) = 0$ and after changing of integration and differentiation we get

$$\begin{aligned}
\pi'(\theta) &= \int \frac{d}{d\theta}\left(\varphi(t)C(\theta)\exp(\theta t)\right)dt & (5.70)\\
&= \int \varphi(t)tC(\theta)\exp(\theta t) + \varphi(t)C'(\theta)\exp(\theta t)\,dt\\
&= \mathsf{E}_\theta\varphi(T)T + \frac{C'(\theta)}{C(\theta)}\mathsf{E}_\theta\varphi(T) = 0. & (5.71)
\end{aligned}$$

For the special test $\psi(t) \equiv \alpha$ we obtain $\mathsf{E}_\theta T + \frac{C'(\theta)}{C(\theta)} = 0$. Substituting this in the expression (5.71) gives

$$\pi'(\theta) = \mathsf{E}_\theta\varphi(T)T - \mathsf{E}_\theta T\mathsf{E}_\theta\varphi(T) = 0.$$

Hence the unbiasedness implies (5.65) and (5.66). Let

$$M = \{(\mathsf{E}_\theta\varphi(T), \mathsf{E}_\theta T\varphi(T)) : \varphi\}.$$

Then M is convex. Setting $\varphi \equiv u$, we see $\{(u, \mathsf{E}_\theta Tu), u \in (0,1)\} \subset M$. From Corollary 5.2 we know that there exist optimal tests for the one-sided test problems with

$$\mathsf{E}_{\theta_0}\varphi_1(T) = \alpha \text{ and } \pi'(\theta_0) > 0$$

and

$$\mathsf{E}_{\theta_0}\varphi_2(T) = \alpha \text{ and } \pi'(\theta_0) < 0.$$

Thus

$$\{(\alpha, u), u > \alpha\mathsf{E}_\theta T\} \subset M \text{ and } \{(\alpha, u), u < \alpha\mathsf{E}_\theta T\} \subset M.$$

Hence

$$\{(\alpha, \alpha\mathsf{E}_\theta T)\} \in int(M).$$

It follows from Theorem 5.6 that there exist constants k_1 and k_2 and an optimal test maximizing the power function:

$$\begin{aligned}
\phi(t) &= \begin{cases} 1 & \text{if } C(\theta_1)\exp(\theta_1 t) > C(\theta_0)(k_1 + k_2 t)\exp(\theta_0 t)\\ 0 & \text{if } C(\theta_1)\exp(\theta_1 t) < C(\theta_0)(k_1 + k_2 t)\exp(\theta_0 t) \end{cases}\\
&= \begin{cases} 1 & \text{if } \exp(bt) > k_1 + k_2 t\\ 0 & \text{if } \exp(bt) < k_1 + k_2 t \end{cases}.
\end{aligned}$$

The critical region $\exp(bt) > k_1 + k_2 t$ is either one-sided or the outside of an interval. If it is one-sided then the test has a monotone power function and therefore it cannot be satisfied that the power function must have a minimum at θ_0. Thus

$$\phi(t) = \begin{cases} 1 & \text{if} \quad C_1 > t \text{ or } C_2 < t \\ 0 & \text{if} \quad C_1 < t < C_2 \end{cases}.$$

That is the test given in (5.64).

□

Corollary 5.3 *Under the conditions of Theorem 5.5 and under the additional assumption that T has a symmetrical distribution around some point a under the null hypothesis, the UMPU α-test is of the form*

$$\varphi(T(\mathbf{x})) = \begin{cases} 1 & \text{if} \quad T(\mathbf{x}) - a > a - C; T(\mathbf{x}) < C \\ \gamma & \text{if} \quad \begin{array}{c} T(\mathbf{x}) - a = a - C \\ T(\mathbf{x}) = C \end{array} \\ 0 & \text{if} \quad C - a < T(\mathbf{x}) - a < a - C \end{cases},$$

with γ, C such that

$$\mathsf{P}_{\theta_0}^T (T - a < C) + \gamma \mathsf{P}_{\theta_0}^T (T = C) = \frac{\alpha}{2}. \tag{5.72}$$

PROOF: T has a symmetrical distribution around some point a; thus

$$\mathsf{P}_{\theta_0}(T < a - u) = \mathsf{P}_{\theta_0}(T < a + u) \text{ and } \mathsf{E}_{\theta_0} T = a.$$

We have to show (5.66), that is

$$\mathsf{E}_{\theta_0} T \varphi(T) = \mathsf{E}_{\theta_0} T \, \mathsf{E}_{\theta_0} \varphi(T).$$

Since $\mathsf{E}_{\theta_0} T = a$ it is equivalent to $\mathsf{E}_{\theta_0}(T - a)(1 - \varphi(T)) = 0$. But that is true, because $T - a$ is symmetric around 0 distributed and $1 - \varphi(t) = \mathbb{1}_{[-d,d]}(t - a)$ for $d = a - C$.

□

Special case 5.5 (Two-sided Z-test) Suppose the normal model $\mathcal{P} = \{\mathsf{N}(\theta, \sigma^2)^{\otimes n}, \theta \in \mathbb{R}\}$ where the variance σ^2 is known. Consider the test problem $H_0 : \theta = \theta_0$ versus $H_1 : \theta \neq \theta_0$. In Special case 2.1 we have seen that $\mathcal{P}$ belongs to a one-parameter exponential family with $\zeta(\theta) = \theta$ and $T(\mathbf{x}) = \frac{1}{\sigma^2} \sum_{i=1}^{n} x_i$. Further we know that under $H_0 : \theta = \theta_0$

$$T(\mathbf{X}) \sim \mathsf{N}\left(\frac{n}{\sigma^2}\theta_0, \frac{n}{\sigma^2}\right).$$

The distribution is symmetric around $a = \frac{n}{\sigma^2}\theta_0$ and $\mathsf{P}_{\theta_0}(T = k) = 0$. From Corollary 5.3 we obtain the following UMPU α-test

$$\varphi(T(\mathbf{x})) = \begin{cases} 1 & \text{if} & \text{otherwise} \\ 0 & \text{if} & C - \frac{n}{\sigma^2}\theta_0 < T(\mathbf{x}) - \frac{n}{\sigma^2}\theta_0 < \frac{n}{\sigma^2}\theta_0 - C \end{cases}.$$

We determine C by

$$\mathsf{E}_{\theta_0}\varphi(\mathbf{X}) = 2\Phi\left(\frac{\sigma}{\sqrt{n}}\left(C - \frac{n}{\sigma^2}\theta_0\right)\right) = \alpha.$$

Thus $C = \frac{\sqrt{n}}{\sigma}z_{\frac{\alpha}{2}} + \frac{n}{\sigma^2}\theta_0$. Remember $T(\mathbf{x})\frac{\sigma^2}{n} = \bar{x}$. Summarizing, the UMPU α-test is the well-known two-sided Z-test:

$$\varphi(T(\mathbf{x})) = \begin{cases} 1 & \text{if} & \text{otherwise} \\ 0 & \text{if} & |Z| \leq z_{1-\frac{\alpha}{2}} \end{cases}, \text{ with } Z = \frac{\sqrt{n}(\bar{x} - \theta_0)}{\sigma}.$$

$\square$

Special case 5.6 (Two-sided variance test) Consider the statistical model for an i.i.d. sample from a normal distribution with known expectation and unknown variances $\mathcal{P} = \{\mathsf{N}(0,\theta)^{\otimes n}, \theta \in \mathbb{R}_+\}$ and the test problem $H_0 : \theta = \theta_0$ versus $H_1 : \theta \neq \theta_0$. From Special case 2.1 we know that $\{\mathsf{N}(0,\theta), \theta \in \mathbb{R}_+\}$ belongs to a one-parameter exponential family with $\zeta(\theta) = -\frac{1}{2}\theta^{-1}$ and $R(x) = x^2$. The distribution of $R(x) = x^2$ is continuous. Because $\zeta(\theta) = \zeta(\theta_0)$ iff $\theta = \theta_0$ the test problem is equivalent to $H_0 : \zeta(\theta) = \zeta(\theta_0)$ versus $H_1 : \zeta(\theta) \neq \zeta(\theta_0)$. Thus from Theorem 5.5 it follows that the following test is UMPU α-test

$$\varphi(\mathbf{x}) = \begin{cases} 1 & \text{if} & T(\mathbf{x}) > k_{\text{upper}}; T(\mathbf{x}) < k_{\text{lower}} \\ 0 & \text{if} & k_{\text{lower}} \leq T(\mathbf{x}) \leq k_{\text{upper}} \end{cases}$$

where $k_{\text{upper}}, k_{\text{lower}}$, such that $\mathsf{E}_{\theta_0}\varphi(\mathbf{X}) = \alpha$ and $\mathsf{E}_{\theta_0}\varphi(\mathbf{X})T(\mathbf{X}) = \mathsf{E}_{\theta_0}T(\mathbf{X})\alpha$. Under H_0 the statistic $\frac{1}{\theta_0}\sum_{i=1}^{n} X_i^2$ is χ^2-distributed with n degrees of freedom. Discuss the first condition

$$\mathsf{E}_{\theta_0}\varphi(\mathbf{X}) = 1 - \mathsf{P}_{\theta_0}\left(\frac{1}{\theta_0}k_{\text{lower}} \leq \frac{1}{\theta_0}\sum_{i=1}^{n} X_i^2 \leq \frac{1}{\theta_0}k_{\text{upper}}\right)$$

$$= 1 - F_{\chi_n^2}\left(\frac{1}{\theta_0}k_{\text{upper}}\right) + F_{\chi_n^2}\left(\frac{1}{\theta_0}k_{\text{lower}}\right) = \alpha. \quad (5.73)$$

Note that $\mathsf{E}_{\theta_0}T(\mathbf{X}) = n\theta_0$; thus the second condition becomes $\mathsf{E}_{\theta_0}\varphi(\mathbf{X})T(\mathbf{X}) = n\theta_0\alpha$ and $\mathsf{E}_{\theta_0}(1 - \varphi(\mathbf{X}))T(\mathbf{X}) = n\theta_0(1 - \alpha)$. This can be rewritten as

$$\int_{\frac{k_{\text{lower}}}{\theta_0}}^{\frac{k_{\text{upper}}}{\theta_0}} \frac{1}{n}yf_n(y)dy = 1 - \alpha, \quad (5.74)$$

where $f_n(y)$ is the density of the χ^2-distribution with n degrees of freedom,

$$f_n(y) = \frac{1}{\Gamma\left(\frac{n}{2}\right)}y^{\frac{n}{2}-1}2^{-\frac{n}{2}}\exp(-y).$$

Note,

$$f_{n+2}(y) = \frac{1}{\Gamma\left(\frac{n}{2}+1\right)} y^{\frac{n}{2}} 2^{\left(-\frac{n+2}{2}\right)} \exp(-y) = f_n(y) y \frac{\Gamma\left(\frac{n}{2}\right)}{2\Gamma\left(\frac{n}{2}+1\right)} = f_n(y) y \frac{1}{n}.$$

So, the condition (5.74) is

$$1 - F_{\chi^2_{n+2}}\left(\frac{k_{\text{upper}}}{\theta_0}\right) + F_{\chi^2_{n+2}}\left(\frac{k_{\text{lower}}}{\theta_0}\right) = \alpha. \tag{5.75}$$

That means we have to solve the nonlinear equation system (5.73), (5.75). For sufficiently large n the equal-tails test given by

$$k_{\text{lower}} = \theta_0 \chi^2_{n;\frac{\alpha}{2}}, \quad k_{\text{upper}} = \theta_0 \chi^2_{n;1-\frac{\alpha}{2}}$$

with

$$1 - F_{\chi^2_n}\left(\frac{k_{\text{upper}}}{\theta_0}\right) = F_{\chi^2_n}\left(\frac{k_{\text{lower}}}{\theta_0}\right) = \frac{\alpha}{2}$$

is a good approximation. Summarizing, the approximative UMPU α-test is the well-known χ^2-squared test for variance in a normal population:

$$\varphi(\mathbf{x}) = \begin{cases} 1 & \text{if} \quad \sum_{i=1}^n x_i^2 < \theta_0 \chi^2_{n;\frac{\alpha}{2}}, \sum_{i=1}^n x_i^2 > \theta_0 \chi^2_{n;1-\frac{\alpha}{2}} \\ 0 & \text{if} \quad \theta_0 \chi^2_{n;\frac{\alpha}{2}} \le \sum_{i=1}^n x_i^2 \le \theta_0 \chi^2_{n;1-\frac{\alpha}{2}} \end{cases}.$$

$\square$

Special case 5.7 (Binomial distribution) Consider the statistical model $\mathcal{P} = \{\text{Bin}(n,\theta) : \theta \in (0,1)\}$ and the test problem $H_0 : \theta = \theta_0$ versus $H_1 : \theta \ne \theta_0$. From Special case 2.2 we know that $\{\text{Bin}(n,\theta) : \theta \in (0,1)\}$ belongs to a one-parameter exponential family with $\zeta(\theta) = \ln(\frac{\theta}{1-\theta})$ and $T(x) = x$. For the test problem $H_0 : \zeta(\theta) = \zeta(\theta_0)$ versus $H_1 : \zeta(\theta) \ne \zeta(\theta_0)$ Theorem 5.5 gives the UMPU α-test

$$\varphi(x) = \begin{cases} 1 & \text{if} \quad x > k_{\text{upper}}; x < k_{\text{lower}} \\ \gamma_1 & \text{if} \quad x = k_{\text{upper}} \\ \gamma_2 & \quad\quad\; x = k_{\text{lower}} \\ 0 & \text{if} \quad k_{\text{lower}} < x < k_{\text{upper}} \end{cases},$$

with $\gamma_1, \gamma_2, k_{\text{upper}}, k_{\text{lower}}$, such that $E_{\theta_0}\varphi(X) = \alpha$ and $E_{\theta_0}\varphi(X)X = E\theta_0 X \alpha$. Because $\zeta(\theta) = \zeta(\theta_0)$ iff $\theta = \theta_0$ the test problems are equivalent. Under H_0 the distribution of X is $\text{Bin}(n,\theta_0)$. Thus the condition $E_{\theta_0}\varphi(X) = \alpha$ becomes

$$\sum_{x=k_{\text{lower}}+1}^{k_{\text{upper}}-1} \text{bin}(n,\theta_0,x) + (1-\gamma_1)\text{bin}(n,\theta_0,k_{\text{upper}})$$
$$+ (1-\gamma_2)\text{bin}(n,\theta_0,k_{\text{lower}})$$
$$= 1 - \alpha,$$

where $\mathrm{bin}(n, \theta, x) = \binom{n}{x}\theta^x(1-\theta)^{n-x}$. Because of $\mathrm{E}_{\theta_0}X = n\theta_0$ and of the relation

$$x\mathrm{bin}(n, \theta_0, x) = n\,\theta_0\,\mathrm{bin}(n-1, \theta_0, x-1)$$

the condition $\mathrm{E}_{\theta_0}\varphi(X)X = \mathrm{E}_{\theta_0}X\alpha$ can be reduced to

$$\sum_{x=k_{\mathrm{lower}}+1}^{k_{\mathrm{upper}}-1} \mathrm{bin}\,(n-1, \theta_0, x-1) + (1-\gamma_1)\mathrm{bin}(n-1, \theta_0, k_{\mathrm{upper}}-1)$$
$$+(1-\gamma_2)\mathrm{bin}(n-1, \theta_0, k_{\mathrm{lower}}-1)$$
$$= 1-\alpha.$$

Thus γ_1, γ_2 and $k_{\mathrm{upper}}, k_{\mathrm{lower}}$ "can be computed from the binomial tables"; see the following Example 5.27. □

Example 5.27 Continuation of Special case 5.7 above. Set $n = 3$ and $\alpha = 0.05$ and $\theta_0 = \frac{1}{4}$. The binomial tables are

x	0	1	2	3
$\mathrm{bin}\left(3, \frac{1}{4}, x\right)$	0.42188	0.42188	0.14063	1.5625×10^{-2}

and

x	0	1	2
$\mathrm{bin}\left(2, \frac{1}{4}, x\right)$	0.5625	0.375	0.0625

We obtain $k_{\mathrm{lower}} = 0$, $k_{\mathrm{upper}} = 3$, $\gamma_1 = 0.8$ and $\gamma_2 = 0.088$. □

5.3.4 Conditional Tests

In this subsection we derive optimal tests in models with multidimensional parameters

$$\mathcal{P} = \left\{\mathrm{P}_\theta : \theta \in \Theta \subseteq \mathbb{R}^k\right\}$$

by using the results above. Suppose P_θ belongs to a k-parameter exponential distribution, that is:

$$p(\mathbf{x}; \theta) = A(\theta)\exp\left(\sum_{j=1}^{k}\zeta_j(\theta)R_j(\mathbf{x})\right)h(\mathbf{x}).$$

Let us denote the natural parameters by $\beta_j = \zeta_j(\theta)$. We are interested in test problems related to the first component of β. Then $\beta_1 = \lambda$ is called the **parameter of interest** and $\vartheta = (\beta_2, .., \beta_k)^T$ is called the **nuisance parameter**. We have

$$p(\mathbf{x}; \theta) = A(\theta) \exp(\lambda U(\mathbf{x}) + \vartheta^T T(\mathbf{x})) h(\mathbf{x}),$$

where $U(\mathbf{x})$ is the sufficient statistic related to the parameter of interest λ and $T(\mathbf{x})$ is the $(k-1)$-dimensional sufficient statistic related to the nuisance parameter ϑ. The joint distribution of (U, T) belongs to a k-parameter exponential family with probability function $p^{(U,T)}(u, t; \theta)$; compare Theorem 3.9. For $\zeta(\theta) = (\lambda, \vartheta)$ and $c_\lambda(t) = \int h(u, t) \exp(\lambda u) h(u, t) du$ it holds

$$\begin{aligned} p^{(U,T)}(u, t; \theta) &= A(\theta) \exp(\lambda u + \vartheta^T t) h(u, t) \\ &= c(\lambda, t) \exp(\lambda u) h(u, t) \, A(\theta) \exp(\vartheta^T t) c(\lambda, t)^{-1} \\ &= p^{U|T=t}(u \mid t; \lambda) \, p^T(t; \theta). \end{aligned}$$

The statistic T has the distribution

$$p^T(t; \theta) = A(\theta) \exp(\vartheta^T t) c(\lambda, t)^{-1}.$$

We see that the conditional distribution

$$p^{U|T=t}(u \mid t; \lambda) = c(\lambda, t) \exp(\lambda u) h(u, t) \qquad (5.76)$$

is independent of ϑ and belongs to a one-parameter exponential family. Thus for each fixed λ' the $(k-1)$-dimensional statistic T is sufficient for $\{(\lambda', \vartheta) : (\lambda', \vartheta) \in Z = \zeta(\Theta)\}$. Consider the conditional models

$$\mathcal{P}_t = \left\{ \mathsf{P}_\lambda^{U|T=t} : \lambda \in A \subseteq \mathbb{R} \right\}, \ t \in \mathbb{R}^{k-1}.$$

The main trick for deviation of optimal tests in the multivariate case is to use the optimal tests for $\mathcal{P}_t$.

First we consider the one-sided test problem:

$$H_0 : \lambda \geq \lambda_0 \text{ versus } H_1 : \lambda < \lambda_0. \qquad (5.77)$$

Note, this test problem describes a decomposition of the k-dimensional parameter space $Z = \zeta(\Theta)$ into $Z_0 = \{(\lambda, \vartheta) : \lambda \geq \lambda_0\}$ and $Z_1 = \{(\lambda, \vartheta) : \lambda < \lambda_0\}$. The boundary set is denoted by $Z_{bound} = \{(\lambda_0, \vartheta) : (\lambda_0, \vartheta) \in Z = \zeta(\Theta)\}$.

Definition 5.13 (α-similar) A test φ is said to be **α-similar** on Z_{bound} iff $E_{(\lambda_0, \vartheta)} \varphi(\mathbf{X}) = \alpha$. φ is similar on Z_{bound} iff there is an α and φ **α-similar** on Z_{bound}.

Definition 5.14 (UMP α-similar) A test φ is **uniform most powerful α-similar** for the test problem $H_0 : \lambda \geq \lambda_0$ versus $H_1 : \lambda < \lambda_0$ iff φ is α-similar on $\mathcal{Z}_{\text{bound}}$ and

$$E_{(\lambda,\vartheta)}\varphi(\mathbf{X}) \geq E_{(\lambda,\vartheta)}\psi(\mathbf{X}), \text{ for all } (\lambda,\vartheta) \in \mathcal{Z}_1$$

and for all α-**similar** tests ψ on $\mathcal{Z}_{\text{bound}}$.

Theorem 5.7 (One-sided conditional test) *Assume that $\mathcal{Z} = \zeta(\Theta)$ is convex and includes a k-dimensional rectangular. The test*

$$\varphi_I(u,t) = \begin{cases} 1 & \text{if } u < c_0(t) \\ \gamma_0(t) & \text{if } u = c_0(t) \\ 0 & \text{if } u > c_0(t) \end{cases},$$

with $\gamma_0(t)$ and $c_0(t)$, such that: $E_{\lambda_0}(\varphi_I(U,T) \mid T = t) = \alpha$ for all t, is an UMP α-similar test for the test problem (5.77).

Corollary 5.4 *Under the conditions of Theorem 5.7 the test $\varphi_I(u,t)$ is an UMPU α-test.*

PROOF OF OF COROLLARY 5.4:* We know $\varphi_I(u,t)$ is better than $\psi(u,t) \equiv \alpha$; thus $E_{(\lambda,\vartheta)}\varphi_I(U,T) \geq \alpha$ for all $(\lambda,\vartheta) \in \mathcal{Z}_1$ and $\varphi_I(u,t)$ is unbiased. Further $\varphi_I(u,t)$ is α-similar, which is $E_{(\lambda,\vartheta)}\varphi_I(U,T) = \alpha$ for all $(\lambda,\vartheta) \in \mathcal{Z}_{\text{bound}}$. Take an arbitrary unbiased test ψ of size at most α. Then $E_{(\lambda,\vartheta)}\psi(U,T) \leq \alpha$ for all $(\lambda,\vartheta) \in \mathcal{Z}_0$ and $E_{(\lambda,\vartheta)}\psi(U,T) \geq \alpha$ for all $(\lambda,\vartheta) \in \mathcal{Z}_1$. From Theorem 2.4 on page 22 it follows that the power function of ψ is continuous. That implies $E_{(\lambda,\vartheta)}\psi(U,T) = \alpha$ for all $(\lambda,\vartheta) \in \mathcal{Z}_{\text{bound}}(\lambda_0)$ and ψ is α-similar on $\mathcal{Z}_{\text{bound}}$. From Theorem 5.7 follows $E_{(\lambda,\vartheta)}\varphi_I(U,T) \geq E_{(\lambda,\vartheta)}\psi(U,T)$ for all $(\lambda,\vartheta) \in \mathcal{Z}_1$.
□

PROOF OF THEOREM 5.7:* From Corollary 5.2 on page 157 it follows that $E_{\lambda_0}(\varphi_I(U,T) \mid T = t) = \alpha$ for all t. By integration we get

$$E_{(\lambda,\vartheta)}(\varphi_I(U,T)) = \alpha \text{ for all } (\lambda,\vartheta) \in \mathcal{Z}_{\text{bound}}. \tag{5.78}$$

Take an arbitrary α-similar test ψ for the test problem (5.77). We have to show

$$E_{(\lambda,\vartheta)}\varphi_I(U,T) \geq E_{(\lambda,\vartheta)}\psi(U,T) \text{ for all } (\lambda,\vartheta) \in \mathcal{Z}_1.$$

Because ψ is α-similar it holds $E_{(\lambda,\vartheta)}\psi(U,T) = \alpha$ for all $(\lambda,\vartheta) \in \mathcal{Z}_{\text{bound}}$. It follows that $E_\vartheta[E_{\lambda_0}(\psi(U,T) \mid T = t) - \alpha] = 0$ for all $(\lambda,\vartheta) \in \mathcal{Z}_{\text{bound}}$. The statistic T is complete on $\mathcal{Z}_{\text{bound}}$. This implies $E_{\lambda_0}(\psi(U,T) \mid T = t) = \alpha$ for all t. Hence $\psi(.,t)$ is a test of size α in $\mathcal{P}_t$. From Corollary 5.2 follows

$$E_\lambda(\varphi_I(U,T) \mid T = t) \geq E_\lambda(\psi(U,T) \mid T = t), \text{ for all } t.$$

By integration we get the statement.

$\qquad\qquad\qquad\qquad\qquad\qquad\qquad\qquad\qquad\qquad\qquad\qquad$ □

Exercise 5.15 Formulate a variant of Theorem 5.7 for the test problem:

$$H_0 : \lambda \le \lambda_0 \text{ versus } H_1 : \lambda > \lambda_0. \qquad (5.79)$$

Special case 5.8 (One-sided t-test) Consider an i.i.d. sample $\mathbf{X} = (X_1, ..., X_n)$ from $N(\mu, \sigma^2)$. Both parameters are unknown: $\theta = (\mu, \sigma^2) \in \Theta = \mathbb{R} \times \mathbb{R}_+$. We are interested in testing:

$$H_0 : \mu \le \mu_0 \text{ versus } H_1 : \mu > \mu_0.$$

The distribution of $\mathbf{X}$ belongs to a two-parameter exponential family with the natural parameters

$$\zeta_1(\theta) = \frac{n}{\sigma^2}\mu, \quad \zeta_2(\theta) = -\frac{1}{2\sigma^2},$$

and the sufficient statistics: $T_1(\mathbf{x}) = \bar{x}$, $T_2(\mathbf{x}) = \sum_{i=1}^{n} x_i{}^2$; compare Special case 2.1. The hypothesis $H_0 : \mu \le \mu_0$ is equivalent to $H_0 : \frac{n}{\sigma^2}\mu \le \frac{n}{\sigma^2}\mu_0$. Thus the parameter of interest is $\lambda = \frac{n}{\sigma^2}\mu$ and the nuisance parameter is $\vartheta = -\frac{1}{2\sigma^2}$. We apply Theorem 5.7 with $U(\mathbf{x}) = \bar{x}$ and $T(\mathbf{x}) = \sum_{i=1}^{n} x_i{}^2$ and obtain the UMP α-similar test

$$\varphi(u, t) = \begin{cases} 1 & \text{if } u > c_0(t) \\ 0 & \text{if } u < c_0(t) \end{cases},$$

where $c_0(t)$ such that $\mathsf{E}_{\mu_0}(\varphi(U, T) \mid T = t) = \alpha$ for all t. Let us rewrite the critical region. Recall

$$s^2 = \frac{1}{n-1}\sum_{i=1}^{n}(x_i - \bar{x})^2 = \frac{1}{n-1}\sum_{i=1}^{n} x_i^2 - \frac{n}{n-1}\bar{x}^2 = \frac{1}{n-1}t - \frac{n}{n-1}u^2.$$

Consider the function

$$F(u) = \frac{\sqrt{n}(u - \mu_0)}{s} = \frac{\sqrt{n}(u - \mu_0)}{\sqrt{\frac{1}{n-1}t - \frac{n}{n-1}u^2}}.$$

For all t, $F(u)$ is nondecreasing. Thus $u > c_0(t) \iff F(U) > F(c_0(t)) = k(t)$. $F(U)$ is t-distributed with $n-1$ degrees of freedom for all (μ_0, σ^2), $\sigma^2 \in \mathbb{R}_+$. Furthermore $F(U(\mathbf{X}))$ and $T(\mathbf{X})$ are independent random variables. Hence

$$\mathsf{E}_{\mu_0}(\varphi(U, T) \mid T = t) = \mathsf{P}_{\mu_0}(F(U(\mathbf{X})) > k(t) \mid T = t) = \mathsf{P}_{\mu_0}(F(U(\mathbf{X})) > k).$$

Let $t_{n-1;1-\alpha}$ be the quantile of the t-distribution with $n-1$ degrees of freedom:

$P(t > t_{n-1;1-\alpha}) = \alpha$. Summarizing the UMP similar α-test is the well-known t-test:

$$\varphi(\mathbf{x}) = \begin{cases} 1 & \text{if} \quad t > t_{n-1,1-\alpha} \\ 0 & \text{if} \quad t \leq t_{n-1,1-\alpha} \end{cases} \quad \text{with } t = \frac{\sqrt{n}(\overline{x} - \mu_0)}{s}.$$

From Corollary 5.4 we obtain that the t-test is also UMPU α-test.   □

Special case 5.9 (A two-sample problem) Consider two independent samples $\mathbf{X} = (X_1, \ldots, X_m)$ i.i.d. from $\mathsf{N}(\mu_1, \sigma^2)$ and $\mathbf{Y} = (Y_1, \ldots, Y_n)$ i.i.d. from $\mathsf{N}(\mu_1 + \mu_2, \sigma^2)$. Consider the test problem

$$H_0 : \sigma^2 \leq \sigma_0^2 \text{ versus } H_1 : \sigma^2 > \sigma_0^2$$

in the statistical model

$$\mathcal{P} = \left\{ \mathsf{N}(\mu_1, \sigma^2)^{\otimes m} \otimes \mathsf{N}(\mu_1 + \mu_2, \sigma^2)^{\otimes n}, \theta = (\mu_1, \mu_2, \sigma^2) \in \mathbb{R}^2 \times \mathbb{R}_+ \right\}. \quad (5.80)$$

The joint sample $\mathbf{Z} = (\mathbf{X}, \mathbf{Y})$ has the density $f(\mathbf{z}; \theta)$:

$$\left(\frac{1}{2\pi\sigma^2}\right)^{\frac{m+n}{2}} \exp\left(-\frac{1}{2\sigma^2} \sum_{i=1}^{m} (x_i - \mu_1)^2 - \frac{1}{2\sigma^2} \sum_{i=1}^{n} (y_i - (\mu_1 + \mu_2))^2\right)$$
$$= A(\theta) \exp\left(\zeta_1(\theta) T_1(\mathbf{z}) + \zeta_2(\theta) T_2(\mathbf{z}) + \zeta_3(\theta) T_3(\mathbf{z})\right) h(\mathbf{z})$$

with

$$A(\theta) = \left(\frac{1}{2\pi\sigma^2}\right)^{\frac{m}{2}} \left(\frac{1}{2\pi\sigma^2}\right)^{\frac{n}{2}} \exp\left(-\frac{m}{2\sigma^2}\mu_1^2 - \frac{n}{2\sigma^2}(\mu_1 + \mu_2)^2\right)$$

and

$$\zeta_1(\theta) = -\frac{1}{2\sigma^2}, \quad T_1(\mathbf{z}) = \sum_{i=1}^{m} x_i^2 + \sum_{i=1}^{n} y_i^2$$

$$\zeta_2(\theta) = \frac{\mu_1 m}{\sigma^2}, \quad T_2(\mathbf{z}) = \frac{1}{m}\sum_{i=1}^{m} x_i = \overline{x}$$

$$\zeta_3(\theta) = \frac{(\mu_1 + \mu_2)n}{\sigma^2}, \quad T_3(\mathbf{z}) = \frac{1}{n}\sum_{i=1}^{n} y_i = \overline{y}.$$

Hence the joint distribution belongs to a three-parameter exponential family. The parameter of interest is $\lambda = -\frac{1}{2\sigma^2}$. The related sufficient statistic is $\sum_{i=1}^{m} x_i^2 + \sum_{i=1}^{n} y_i^2$. The nuisance parameters are $\frac{\mu_1 m}{\sigma^2}, \frac{(\mu_1 + \mu_2)n}{\sigma^2}$. The test problem can be reformulated as a test problem with respect to the parameter of interest, because

$$\sigma^2 \leq \sigma_0^2 \Leftrightarrow \lambda \leq \lambda_0, \text{ with } \lambda_0 = -\frac{1}{2\sigma_0^2}.$$

Note, U has a continuous distribution. Applying the conditional test of Theorem 5.7, we obtain the UMP α-similar test

$$\varphi(u, t) = \begin{cases} 1 & \text{if } u > c(t) \\ 0 & \text{if } u \leq c(t) \end{cases},$$

with

$$u = \sum_{i=1}^{m} x_i^2 + \sum_{i=1}^{n} y_i^2, \ t = (\bar{x}, \bar{y})$$

and $c(t)$ such that $\mathsf{E}_{\lambda_0}(\varphi(U, T) \mid T = t) = \alpha$ for all t. It holds

$$s_x^2 = \frac{1}{m-1} \sum_{i=1}^{m} (x_i - \bar{x})^2, \ \sum_{i=1}^{m} x_i^2 = s_x^2(m-1) + m\bar{x}^2.$$

Thus $u = s_x^2(m-1) + m\bar{x}^2 + s_y^2(n-1) + n\bar{y}^2$ and $u > c(t) \Leftrightarrow s_{(x,y)}^2 > k(t)$ with

$$s_{(x,y)}^2 = \frac{1}{\sigma_0^2} s_x^2(m-1) + \frac{1}{\sigma_0^2} s_y^2(n-1).$$

Under $\sigma = \sigma_0$ the distribution of $s_{(x,y)}^2 \sim \chi_{m+n-2}^2$ is independent of t. Hence the UMP α-similar test is

$$\varphi(u, t) = \begin{cases} 1 & \text{if } s_{(x,y)}^2 > \chi_{m+n-2;1-\alpha}^2 \\ 0 & \text{if } s_{(x,y)}^2 \leq \chi_{m+n-2;1-\alpha}^2 \end{cases}.$$

$\square$

Consider now the two-sided test problem

$$H_0 : \lambda = \lambda_0 \text{ versus } H_1 : \lambda \neq \lambda_0. \tag{5.81}$$

This test problem describes a decomposition of the k-dimensional parameter space $\mathcal{Z} = \zeta(\Theta)$ into $\mathcal{Z}_{\text{bound}}$ and $\mathcal{Z}_1 = \{(\lambda, \vartheta) : \lambda \neq \lambda_0\}$. We have the result:

Theorem 5.8 (Two-sided conditional test) *The test*

$$
\varphi_{II}(u,t) = \begin{cases} 1 & if \quad u < c_1(t), u > c_2(t) \\ \gamma_1(t) & \quad\quad u = c_1(t) \\ \gamma_2(t) & if \quad\quad u = c_2(t) \\ 0 & if \quad c_1(t) < u < c_2(t) \end{cases},
$$

with $\gamma_i(t)$ *and* $c_i(t)$ *such that* : $\mathsf{E}_{\lambda_0}(\varphi_{II}(U,T) \mid T = t) = \alpha$ *for all* t *and*

$$
\mathsf{E}_{\lambda_0}(U\varphi_{II}(U,T) \mid T = t) = \alpha\mathsf{E}_{\lambda_0}(U \mid T = t) \tag{5.82}
$$

is an UMPU α*-test for the test problem (5.81).*

PROOF: * For all t the test $\varphi_{II}(.,t)$ is the UMPU test for $H_0 : \lambda = \lambda_0$ versus $H_1 : \lambda \neq \lambda_0$ in $\mathcal{P}_t$, because of Theorem 5.5. First we show that φ_{II} is unbiased α-test in $\mathcal{P}$. By integration it follows from $\mathsf{E}_{\lambda_0}(\varphi_{II}(U,T) \mid T = t) = \alpha$ that $\mathsf{E}_{\lambda_0,\vartheta}(\varphi_{II}(U,T)) = \alpha$, for all $(\lambda,\vartheta) \in \mathcal{Z}_{\text{bound}}$. Thus φ_{II} is α-test in $\mathcal{P}$. For all t we have $\mathsf{E}_{\lambda}(\varphi_{II}(U,T) \mid T = t) \geq \alpha$ for all $\lambda \neq \lambda_0$, because $\varphi_{II}(.,t)$ is an unbiased test in $\mathcal{P}_t$. By integration we get $\mathsf{E}_{\lambda,\vartheta}(\varphi_{II}(U,T)) \geq \alpha$ for all $(\lambda,\vartheta) \in \mathcal{Z}$. Hence $\varphi_{II}(u,t)$ is an unbiased test in $\mathcal{P}$.

Now we show that the set of all unbiased α-tests in $\{\mathcal{P}_t : t \in \mathbb{R}^{k-1}\}$ coincides with the set $\Psi_{un,\alpha}$ of all unbiased α-tests in $\mathcal{P}$. For all $t \in \mathbb{R}^{k-1}$ let $\psi(.,t)$ be an α-test in $\mathcal{P}_t$. Then $\mathsf{E}_{\lambda_0}(\psi(U,T) \mid T = t) = \alpha$. By integration we obtain $\mathsf{E}_{\lambda_0,\vartheta}(\psi(U,T)) = \alpha$, for all $(\lambda,\vartheta) \in \mathcal{Z}_{\text{bound}}$; thus ψ is an α-test in $\mathcal{P}$. The property, unbiased α-test, in $\mathcal{P}_t$ implies

$$
\mathsf{E}_{\lambda_0}(\psi(U,T) \mid T = t) = \alpha \text{ and } \frac{\partial}{\partial\lambda}\mathsf{E}_{\lambda}(\psi(U,T) \mid T = t) = 0 \text{ at } \lambda_0.
$$

From Theorem 2.4 on page 22 it follows that

$$
\frac{\partial}{\partial\lambda}\mathsf{E}_{\lambda}(\psi(U,T) \mid T = t) = \mathsf{E}_{\lambda}(U\psi(U,T) \mid T = t) - \mathsf{E}_{\lambda_0}(U \mid T = t)\alpha.
$$

Thus under (5.82) $\frac{\partial}{\partial\lambda}\mathsf{E}_{\lambda}(\psi(U,T) \mid T = t) = 0$ for all t at λ_0. By integration we get $\mathsf{E}_{\lambda,\vartheta}\left(\frac{\partial}{\partial\lambda}\psi(U,T)\right) = 0$ and Theorem 2.4 implies

$$
\frac{\partial}{\partial\lambda}\mathsf{E}_{\lambda,\vartheta}(\psi(U,T)) = 0 \text{ for all } (\lambda,\vartheta) \in \mathcal{Z}_{\text{bound}}.
$$

Thus the power function of ψ is minimal on $\mathcal{Z}_{\text{bound}}$ and $\mathsf{E}_{\lambda,\vartheta}(\psi(U,T)) \geq \alpha$ for all $(\lambda,\vartheta) \in \mathcal{Z}$. Hence ψ is an unbiased test in $\mathcal{P}$. Because of Theorem 5.5, the test $\varphi_{II}(.,t)$ is the UMPU test in $\mathcal{P}_t$ for all t. That means

$$
\mathsf{E}_{\lambda}(\varphi_{II}(U,T) \mid T = t) \geq \mathsf{E}_{\lambda}(\psi(U,T) \mid T = t), \text{ for all } \psi \in \Psi_{un,\alpha}.
$$

By integration we obtain the statement of Theorem 5.8.

□

Special case 5.10 (Two-sided variance test) Consider
an i.i.d. sample $\mathbf{X} = (X_1, ..., X_n)$ from $N(\mu, \sigma^2)$. Both parameters are unknown $\theta = (\mu, \sigma^2) \in \Theta = \mathbb{R} \times \mathbb{R}_+$. We are
interested in testing:

$$H_0 : \sigma^2 = \sigma_0^2 \text{ versus } H_1 : \sigma^2 \neq \sigma_0^2.$$

The distribution of $\mathbf{X}$ belongs to a two-parameter exponential family with the
natural parameters

$$\zeta_1(\theta) = \frac{n}{\sigma^2} \mu, \quad \zeta_2(\theta) = -\frac{1}{2\sigma^2}$$

and the sufficient statistics $T_1(\mathbf{x}) = \bar{x}$ and $T_2(\mathbf{x}) = \sum_{i=1}^{n} x_i^2$; compare Special
case 2.1 on page 15. Both statistics have a continuous distribution. The hy-
pothesis $H_0 : \sigma^2 = \sigma_0^2$ is equivalent to $H_0 : -\frac{1}{2\sigma^2} \neq -\frac{1}{2\sigma_0^2}$. Thus the parameter
of interest is $\lambda = -\frac{1}{2\sigma^2}$ and the nuisance parameter is $\vartheta = \frac{n}{\sigma^2} \mu$. We apply
Theorem 5.8 with $u(\mathbf{x}) = \sum_{i=1}^{n} x_i^2$ and $T(\mathbf{x}) = \bar{x}$, and obtain the UMPU
α-test:

$$\varphi_{II}(u, t) = \begin{cases} 1 & \text{if} \quad u < c_1(t), u > c_2(t) \\ 0 & \text{if} \quad c_1(t) \leq u \leq c_2(t) \end{cases},$$

with $c_i(t)$ such that $\mathsf{E}_{\sigma_0^2}(\varphi_{II}(U, T) \mid T = t) = \alpha$ for all t and

$$\mathsf{E}_{\sigma_0^2}(U\varphi_{II}(U, T) \mid T = t) = \alpha \mathsf{E}_{\sigma_0^2}(U \mid T = t)$$

for all t. Consider the region $c_1(t) < u < c_2(t)$. It holds

$$s^2 = \frac{1}{n-1} \sum_{i=1}^{n} (x_i - \bar{x})^2 = \frac{1}{n-1} (u - n t^2).$$

Thus $c_1(t) < u < c_2(t)$ is equivalent to $(c_1(t) - nt^2) < s^2(n-1) < (c_2(t) - nt^2)$. Define $k_i(t) = c_i(t) - nt^2$. The statistic $\frac{S^2(n-1)}{\sigma_0^2}$ is χ^2-distributed
with $n-1$ degrees of freedom and is independent of $\bar{X}$. We have to determine
$k_i = k_i(t)$, such that

$$\mathsf{P}_{\sigma_0^2}(k_1 < s^2(n-1) < k_2) = 1 - \alpha \text{ and } \mathsf{E}_{\sigma_0^2}\varphi_{II}(\mathbf{X})s^2(n-1) = (n-1)\sigma_0^2\alpha.$$

Using the calculations in Special case (5.6) an approximative solution is
$\frac{1}{\sigma_0^2} k_1 = \chi^2_{n-1;\frac{\alpha}{2}}$ and $\frac{1}{\sigma_0^2} k_2 = \chi^2_{n-1;1-\frac{\alpha}{2}}$. Summarizing, we get the approximative UMPU α-test is the well-known variance test:

$$\varphi(\mathbf{x}) = \begin{cases} 1 & \text{if} & \text{otherwise} \\ 0 & \text{if} \quad \chi^2_{n-1;\frac{\alpha}{2}} \leq \frac{s^2(n-1)}{\sigma_0^2} \leq \chi^2_{n-1;1-\frac{\alpha}{2}} \end{cases}.$$

□

5.4   List of Problems

1. Consider the following rejection regions: $C_1 = \{|\bar{x} - 10| > 0.5\}$ and $C_1 = \{|\bar{x} - 10| > 0.8\}$. Which one has the larger α? (Lindgren (1962), Problem 9-21.)

2. Consider two Neyman–Pearson tests

$$\varphi_1(x) = \begin{cases} 1 & \text{if } p_0(x) < k\,p_1(x) \\ 0 & \text{if } p_0(x) \geq k\,p_1(x) \end{cases} , \quad \varphi_2(x) = \begin{cases} 1 & \text{if } p_0(x) \leq k\,p_1(x) \\ 0 & \text{if } p_0(x) > k\,p_1(x) \end{cases} .$$

Let the sizes of φ_i be α_i and β_i. Show that for the Neyman–Pearson tests with $\alpha = \lambda\alpha_1 + (1 - \lambda)\alpha_2$, $0 \leq \lambda \leq 1$ it holds $\beta = \lambda\beta_1 + (1 - \lambda)\beta_2$.

3. Let (X_1, X_2) be i.i.d. $X \sim \text{Ber}(p)$. Consider testing $p = 0.5$ against $p = 0.8$. a) List all possible critical regions. b) Find all (α, β) for nonrandomized tests. Plot them. c) Plot the set of $\alpha\beta$-representations. d) Derive the Neymann–Pearson test with size $\alpha = 0.05$. (Lindgren (1962), Problems 9-20, 9-55.)

4. Consider the critical region $X > k$ for a single observation, to test $H_0 : U[0, 1]$ versus $H_1 : f(x) = 2x$, for $0 < x < 1$. a) Find α and β as functions of k. b) Plot the $\alpha\beta$-curve of a). c) Find k such that $\alpha = 4\beta$. (Lindgren (1962), Problem 9-57.)

5. Let two discrete probability distributions be defined by

x	2	3	4	5	6	7	8
$p_0(x)$	0.05	0.02	0.33	0.1	0.2	0.1	0.2
$p_1(x)$	0.01	0.3	0.01	0.18	0.2	0.2	0.1

a) Give the Neyman–Pearson test for $\alpha = 0.02$. b) Give the Neyman–Pearson test for $\alpha = 0.05$. c) Calculate the sizes of error of second type for both tests. d) Give an alternative alpha test for $\alpha = 0.05$. e) Compare your test in d) with the Neyman–Pearson test in b).

6. Let $(X_1, \ldots, X_n)$ be an i.i.d. sample from $\text{Poi}(\lambda)$. Find the most powerful test for $H_0 : \lambda = 1$ versus $H_1 : \lambda = 2$.

7. Let z be an observation from one of the four-points-distributions given in the following table:

	z_1	z_2	z_3	z_4
θ_1	.2	.3	.1	.4
θ_2	.5	.1	.2	.2
θ_3	.3	.0	.4	.3

Consider the test problem $H_0 : \theta_1$ versus $H_1 : \{\theta_2, \theta_3\}$. a) For each value z give the likelihood ratio: $\Lambda^* = \frac{\sup_{\theta \in \Theta_0} L(\theta)}{\sup_{\theta \in \Theta_1} L(\theta)}$. b) Determine the critical regions defined by $\Lambda^* < K$. c) Find α for each test in b). d) Compare the tests with respective critical regions $\{z_1\}$ and $\{z_2, z_3\}$ when the "true" parameter is θ_2. (Lindgren (1962), Problem 9-23.)

8. Show that the exponential distribution has a monotone ratio. Find the UMP test for a one-sided hypothesis.

9. Let $(X_1, X_2, \ldots, X_n)$ be an i.i.d. sample from $\mathsf{N}(\mu, 1)$. a) Find the UMP test for $H_0 : \mu \leq \mu_0$ versus $H_1 : \mu > \mu_0$. b) Plot the power function of the test. c) Find the UMP test for $H_0 : \mu \geq \mu_0$ versus $H_1 : \mu < \mu_0$. d) Find the UMPU α-test for $H_0 : \mu = \mu_0$ versus $H_1 : \mu \neq \mu_0$.

10. Consider an i.i.d. sample $(X_1, X_2, \ldots, X_n)$ of an exponential distribution $\mathsf{Exp}(\lambda)$ with density : $f(x; \lambda) = \lambda \exp(-\lambda x)$ for $x > 0$; $f(x; \lambda) = 0$ otherwise. a) Find the statistic T for testing $H_0 : \lambda = \lambda_0$ versus $H_1 : \lambda \neq \lambda_0$. b) Define a critical region.

11. Consider a sample $\mathbf{X} = (X_1, \ldots, X_n)$ of independent observations with $X_i \sim \mathsf{Poi}(n_i \lambda)$ with known n_i and with

$$p_i(x) = \frac{(n_i \lambda)^x}{x!} \exp(-n_i \lambda).$$

a) Does the distribution of $\mathbf{X}$ belong to an exponential family? b) Derive the sufficient statistic. c) Derive the MLE for λ. d) Has the distribution a monotone likelihood quotient? e) Which properties has the uniform most powerful test of size α for $H_0 : \lambda \geq 1$ versus $H_1 : \lambda < 1$? f) Derive the most powerful test for $H_0 : \lambda \geq 1$ versus $H_1 : \lambda < 1$.

12. Let $(X_1, \ldots, X_n)$ be an i.i.d. sample from $\mathsf{N}(\mu, \sigma^2)$, where σ^2 is unknown. Find the UMPU α-test for $H_0 : \mu = \mu_0$ versus $H_1 : \mu \neq \mu_0$. (Hint: It is the two-sided t-test.)

13. Let $X_1, \ldots, X_n$ be an i.i.d. sample, $X_i \sim \mathsf{N}(\mu_1, \sigma^2)$, and let $Y_1, \ldots, Y_n$ be an i.i.d sample, $Y_j \sim \mathsf{N}(\mu_2, 2\sigma^2)$, independent of the first sample. Consider the joint sample $\mathbf{Z} = (X_1, \ldots, X_n, Y_1, \ldots, Y_n)$ with the unknown parameters (μ_1, μ_2, σ^2). a) Show that $\mathbf{Z}$ belongs to a three-parameter exponential family. b) Give the sufficient statistics. c) Determine the parameter of interest and the nuisance parameters for $H_0 : \sigma^2 \leq \Delta_0$ versus $H_1 : \sigma^2 > \Delta_0$. d) Transform the hypotheses. e) Derive the UMP α-similar test for $H_0 : \sigma^2 \leq \Delta_0$ versus $H_1 : \sigma^2 > \Delta_0$. (Hint: Use: $(n-1)S_x^2 + (n-1)\frac{1}{2}S_y^2 \sim \chi_{2n-2}^2 \sigma^2$.)

5.5   Further Reading

The concept of assessing evidence, where only the null hypothesis is specified, goes back to Fisher (1890–1962). For an interesting discussion, from a historical point of view, see the paper of Lehmann (1993) and the book of Heyde and Seneta (2001).

In this textbook goodness-of-fit tests are not considered. For checking whether the underlying distribution is equal to a prespecified continuous distribution there are tests based on the distance between the empirical distribution and the hypothetical distribution. The derivation of such tests for various weighted distances is given in Shorack and Wellner (1986).

For testing the goodness-of-fit of a discrete distribution Chi-squared tests are appropriate. Greenwood and Nikulin (1996) provide the basic material on Chi-squared testing. They complement the theory by the application of the proposed test procedure to special problems.

Computer-intensive methods for carrying out tests are only mentioned in this textbook. For a further study the reader is referred to Efron and Tibshirani (1993) and Davison and Hinkley (2003). In the book of Zhu (2005) different Monte Carlo algorithms are presented and their justifications are proved.

The likelihood ratio test as an extension of tests of Neyman–Pearson type and related asymptotic test procedures are presented in Lehmann (1999). In Lehmann (1998) rank tests are considered. These tests are useful to compare distributions in cases where no parametric model can be justified. Moreover, they are applied to test for independence and randomness.

In Section 5.2 on page 126 the p-value is defined. Different ways for defining a p-value for composite null hypotheses are given by Bayarri and Berger (2000). An interesting approach to p-values is provided by Schervish (1995). Here also the connection between p-values and testing from the Bayesian point of view is considered. For a further discussion of p-values we refer to Garthwaite et al. (2002).

A rigorous mathematical treatment of testing hypotheses is given by Lehmann and Romano (2005). The authors present an optimality theory for hypothesis testing and confidence sets—small sample properties as well as asymptotic properties are investigated. Moreover, an introduction in the theory of resampling methods for testing is developed.

A classical monograph which embeds test theory procedures in a general decision theory and which also gives the relationship to game theory is the book of Ferguson (1967).

A modern decision theory and their systematic applications to the fields of testing hypotheses and selection populations are included in the book of Liese and Miescke (2008).

Chapter 6

Linear Model

6.1 Introduction

In earlier chapters we considered inference methods. They were the basis to characterize the properties of estimators proposed in Chapter 4 to estimate unknown parameters of a distribution. In Chapter 5 general approaches to testing hypotheses were discussed. Now let us apply these tools provided in the previous chapters to examine the relationship between an outcome variable and some explanatory variables. We will assume that the functions describing such relationships are of a specific form, including unknown parameters. Thus, the determination of the relationship between the considered variables is a problem of estimating unknown parameters. Moreover, the problem of finding an appropriate function leads to the problem of testing hypotheses about parameters.

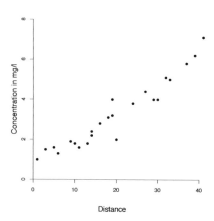

Figure 6.1: Nitrate concentration in rivers depending on distance from the source.

Example 6.1 (Nitrate concentration in rivers) The concentration of nitrate in the river depends on the distance to the source. Figure 6.1 shows a scatterplot of data (x_i, y_i) taken at measuring points at 24 rivers. Here y_i is the concentration of nitrate (in mg/l) and x_i is the distance of the measuring point from the source (in km), $i = 1, \ldots, 24$ (data are taken from Rudolf and Kuhlisch (2008)).   □

x	1	6	3	11	5	13	10	9	14	14	19	18
y	1.0	1.3	1.5	1.6	1.6	1.8	1.8	1.9	2.2	2.4	3.2	3.1
x	16	24	27	33	32	37	39	41	19	20	29	30
y	2.8	3.8	4.4	5.0	5.1	5.8	6.2	7.1	4.0	2.0	4.0	4.0

Table 6.1: Nitrate concentration y depending on the distance from the source x

The plot in Figure 6.1 shows a nearly linear dependence of the concentration on the distance. But the presence of randomness leads the data points to scatter about the straight line. So we consider the observation y_i as a value of a r.v. Y_i satisfying the equation

$$Y_i = \beta_1 + \beta_2 x_i + \varepsilon_i, \qquad i = 1, \ldots, n, \tag{6.1}$$

where the ε_i's are random errors.

The variable Y_i, $i = 1, \ldots, n$, is called **response variable** or output, and x_i is called the covariate, the explanatory variable, the input or the design point. Model (6.1) is a very simple example of a linear regression model, the so-called *simple linear regression*. It can be generalized as follows: We consider not only a linear dependence between input and output, but also other types of functional relationships. An important example is the polynomial model. Furthermore, we can extend the simple linear regression model by including more **explanatory variables**. That is, we consider the dependence of Y_i on a vector $x_i = (x_{i1}, \ldots, x_{ip})^\mathsf{T}$.

The relationship between the response Y_i and the explanatory variables x_{ij} is not completely predictable. Very often there are unavoidable, uncontrolled factors that result in different responses for the same value of x_i. Such factors include errors in the measurements and changes in experimental conditions, but also some randomness inherent to the phenomenon itself, for example the biological variability. These **unobservable errors** are described by the random variable ε_i.

The aim of the statistical analysis is to sort out the underlying relationship from the "noise," the random component in the Y_i.

In the following chapter we consider only real-valued continuous outputs Y_i. Multivariate response variables are the subject of multivariate analysis; discrete Y_i's are modeled, for example, by logistic regression models.

6.2   Formulation of the Model

Suppose we have observations y_i at covariates $x_{i1}, \ldots, x_{ip}$, $i = 1, \ldots, n$. We consider these observations as realizations of r.v.'s Y_i satisfying the equations

$$Y_i = x_{i1}\beta_1 + \cdots + x_{ip}\beta_p + \varepsilon_i, \qquad i = 1, \ldots, n. \qquad (6.2)$$

The β_j's are the unknown coefficients and the ε_i's are the unobservable random errors with expectation zero. We assume that the variables x_{ij} are fixed. They are actually known.

It is convenient to write the model (6.2) in matrix form

$$\mathbf{Y} = X\beta + \varepsilon, \qquad (6.3)$$

where

- $\mathbf{Y}$ is the $n \times 1$ vector whose i-th component is Y_i,
- X is the fixed design matrix, i.e., the $n \times p$ matrix whose i-th row is the vector $x_i^{\mathsf{T}} = (x_{i1}, \ldots, x_{ip})$,
- $\beta = (\beta_1, \ldots, \beta_p)^{\mathsf{T}} \in \mathbb{R}^p$ is the parameter of unknown coefficients and
- ε is the $n \times 1$ vector of errors.

Let us give some examples for the explanatory variables:

Special case 6.1 (Polynomial regression) Consider a single covariate, say z. Suppose the relationship between the output y and z can be described by a polynomial of degree $p - 1$. Thus we have

$$Y_i = \beta_1 + \beta_2 z_i + \cdots + \beta_p z_i^{p-1} + \varepsilon_i, \qquad i = 1, \ldots, n,$$

and model (6.2) holds with

$$x_{i1} = 1, \qquad \text{and} \qquad x_{ij} = z_i^{j-1}, \quad j = 2, \ldots, p.$$

This model is called **polynomial regression**. For $p = 2$ we obtain the simple linear model, that is $\beta_1 + \beta_2 z$, the straight-line regression with slope β_2 and intercept β_1. $\square$

Example 6.2 (Regression with interactions) Let us investigate the relationship between an output and two covariates. For example, consider the product yield of a certain chemical reaction depending on the reaction temperature z_1 and the pressure z_2. The theory suggests that in average the yield will decrease when the pressure increases. The decline depends also on the temperature—it is more rapid for a higher than for a lower one. Thus the chemist would include in the model not only z_1 and z_2 separately, but a third variable describing the **interaction** between both, for example $z_1 \cdot z_2$. Hence we have the following model

$$Y_i = \beta_1 + \beta_2 z_{i1} + \beta_3 z_{i2} + \beta_4 z_{i1} z_{i2} + \varepsilon_i, \qquad i = 1, \ldots, n,$$

or in the matrix form (6.3) with $p = 4$

$$x_{i1} = 1, \quad x_{i2} = z_{i1}, \quad x_{i3} = z_{i2}, \quad x_{i4} = z_{i1} z_{i2}.$$

$\square$

Let us summarize: In general we consider the dependence of a random response Y on r covariates $z_1, \ldots, z_r$, say $z^{\mathsf{T}} = (z_1, \ldots, z_r)$. Copies of Y and z are observed and the functional relationship is described by model (6.2) with

$$x_{ij} = f_j(z_{i1}, \ldots, z_{ir}) = f_j(z_i),$$

where the f_j's are known functions. In other words, we have for $i = 1, \ldots, n$

$$Y_i = \sum_{j=1}^{p} x_{ij}\beta_j + \varepsilon_i = \sum_{j=1}^{p} f_j(z_i)\beta_j + \varepsilon_i.$$

The function $m : \mathbb{R}^r \to \mathbb{R}$ defined by

$$m(z) = \sum_{j=1}^{p} f_j(z)\beta_j$$

is called a **regression function**.

OBS! Note that the notation *linear* model means linearity in the parameter β, not linearity between Y and the covariate z.

The following models are examples for **nonlinear regression**:

$$Y_i = \alpha \, x_{i1}^{\beta} \, x_{i2}^{\gamma} \, \eta_i, \quad \text{where } \eta_i \text{ denotes the error,} \tag{6.4}$$

and

$$Y_i = \alpha(1 - \exp(-x_i/\beta)) + \varepsilon_i. \tag{6.5}$$

Model (6.4) has a multiplicative structure. Taking the logarithm we get a linear model:

$$\log Y_i = \log \alpha + \beta \log x_{i1} + \gamma \log x_{i2} + \log \eta_i.$$

Model (6.5) cannot be transformed into a linear regression model.

Up to now we have only considered the inclusion of quantitative variables. By using a simple numerical coding, qualitative variables can be incorporated into the model. These variables are called **dummy variables**. In the simplest case these variables take the values 1 or 0.

Example 6.3 (Production process) Consider a production process, where the product yield depends on certain quantitative factors, as temperature, pressure and others. But in addition we know that the production is carried out at two different production lines with two different working teams. Suppose we have 10 observations from line 1 with team A, 8 observations from line 1 with team B, 12 observations from line 2 with team A and 6 observations from line 2 with team B. Taking into account these qualitative factors one could define the matrix X as follows: The i-th row of the (36×5) matrix is given by $(1, x_{i2}, x_{i3}, x_{i4}, x_{i5})$, where x_{i2} and x_{i3} are the values of the temperature and the pressure, respectively, for the i-th observation and

$$x_{i4} = \begin{cases} 1 & \text{if the } i\text{-th product is produced at line 1} \\ 0 & \text{if the } i\text{-th product is produced at line 2} \end{cases}$$

and

$$x_{i5} = \begin{cases} 1 & \text{if the } i\text{-th product is produced by team A} \\ 0 & \text{if the } i\text{-th product is produced by team B} \end{cases}.$$

The equation is

$$y_i = \beta_1 + \beta_2 x_{i2} + \beta_3 x_{i3} + \beta_4 x_{i4} + \beta_5 x_{i5} + \varepsilon_i.$$

Suppose the parameter β is estimated by $\hat{\beta}$. Then, for example, the expected output at line 1 with team B is estimated by

$$\widehat{\text{yield}} = (\hat{\beta}_1 + \hat{\beta}_4) + \hat{\beta}_2 \,\text{temperature} + \hat{\beta}_3 \,\text{pressure}.$$

□

This textbook does not include the theory of analysis of variances. Nevertheless, we will demonstrate that the classification models of the variance analysis can be treated as linear models.

Example 6.4 (Crop yield of wheat) An experiment was carried out to examine the effect of four brands of wheat on the crop yield. To that aim, on 20 plots of equal size (under homogenous conditions) these four types of wheat were planted. Table 6.2 gives the yield. We formulate the following model: The response variable "crop yield" depends on the factor brand. This factor takes four levels. We can consider these observations as values of four samples of i.i.d. r.v.'s. $\mathbf{Y}_1, \ldots, \mathbf{Y}_4$, where $\mathbf{Y}_i = (Y_{i1}, \ldots, Y_{i5})$ is the sample of the yield of wheat of type i. Let μ_i be the expected yield of brand i, then

$$Y_{ij} = \mu_i + \varepsilon_{ij}, \qquad i = 1, \ldots, 4, \quad j = 1, \ldots, 5, \tag{6.6}$$

Brand	Yield				
1	82	94	100	84	92
2	83	78	68	80	86
3	96	110	107	104	106
4	88	98	82	90	96

Table 6.2: Yield depending on brand of wheat

where ε_{ij} are error terms. To describe the effect of the factor brand let us introduce the *grand mean* μ, which can be interpreted as mean of all Y_{ij}'s. The mean crop yield μ_i of wheat i will be higher or less than μ. Thus the effect of brand i is characterized by the difference $\alpha_i = \mu - \mu_i$. This will result by the following model

$$Y_{ij} = \mu + \alpha_i + \varepsilon_{ij}.$$

□

Special case 6.2 (One-Way ANOVA-Models) The general form of the one-way-ANOVA is given by the equality

$$Y_{ij} = \mu + \alpha_i + \varepsilon_{ij}, \qquad i = 1, \dots, I, \quad j = 1, \dots, n_i. \qquad (6.7)$$

Here we have one factor with I levels. On each level there are n_i observations. If $n_i = N$ for all i, then we have a *balanced* model. We will consider only balanced models. Furthermore we assume

$$\sum_{i=1}^{I} \alpha_i = 0. \qquad (6.8)$$

Condition (6.8) can be interpreted as follows: Let μ_i be the expectation of the i-th population, which is decomposed into the grand mean μ and the effect α_i. Then $\frac{1}{I}\sum_i \mu_i = \frac{1}{I}\sum_i(\mu + \alpha_i)$.

We can write equation (6.7) in the form (6.3): For $n = N \cdot I$ define the n-dimensional vectors

$$\mathbf{Y} = (Y_{11}, \dots, Y_{1N}, Y_{21}, \dots, Y_{2N}, \dots, Y_{I1}, \dots, Y_{IN})^{\mathsf{T}}$$

and

$$\varepsilon = (\varepsilon_{11}, \dots, \varepsilon_{1N}, \varepsilon_{21}, \dots, \varepsilon_{2N}, \dots, \varepsilon_{I1}, \dots, \varepsilon_{IN})^{\mathsf{T}}.$$

Because of condition (6.8) there are only I unknown parameters: $\mu, \alpha_1, \dots, \alpha_{I-1}$. Now, let us define the $n \times I$ matrix X: Denote with $\mathbb{1}_N$ the vector of dimension N consisting only of ones's and with $\mathbb{0}_N$ that consisting only of zero's. Then we can write

$$\mathbf{Y} = \begin{pmatrix} \mathbb{1}_N & \mathbb{1}_N & \mathbb{O}_N & \mathbb{O}_N & \cdots & \mathbb{O}_N & \mathbb{O}_N \\ \mathbb{1}_N & \mathbb{O}_N & \mathbb{1}_N & \mathbb{O}_N & \cdots & \mathbb{O}_N & \mathbb{O}_N \\ \mathbb{1}_N & \mathbb{O}_N & \mathbb{O}_N & \mathbb{1}_N & \cdots & \mathbb{O}_N & \mathbb{O}_N \\ \vdots & \vdots & \vdots & \vdots & \cdots & \vdots & \vdots \\ \mathbb{1}_N & \mathbb{O}_N & \mathbb{O}_N & \mathbb{O}_N & \cdots & \mathbb{1}_N & \mathbb{O}_N \\ \mathbb{1}_N & \mathbb{O}_N & \mathbb{O}_N & \mathbb{O}_N & \cdots & \mathbb{O}_N & \mathbb{1}_N \\ \mathbb{1}_N & -\mathbb{1}_N & -\mathbb{1}_N & -\mathbb{1}_N & \cdots & -\mathbb{1}_N & -\mathbb{1}_N \end{pmatrix} \begin{pmatrix} \mu \\ \alpha_1 \\ \vdots \\ \alpha_{I-1} \end{pmatrix} + \varepsilon.$$

□

Let us now formulate the assumptions about the distribution—or in other words, let us state the statistical model: Without specifying the type of the distribution of the errors we assume only that the errors ε_i have zero mean and a finite variance. Then we obtain from (6.3):

$$\mathsf{E}_\theta \mathbf{Y} = X\beta + \mathsf{E}_\theta \varepsilon = X\beta \qquad \text{and} \qquad \mathsf{Cov}_\theta \mathbf{Y} = \mathsf{Cov}_\theta \varepsilon = C,$$

where $C = (c_{ij})$ is a positive semidefinite $n \times n$ matrix.
The parameter θ consists of β, C and a parameter κ characterizing the distribution of $\mathbf{Y}$ up to the first and second moments.
Let us summarize:

Definition 6.1 (Linear model) The linear model is given by

$$\mathbf{Y} = X\beta + \varepsilon.$$

The distribution of $\mathbf{Y}$ belongs to the class

$$\mathcal{P} = \{\mathsf{P}_\theta : \mathsf{E}_\theta \mathbf{Y} = X\beta, \quad \mathsf{Cov}_\theta \mathbf{Y} = C,$$
$$\theta = (\beta, C, \kappa) \in \Theta \subseteq \mathbb{R}^p \times \mathcal{M}^{\geq} \times \mathcal{K}\}.$$

If one can assume that the observations Y_i are uncorrelated, then the covariance matrix has a diagonal form, where $c_{ii} = \mathsf{Var}_\theta Y_i$ and $c_{ij} = 0$ for $i \neq j$. Moreover, if the observations are uncorrelated with equal variances, then $C = \sigma^2 I_n$, where $\sigma^2 = \mathsf{Var}_\theta Y_i$ and I_n is the identity matrix. The distribution of a linear model with uncorrelated errors having the same variance is given by

$$\mathcal{P} = \{\mathsf{P}_\theta : \mathsf{E}_\theta \mathbf{Y} = X\beta, \quad \mathsf{Cov}_\theta \mathbf{Y} = \sigma^2 I_n, \tag{6.9}$$
$$\theta = (\beta, \sigma^2, \kappa) \in \Theta \subseteq \mathbb{R}^p \times \mathbb{R}_+ \times \mathcal{K}\}.$$

Another important case is the one with $C = \sigma^2 \Sigma$, where Σ is a known positive definite matrix and $\sigma^2 > 0$ is unknown. Then, again, the parameter is $\theta = (\beta, \sigma^2, \kappa)$. In this case the data can be transformed by the matrix $\Sigma^{-1/2}$ as follows:

$$\widetilde{\mathbf{Y}} = \Sigma^{-1/2} \mathbf{Y} = \Sigma^{-1/2} X\beta + \Sigma^{-1/2} \varepsilon = \widetilde{X}\beta + \widetilde{\varepsilon}. \qquad (6.10)$$

For the transformed model we obtain

$$\mathsf{E}_\theta \widetilde{\mathbf{Y}} = \widetilde{X}\beta, \quad \mathsf{Cov}_\theta \widetilde{\mathbf{Y}} = \sigma^2 I_n.$$

If one can justify that the errors are normally distributed, then we have a parametric model, the so-called **normal linear model**:

$$\mathcal{P} = \{\mathsf{N}_n(X\beta, C) : (\beta, C) \in \mathbb{R}^k \times \mathcal{M}^{\geq}\},$$

where $\mathcal{M}^{\geq}$ denotes the set of all positive definite matrices, or shortly

$$\mathbf{Y} \sim \mathsf{N}_n(X\beta, C).$$

And also here we will consider the special case $C = \sigma^2 I_n$.

Remark 6.1 We assume the covariates as nonrandom variables, i.e., we have a so-called **fixed design model**. Such an approach is always justified when the x_i's are known before the Y_i are measured. But there are situations where the inputs have to be modeled as random variables, i.e., we have a **random design model**. In this case the observations (Y_i, X_i), $i = 1, \ldots, n$, form a sample of independent and *identically* distributed r.v.'s. Then the regression function m is the conditional expectation of the response given the covariates. The statistical methods proposed in the following sections can be applied also to the random design model. But we have to be careful—all properties are "conditional" properties; for example unbiasedness of a parameter estimator is then conditional unbiasedness, i.e. the conditional expectation of the estimator given the covariate $\mathbf{X} = \mathbf{x}$ is equal to the parameter.

Before we begin with the statistical investigation of the model (6.3) let us consider the model from another point of view. With $\mu = X\beta$ we can write [1]

$$\mathbf{Y} = \mu + \varepsilon, \qquad (6.11)$$

where the $n \times 1$ vector μ lies in a subspace

$$\mathcal{R}[X] = \{z \in \mathbb{R}^n \,|\, z = X\beta, \ \beta \in \mathbb{R}^p\}$$

of $\mathbb{R}^n$. The dimension of this subspace is the rank of the matrix X, usually denoted by q. The components of the vector μ are

$$\mu_i = x_{i1}\beta_1 + \cdots + x_{ip}\beta_p = x_i^{\mathsf{T}}\beta.$$

---

[1] We use here the same notation as for the (scalar) grand mean in the models of the variance analysis. We think that it is clear from the context which μ is meant.

Since $\mathsf{E}_\theta \mathbf{Y} = \mu$, the linear model (6.11) can be considered as a generalization of a sample of i.i.d. r.v.'s with expectation $\mu_i = \mu \in \mathbb{R}$ for all $i = 1, \ldots, n$ to nonidentical and possible dependent variables.

The formulation of the model in the form (6.11) will give us some additional interpretations of the geometrical background of the estimation and test procedures presented in the next sections. [2]

6.3 The Least Squares Estimator

6.3.1 The Model with Uncorrelated Errors

We consider the linear model (6.3) with $\mathsf{Cov}_\theta \varepsilon = \sigma^2 I_n$. That is the errors are uncorrelated r.v.'s with the same variance. The first problem is to estimate the unknown parameter β. To do this we apply the method of least squares. The idea is to estimate the coefficients β_j by those values for which the sum of the squared differences

$$\sum_{i=1}^{n} (y_i - (x_{i1}\beta_1 + \cdots + x_{ip}\beta_p))^2$$

is minimized. These differences are the deviations of the observations from the expected values $\mathsf{E}_\theta Y_i$. In matrix notation this is equivalent to the minimization of the quadratic form

$$(\mathbf{y} - X\beta)^{\mathsf{T}}(\mathbf{y} - X\beta). \tag{6.12}$$

Differentiating (6.12) we get

$$\frac{\partial}{\partial \beta}(\mathbf{y} - X\beta)^{\mathsf{T}}(\mathbf{y} - X\beta) = 2(X^{\mathsf{T}}X\beta - X^{\mathsf{T}}\mathbf{y})$$

and

$$\frac{\partial^2}{\partial \beta \partial \beta^{\mathsf{T}}}(\mathbf{y} - X\beta)^{\mathsf{T}}(\mathbf{y} - X\beta) = 2X^{\mathsf{T}}X \succeq 0.$$

Thus, the determination of an estimate minimizing the sum of squared differences leads to the following system of equations:

$$X^{\mathsf{T}}X\beta = X^{\mathsf{T}}\mathbf{y}. \tag{6.13}$$

Often, in literature the system of equations (6.13) is named **normal equations**. We define:

Definition 6.2 (Least squares estimator (LSE)) A solution of the system of equations (6.13) is called least squares estimate (**LSE**) for β.

---

[2]Sometimes the model $\mathbf{Y} = \mu + \varepsilon$ is called a linear model and $\mathbf{Y} = X\beta + \varepsilon$ is named a linear regression model. We prefer the name linear model for the latter because it is also a useful tool in variance analysis.

Assuming that the matrix X has full rank, that is $q = p$, then the normal equations have a unique solution, namely

$$\hat{\beta} = (X^\mathsf{T} X)^{-1} X^\mathsf{T} \mathbf{y}. \tag{6.14}$$

In this case the LSE has the following properties:

Theorem 6.1 *Assume model (6.9) with* $\mathrm{rank}(X) = p$. *Then the normal equations (6.13) have the unique solution* $\hat{\beta}$ *given in (6.14). Furthermore the estimator* $\hat{\beta} = (X^\mathsf{T} X)^{-1} X^\mathsf{T} \mathbf{Y}$ *has the following properties:*

1. $\hat{\beta}$ *is linear in* $\mathbf{Y}$.
2. $\hat{\beta}$ *is unbiased, i.e.,* $\mathsf{E}_\theta \hat{\beta} = \beta$ *for all* θ.
3. *The covariance is given by* $\mathsf{Cov}_\theta \hat{\beta} = \sigma^2 (X^\mathsf{T} X)^{-1}$.

PROOF: We have

$$\mathsf{E}_\theta \hat{\beta} = \mathsf{E}_\theta (X^\mathsf{T} X)^{-1} X^\mathsf{T} \mathbf{Y} = (X^\mathsf{T} X)^{-1} X^\mathsf{T} \mathsf{E}_\theta \mathbf{Y} = (X^\mathsf{T} X)^{-1} X^\mathsf{T} X \beta = \beta.$$

The covariance of $\hat{\beta}$ is

$$\mathsf{Cov}_\theta \hat{\beta} = (X^\mathsf{T} X)^{-1} X^\mathsf{T} (\mathsf{Cov} \mathbf{Y}) X (X^\mathsf{T} X)^{-1} = \sigma^2 (X^\mathsf{T} X)^{-1} X^\mathsf{T} I_n X (X^\mathsf{T} X)^{-1}$$
$$= \sigma^2 (X^\mathsf{T} X)^{-1}.$$

$\square$

A remark about the covariance: Set $V = (X^\mathsf{T} X)^{-1}$. Then the third statement of Theorem 6.1 means:

$$\mathsf{Var}_\theta \hat{\beta}_j = \sigma^2 V_{jj} \quad \text{and} \quad \mathsf{Cov}_\theta(\hat{\beta}_j, \hat{\beta}_k) = \sigma^2 V_{jk}.$$

Replacing here the unknown σ^2 by an estimator (see Section 6.3.4) and taking the square root one obtains the standard error of the estimators for the components β_j.

Moreover, we see that in general the estimators $\hat{\beta}_j$ are correlated.

Consider the linear model (6.11). Here the least squares estimate of μ is defined by the minimizer of the squared distance

$$(\mathbf{y} - \mu)^\mathsf{T} (\mathbf{y} - \mu) = \|\mathbf{y} - \mu\|^2$$

over $\mu \in \mathcal{R}[X]$ ($\|z\|$ denotes the Euclidean distance of the vector z).

Theorem 6.2 *The solution*

$$\hat{\mu}(\mathbf{y}) = \arg \min_{\mu \in \mathcal{R}[X]} ||\mathbf{y} - \mu||^2$$

is the orthogonal projection of $\mathbf{y}$ *on* $\mathcal{R}[X]$, *say* $\hat{\mu}(\mathbf{y}) = P\mathbf{y}$, *where* P *is the projection matrix.*

PROOF: Consider the squared distance

$$||\mathbf{y} - \mu||^2 = ||\mathbf{y} - P\mathbf{y} + P\mathbf{y} - \mu||^2$$
$$= ||\mathbf{y} - P\mathbf{y}||^2 + ||P\mathbf{y} - \mu||^2 + 2(\mathbf{y} - P\mathbf{y})^{\mathsf{T}}(P\mathbf{y} - \mu). \quad (6.15)$$

Since the projection matrix P is idempotent, i.e., $P^2 = P$, the third term in (6.15) is equal to

$$(\mathbf{y} - P\mathbf{y})^{\mathsf{T}}(P\mathbf{y} - \mu) = \mathbf{y}^{\mathsf{T}} P\mathbf{y} - \mathbf{y}^{\mathsf{T}}\mu - \mathbf{y}^{\mathsf{T}} P\mathbf{y} + \mathbf{y}^{\mathsf{T}} P\mu = -\mathbf{y}^{\mathsf{T}}\mu + \mathbf{y}^{\mathsf{T}}\mu = 0.$$

Therefore, the r.h.s. of (6.15) is minimized by $\hat{\mu}(\mathbf{y}) = P\mathbf{y}$ and

$$\min_{\mu \in \mathcal{R}[X]} ||\mathbf{y} - \mu||^2 = ||\mathbf{y} - P\mathbf{y}||^2.$$

□

The projection matrix has the form $P = X(X^{\mathsf{T}}X)^- X^{\mathsf{T}}$, where the matrix $(X^{\mathsf{T}}X)^-$ is the generalized inverse of the matrix $X^{\mathsf{T}}X$.

Remark 6.2 (g-inverse) Let A be an arbitrary $p \times q$ matrix. A $q \times p$ matrix A^- is called generalized inverse of A, if $AA^- A = A$. If A is nonsingular, then $A^- = A^{-1}$. If X is an arbitrary $n \times k$ matrix, then $X(X^{\mathsf{T}}X)^- X^{\mathsf{T}}$ is independent of the choice of $(X^{\mathsf{T}}X)^-$. Moreover, $X(X^{\mathsf{T}}X)^- X^{\mathsf{T}}X = X$ and $X^{\mathsf{T}}X(X^{\mathsf{T}}X)^- X^{\mathsf{T}} = X^{\mathsf{T}}$.

Remark 6.3 Since $P\mathbf{y} = X\hat{\beta} = \hat{\mathbf{y}}$ the matrix P is often called *hat matrix* and denoted by H. We prefer the symbol P, because it is a projection matrix.

The properties of the estimator $\hat{\mu}(\mathbf{Y})$ are given in the following theorem:

Theorem 6.3 *The solution* $\hat{\mu}$ *of* $\min_{\mu \in \mathcal{R}[X]} ||\mathbf{y} - \mu||^2$ *is unique and satisfies:*

1. $\hat{\mu}(\mathbf{y})$ *is linear in* $\mathbf{y}$.
2. *The estimator* $\hat{\mu}(\mathbf{Y})$ *is unbiased, i.e.,* $\mathsf{E}_\theta \hat{\mu}(\mathbf{Y}) = \mu$.
3. *The covariance is given by* $\mathsf{Cov}_\theta \hat{\mu}(\mathbf{Y}) = \sigma^2 P$.

PROOF: We have

$$\mathsf{E}_\theta\, \hat{\mu}(\mathbf{Y}) \;=\; \mathsf{E}_\theta\, P\mathbf{Y} \;=\; P\mathsf{E}_\theta\mathbf{Y} \;=\; P\mu \;=\; \mu$$

and the idempotence of P implies

$$\mathsf{Cov}_\theta\, \hat{\mu}(\mathbf{Y}) \;=\; \mathsf{Cov}_\theta\, P\mathbf{Y} \;=\; P(\mathsf{Cov}_\theta\mathbf{Y})P \;=\; \sigma^2 P.$$

$\square$

Figure 6.2 illustrates the geometry of the least squares estimation.

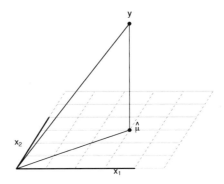

Figure 6.2: The horizontal plane represents the p-dimensional space in which $\mu = X\beta$ lies. The observation $\mathbf{y}$ lies in the space spanned by all three axes. The value $\hat{\mu}(\mathbf{y})$, that gives the minimum of the squared distance, lies vertically below the observation $\mathbf{y}$. It corresponds to the orthogonal projection of $\mathbf{y}$ into the p-dimensional subspace spanned by the columns of X, i.e., $\hat{\mu} = P\mathbf{y}$; the fitted value $\hat{\mu}$ is the point closest to y in that subspace.

Now let us come back to the problem of estimating β in the model (6.3). The estimate $\hat{\beta} = \hat{\beta}(\mathbf{y})$ is a solution of the normal equation if and only if

$$\hat{\mu} \;=\; X\hat{\beta} \;=\; P\mathbf{y} \;=\; X(X^\mathsf{T}X)^- X^\mathsf{T}\mathbf{y}. \qquad (6.16)$$

If the matrix X is of full rank $q = p$, the equation (6.16) can be solved with respect to β:

$$\hat{\mu} \;=\; X\hat{\beta} = X(X^\mathsf{T}X)^{-1}X^\mathsf{T}\mathbf{y} \quad \Leftrightarrow \quad \hat{\beta} = (X^\mathsf{T}X)^{-1}X^\mathsf{T}\mathbf{y}.$$

If $q < p$, the system of equations (6.16) has no unique solution for β. Equivalently, the normal equations have no unique solution. For each generalized inverse matrix $(X^\mathsf{T}X)^-$ the vector $\hat{\beta} = (X^\mathsf{T}X)^-X^\mathsf{T}y$ is a solution of (6.13). Furthermore, note that an estimator $\hat{\beta} = (X^\mathsf{T}X)^-X^\mathsf{T}\mathbf{Y}$ is not unbiased. But the projection is still unique! Compare the following example for illustration, and see also Figure 6.3.

Example 6.5 For simplicity of illustration we suppose that $n = 4$ and $p = 3$. The matrix X is given by

$$X = \begin{pmatrix} 1 & 1 & 0 \\ 1 & 1 & 0 \\ 1 & 0 & 1 \\ 1 & 0 & 1 \end{pmatrix}.$$

One can see that the column vectors are linearly dependent. Generalized inverse matrices of the matrix

$$X^\mathsf{T}X = \begin{pmatrix} 4 & 2 & 2 \\ 2 & 2 & 0 \\ 2 & 0 & 2 \end{pmatrix}$$

are, for example, given by (values are rounded)

$$G_1 = (X^\mathsf{T}X)^- = \begin{pmatrix} 0.11111 & 0.05556 & 0.05556 \\ 0.05556 & 0.27778 & -0.22222 \\ 0.05556 & -0.22222 & 0.27778 \end{pmatrix}$$

and

$$G_2 = (X^\mathsf{T}X)^- = \begin{pmatrix} 0.25000 & 0 & 0 \\ 0 & 0.25000 & -0.25000 \\ 0 & -0.25000 & 0.25000 \end{pmatrix}.$$

For observations $\mathbf{y} = (2, 3, 6, 8)^\mathsf{T}$ the resulting estimators are

$$\hat{\beta}^\mathsf{T}_{(1)} = (3.1667, -0.6667, 3.8333) \quad \text{and} \quad \hat{\beta}^\mathsf{T}_{(2)} = (4.75, -2.25, 2.25).$$

These are only two elements of the set of the solutions of the normal equations. But note, even if the generalized inverses are different, the projection matrix satisfies

$$XG_{(1)}X^\mathsf{T} = XG_{(2)}X^\mathsf{T} = \begin{pmatrix} 0.500 & 0.500 & 0 & 0 \\ 0.500 & 0.500 & 0 & 0 \\ 0 & 0 & 0.500 & 0.500 \\ 0 & 0 & 0.500 & 0.500 \end{pmatrix}.$$

Compare also Figure 6.3. □

Exercise 6.1 Show that G_1 and G_2 are generalized inverses of $X^\mathsf{T}X$.

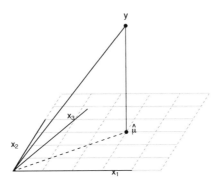

Figure 6.3: The vectors x_1, x_2 and x_3 are linearly dependent. Two vectors are enough to span the subspace in which the observation $\mathbf{y}$ is orthogonally projected. One sees that $\hat{\mu}$ can be represented as $\hat{\mu} = \hat{\beta}_1 x_1 + \hat{\beta}_2 x_2$ or $\hat{\mu} = \tilde{\beta}_1 x_1 + \tilde{\beta}_3 x_3$. Thus we have at least two different betas to get the unique projection $\hat{\mu}$.

Let us now apply the least squares method to the special cases:

Special case 6.3 (Simple linear regression) Let us consider the regression straight line defined by $m(x) = \beta_1 + \beta_2 x$, that is, we have the model (6.1). For simplicity we write $x_{i1} = 1$ and $x_{i2} = x_i$. The solution of the normal equations is given by

$$\hat{\beta}_1 = \bar{y} - \hat{\beta}_2 \bar{x} \quad \text{and} \quad \hat{\beta}_2 = \frac{\sum_{i=1}^{n}(x_i - \bar{x})(y_i - \bar{y})}{\sum_{i=1}^{n}(x_i - \bar{x})^2} \tag{6.17}$$

with $\bar{y} = \frac{1}{n}\sum_{i=1}^{n} y_i$. The covariance matrix is given by

$$\sigma^2 (X^\mathsf{T} X)^{-1} = \frac{\sigma^2}{\sum_{i=1}^{n}(x_i - \bar{x})^2} \begin{pmatrix} \frac{1}{n}\sum_{i=1}^{n} x_i^2 & -\bar{x} \\ -\bar{x} & 1 \end{pmatrix}. \tag{6.18}$$

Note that the estimated regression line $\widehat{m}(x) = \hat{\beta}_1 + \hat{\beta}_2 x$ can be written in the form

$$\frac{\widehat{m}(x) - \bar{y}}{s_Y} = r \frac{x - \bar{x}}{s_x}, \tag{6.19}$$

where s_Y^2 is the sample variance of the y_i's, s_x^2 is the sample variance of the x_i's and r is the empirical correlation:

$$s_Y^2 = (n-1)^{-1} \sum_{i=1}^{n}(y_i - \bar{y})^2, \qquad s_x^2 = (n-1)^{-1} \sum_{i=1}^{n}(x_i - \bar{x})^2$$

and

$$r = \frac{\sum_{i=1}^{n}(x_i - \bar{x})(y_i - \bar{y})}{\sqrt{\sum_{i=1}^{n}(y_i - \bar{y})^2}\,\sqrt{\sum_{i=1}^{n}(x_i - \bar{x})^2}}.$$

Equation (6.19) gives the name "**regression**." Since $|r| < 1$ (except the cases where all pairs (y_i, x_i) lie on a straight line) the normalized deviation of the "estimated response $\hat{m}(x)$" from the average $\bar{y}$ is smaller than the normalized covariate x from the average $\bar{x}$; in that sense we have a regression.

Remark 6.4 Note that in our approach the x_i's are considered as fixed quantities, and the y_i's are values of random variables. Thus, in a rigorous mathematical sense r is not an estimate of the correlation between two random variables!

Sometimes a linear relationship is formulated in the form $y_i = a + b(x_i - \bar{x}) + \varepsilon_i$. Then it is obvious from (6.17) that the estimates for a and b are given by

$$\hat{a} = \bar{y} \qquad \text{and} \qquad \hat{b} = \hat{\beta}_2.$$

Exercise 6.2 Compute the covariance between the estimators $\hat{a}$ and $\hat{b}$.

$\square$

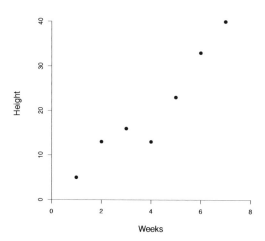

Figure 6.4: Height of soybeans versus weeks with fitted line.

Example 6.6 (Soybeans) Consider the data of Example 2.5 with the model given in Example 2.13 on page 11. The R-code `lm(height~weeks)` gives the following printout:

```
Coefficients:
              Estimate    Std. Error  t value    Pr(>|t|)
(Intercept)   -1.29             3.60    -0.36      0.7354
x              5.43             0.80     6.75      0.0011
```

Thus, the estimated regression line is given by

$$\widehat{m}(x) = -1.29 + 5.43x.$$

The column "Std. Error" contains estimates for the standard deviations of the parameter estimators according to formula (6.18). The data points and the fitted line are shown in Figure 6.4.   □

Let us continue with some examples with more than one covariate:

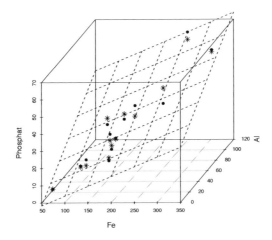

Figure 6.5: The fitted plane for Example 6.7. The points are the observations; the stars are the fitted points $\hat{y}_i = \widehat{m}(x_{i1}, x_{i2})$.

Example 6.7 (Phosphate adsorption) The following example is taken from Devore and Peck (1993). An article reported on a regression analysis

No.	x_1	x_2	y	No.	x_1	x_2	y
1	61	13	4	8	169	61	30
2	175	21	18	9	160	39	28
3	111	24	14	10	244	71	36
4	124	23	18	11	257	112	65
5	130	64	26	12	333	88	62
6	173	38	26	13	199	54	40
7	169	33	21				

Table 6.3: Phosphate adsorption index

with dependent variable y = phosphate adsorption index of a sediment and two predictor variables, x_1 = amount of extractable iron, and x_2 = amount of extractable aluminum. The analysis was based on the $n = 13$ data given in Table 6.3. Each observation consists of a triple (x_1, x_2, y).

In this example we have two predictor variables. The "fitted line of simple linear regression" becomes a plane, i.e., the points $(\widehat{m}(x_1, x_2), x_1, x_2)$ lie in a plane in $\mathbb{R}^3$. Compare Figure 6.5.   □

Example 6.8 The following example is taken from Davison (2003). "Norman Miller of the University of Wisconsin wanted to see how seat height, tire pressure and the use of a dynamo affected the time taken to ride his bicycle up a hill. He decided to collect data at each combination of two seat heights, 26 and 30 inches from the center of crank, two tire pressures, 40 and 55 pounds per square inch and with dynamo on and off, giving eight combinations in all. The times were expected to be quite variable, and in order to get more accurate results he decided to make two timings for each combination. He wrote each of the eight combinations on two pieces of card, and then drew the sixteen from a box in a random order...." The following table gives timings (in seconds) obtained with his wristwatch. Let us model the influence of the quantities "seat height," "tire pressure" and "dynamo" by a linear model and estimate the parameters. With

	Seat height	Dynamo	Tire pressure
−	26	Off	40
+	30	On	55

the matrix X is defined accordingly to the data table:

Figure 6.6: Norman Miller.

Setup	Height	Dynamo	Tire	Time
1	-	-	-	51
2	-	-	-	54
3	+	-	-	41
4	+	-	-	43
5	-	+	-	54
6	-	+	-	60
7	+	+	-	44
8	+	+	-	43
9	-	-	+	50
10	-	-	+	48
11	+	-	+	39
12	+	-	+	39
13	-	+	+	53
14	-	+	+	51
15	+	+	+	41
16	+	+	+	44

$$X = \begin{pmatrix} 1 & -1 & -1 & -1 \\ 1 & -1 & -1 & -1 \\ 1 & 1 & -1 & -1 \\ 1 & 1 & -1 & -1 \\ 1 & -1 & 1 & -1 \\ 1 & -1 & 1 & -1 \\ 1 & 1 & 1 & -1 \\ 1 & 1 & 1 & -1 \\ 1 & -1 & -1 & 1 \\ 1 & -1 & -1 & 1 \\ 1 & 1 & -1 & 1 \\ 1 & 1 & -1 & 1 \\ 1 & -1 & 1 & 1 \\ 1 & -1 & 1 & 1 \\ 1 & 1 & 1 & 1 \\ 1 & 1 & 1 & 1 \end{pmatrix}$$

The least squares estimates in the model $\mathbf{Y} = X\beta + \varepsilon$ are:

Coefficients:	(Intercept)	Seat Height	Dynamo	Tire Pressure
	47.188	−5.438	1.562	−1.563

We can interpret these results as follows: The overall mean time is 47.188 (intercept); the effect of increasing seat height from 26 to 30 inches is $2 * \hat{\beta}_1 \approx -10.88$, the effect of switching the dynamo on is 3.12, and the effect of increasing the tire pressure is −3.13.   □

6.3.1.1 Estimable Parameter

Let us consider the problem of estimating a linear parameter $\gamma = F\beta$, where F is a known $m \times p$ matrix.
There are cases where an unbiased estimator for γ exists, although $\hat{\beta}$ is not unique. To describe such situations we introduce the notation of an estimable parameter.

Definition 6.3 (Estimability) The parameter $\gamma = F\beta$ is **estimable**, if there exists a linear unbiased estimator for γ. That is, if there exists a $n \times m$ matrix L such that

$$E_\theta L^T Y = F\beta \qquad \text{for all } \theta = (\beta, \sigma^2, \kappa).$$

The condition of unbiasedness implies

$$E_\theta L^T Y = L^T X \beta = F\beta \qquad \text{for all } \beta.$$

Hence the parameter $F\beta$ is estimable if and only if there exists a matrix L such that

$$L^T X = F. \tag{6.20}$$

Corollary 6.1 If the rank of the matrix X is equal to p, then all linear parameters $\gamma = F\beta$ are estimable.

PROOF: If X has full rank, then $L = X(X^T X)^{-1}F^T$ satisfies equation (6.20). □

Example 6.9 Let us continue Example 6.5. The parameter β is not estimable since X is not of full rank. However $\gamma = 2\beta_1 + \beta_2 + \beta_3$ is estimable. With $L^T = (1, 0, 0, 1)$ we obtain $F = L^T X = (2, 1, 1)$. Note that also the component β_1 is not estimable. □

The least squares estimator for an estimable parameter $\gamma = F\beta$ is derived as follows: Equation (6.20) implies for estimable parameters:

$$\gamma = F\beta = L^T X \beta = L^T \mu.$$

Hence estimable parameters are linear combinations of the components of μ. Since $\hat{\mu}(Y)$ is a linear unbiased estimator, μ is estimable. Thus, for an estimable parameter $\gamma = F\beta$ we choose $\hat{\gamma} = F\hat{\beta}$, where $\hat{\beta}$ is a solution of the normal equation, i.e., $\hat{\beta} = (X^{\mathsf{T}}X)^{-}X^{\mathsf{T}}\mathbf{Y}$ for $q < p$, and $\hat{\beta} = (X^{\mathsf{T}}X)^{-1}X^{\mathsf{T}}\mathbf{Y}$ if X is of full rank. Although the generalized inverse is not unique, the matrix $F(X^{\mathsf{T}}X)^{-}X^{\mathsf{T}}$ with F satisfying (6.20) is unique, and we have (compare Remark 6.2 on page 193):

$$\mathsf{E}_{\theta}\hat{\gamma}(\mathbf{Y}) = F(X^{\mathsf{T}}X)^{-}X^{\mathsf{T}}\mathsf{E}_{\theta}\mathbf{Y} = F(X^{\mathsf{T}}X)^{-}X^{\mathsf{T}}X\beta. \qquad (6.21)$$

Condition (6.20) implies $L^{\mathsf{T}}X = F$ for some L. Furthermore, since $X(X^{\mathsf{T}}X)^{-}X^{\mathsf{T}}X = X$ we get for the r.h.s. of (6.21)

$$L^{\mathsf{T}}X(X^{\mathsf{T}}X)^{-}X^{\mathsf{T}}X\beta = L^{\mathsf{T}}X\beta = F\beta.$$

Thus, the estimator $\hat{\gamma}(\mathbf{Y}) = F\hat{\beta}(\mathbf{Y})$ is an unbiased estimator. Further we obtain
$$\mathsf{Cov}_{\theta}\hat{\gamma}(\mathbf{Y}) = \sigma^{2}F(X^{\mathsf{T}}X)^{-}F^{\mathsf{T}}.$$

6.3.2   Generalized Least Squares Estimator

Assume now a model where dependent errors ε_i are allowed. However, we suppose that the variances and covariances of the observations are known up to a common factor σ^2, i.e., $\mathsf{Cov}_{\theta}\mathbf{Y} = \sigma^2\Sigma$, where Σ is a known positive definite matrix.

As described above on page 190, we can transform our model into the model (6.10). The LSE for $\tilde{\mu} = \tilde{X}\beta$ is defined by the orthogonal projection on the space $\mathcal{R}[\tilde{X}]$, which is given by

$$\tilde{P} = \tilde{X}(\tilde{X}^{\mathsf{T}}\tilde{X})^{-}\tilde{X}^{\mathsf{T}},$$

and the corresponding LSE for β in the model (6.10) is

$$\hat{\beta} = (\tilde{X}^{\mathsf{T}}\tilde{X})^{-}\tilde{X}^{\mathsf{T}}\tilde{\mathbf{y}}.$$

The re-transformation leads to the following estimates for $\mu = X\beta$ and β:

$$\hat{\mu}_{\mathrm{GLSE}} = X(X^{\mathsf{T}}\Sigma^{-1}X)^{-}X^{\mathsf{T}}\Sigma^{-1}\mathbf{y}, \qquad (6.22)$$
$$\hat{\beta}_{\mathrm{GLSE}} = (X^{\mathsf{T}}\Sigma^{-1}X)^{-}X^{\mathsf{T}}\Sigma^{-1}\mathbf{y}. \qquad (6.23)$$

Note that $\tilde{P}\tilde{\mathbf{y}}$ is the minimizer of a weighted squared distance:

$$\min_{\mu\in\mathcal{R}[\tilde{X}]} (\tilde{\mathbf{y}} - \mu)^{\mathsf{T}}(\tilde{\mathbf{y}} - \mu).$$

Moreover

$$\min_{\mu\in\mathcal{R}[\tilde{X}]} (\tilde{\mathbf{y}} - \mu)^{\mathsf{T}}(\tilde{\mathbf{y}} - \mu) = \min_{\mu\in\mathcal{R}[X]} (\mathbf{y} - \mu)^{\mathsf{T}}\Sigma^{-1}(\mathbf{y} - \mu).$$

The weights are given by the inverse of the matrix Σ. This leads to the following definition.

Definition 6.4 (Generalized least squares estimator (GLSE))
Consider the linear model $\mathsf{E}_\theta \mathbf{Y} = X\beta$ and $\mathsf{Cov}_\theta \mathbf{Y} = \sigma^2 \Sigma$, where Σ is a known positive definite matrix. Any solution $\hat\beta$ of the minimization problem

$$\min_\beta (\mathbf{y} - X\beta)^\mathsf{T} \Sigma^{-1} (\mathbf{y} - X\beta)$$

is called **generalized least squares estimate (GLSE)** for β.

For an estimable parameter $\gamma = F\beta$ we define

$$\hat\gamma_{\mathrm{GLSE}} = F\hat\beta = F(X^\mathsf{T}\Sigma^{-1}X)^- X^\mathsf{T}\Sigma^{-1}\mathbf{y}.$$

Exercise 6.3 a) Show that $\hat\gamma_{\mathrm{GLSE}}$ defines an unbiased estimator for an estimable parameter $\gamma = F\beta$. Further, derive its covariance.
b) Suppose that X is of full rank. Derive the GLSE for β. Write down its covariance.

Remark 6.5 Sometimes it also makes sense to take arbitrary weights for the least squares approach, for instance if the covariance matrix is unknown and there is some prior knowledge about the variation at different design points. Then a weighted least squares estimator can be used. It is defined as the solution of

$$\min_\beta (\mathbf{y} - X\beta)^\mathsf{T} W (\mathbf{y} - X\beta),$$

where W is a suitable nonnegative definite matrix.

6.3.3   Gauss–Markov Theorem

The following theorem says that the GLSE are not only unbiased, but best estimators in the class of all linear unbiased estimators.

Theorem 6.4 (Gauss–Markov) *Consider the model* $\mathsf{E}_\theta \mathbf{Y} = X\beta$ *with* $\mathsf{Cov}_\theta \mathbf{Y} = \sigma^2 \Sigma$, *where* Σ *is a known positive definite* $n \times n$ *matrix. Then any linear unbiased estimator* $\tilde\gamma = \tilde{L}^\mathsf{T}\mathbf{Y}$ *for the estimable parameter* $\gamma = F\beta$ *satisfies*

$$\mathsf{Cov}_\theta \tilde\gamma(\mathbf{Y}) \succeq \mathsf{Cov}_\theta \hat\gamma(\mathbf{Y}) \qquad \textit{for all} \quad \theta,$$

where $\hat\gamma(\mathbf{Y}) = F(X^\mathsf{T}\Sigma^{-1}X)^- X^\mathsf{T}\Sigma^{-1}\mathbf{Y}$. *That is, the GLSE for an estimable parameter is BLUE (Best Linear Unbiased Estimator).*

PROOF: We have already shown that the estimator $\hat\gamma$ is linear in $\mathbf{Y}$ and unbiased. Let us now compare the covariance of $\hat\gamma$ with that of an arbitrary linear unbiased estimator $\tilde{\gamma} = \tilde{L}^{\mathsf{T}}\mathbf{Y}$. For simplicity of notation set $\hat{\gamma} = L_0^{\mathsf{T}}\mathbf{Y}$ with $L_0^{\mathsf{T}} = F(X^{\mathsf{T}}\Sigma^{-1}X)^- X^{\mathsf{T}}\Sigma^{-1}$. We have

$$\mathsf{Cov}_\theta\hat{\gamma} = \sigma^2 L_0^{\mathsf{T}}\Sigma L_0 \quad \text{and} \quad \mathsf{Cov}_\theta\tilde{\gamma} = \sigma^2 \tilde{L}^{\mathsf{T}}\Sigma\tilde{L}.$$

With $\tilde{L} = L_0 + (\tilde{L} - L_0)$ we obtain

$$\tilde{L}^{\mathsf{T}}\Sigma\tilde{L} = L_0^{\mathsf{T}}\Sigma L_0 + (\tilde{L}-L_0)^{\mathsf{T}}\Sigma L_0 + L_0^{\mathsf{T}}\Sigma(\tilde{L}-L_0) + (\tilde{L}-L_0)^{\mathsf{T}}\Sigma(\tilde{L}-L_0). \quad (6.24)$$

Since both estimators are unbiased, we have $\tilde{L}^{\mathsf{T}}X = F$ and $L_0^{\mathsf{T}}X = F$. This implies

$$\begin{aligned} L_0^{\mathsf{T}}\Sigma(\tilde{L} - L_0) &= F(X^{\mathsf{T}}\Sigma^{-1}X)^- X^{\mathsf{T}}\Sigma^{-1}\Sigma(\tilde{L} - L_0) \\ &= F(X^{\mathsf{T}}\Sigma^{-1}X)^- X^{\mathsf{T}}(\tilde{L} - L_0) = 0. \end{aligned}$$

Thus, the second and the third term on the r.h.s. of (6.24) is equal to zero. We obtain

$$\mathsf{Cov}_\theta\tilde{\gamma} = \sigma^2 \tilde{L}^{\mathsf{T}}\Sigma\tilde{L} = \mathsf{Cov}_\theta\hat{\gamma} + \sigma^2(\tilde{L} - L_0)^{\mathsf{T}}\Sigma(\tilde{L} - L_0) \succeq \mathsf{Cov}_\theta\hat{\gamma}.$$

$\square$

Remark 6.6 An unweighted least squares estimator is called an **ordinary least squares estimator (OLSE)**:

$$\hat{\beta}_{\text{OLSE}} = (X^{\mathsf{T}}X)^- X^{\mathsf{T}}\mathbf{Y} \quad \text{and} \quad \hat{\mu}_{\text{OLSE}} = X(X^{\mathsf{T}}X)^- X^{\mathsf{T}}\mathbf{Y}.$$

Applying the OLSE to the model with covariance matrix $\sigma^2\Sigma$, where Σ is known, we get a worse result. That is, suppose $\gamma = F\beta$ is estimable and set

$$\hat{\gamma}_{\text{OLSE}}(\mathbf{Y}) = F(X^{\mathsf{T}}X)^- X^{\mathsf{T}}\mathbf{Y} \quad \text{and} \quad \hat{\gamma}_{\text{GLSE}}(\mathbf{Y}) = F(X^{\mathsf{T}}\Sigma^{-1}X)^- X^{\mathsf{T}}\Sigma^{-1}\mathbf{Y}.$$

Then

$$\mathsf{E}_\theta\hat{\gamma}_{\text{OLSE}}(\mathbf{Y}) = \mathsf{E}_\theta\hat{\gamma}_{\text{GLSE}}(\mathbf{Y}) = F\beta,$$

but

$$\mathsf{Cov}_\theta\hat{\gamma}_{\text{OLSE}}(\mathbf{Y}) = \sigma^2 F(X^{\mathsf{T}}X)^- X^{\mathsf{T}}\Sigma X(X^{\mathsf{T}}X)^- F^{\mathsf{T}} \succeq \sigma^2 F(X^{\mathsf{T}}\Sigma^{-1}X)^- F^{\mathsf{T}}.$$

Exercise 6.4 Suppose the model $\mathsf{E}_\theta\mathbf{Y} = X\beta$ with $\mathsf{Cov}_\theta\mathbf{Y} = \sigma^2\Sigma$, where X is of full rank and the matrix Σ is known. Compare the covariance of the OLSE and the covariance of the GLSE for β.

6.3.4 Estimation of the Variance

Consider model (6.3) with $\mathrm{Cov}_\theta \varepsilon = \sigma^2 I_n$. The aim of this section is to derive an estimator for the parameter σ^2, i.e., for the variance of the errors. Since the errors are unobservable, we use estimates of these errors for the construction of the estimator of σ^2. The error $\varepsilon_i = Y_i - x_i^{\mathsf{T}}\beta$ is estimated by the i-th residual $\hat{\varepsilon}_i = y_i - x_i^{\mathsf{T}}\hat{\beta}$. In vector terms,

$$\hat{\varepsilon} = \mathbf{y} - X\hat{\beta} = (I_n - P)\mathbf{y}.$$

Note that $\hat{\varepsilon}$ is well defined, independent of the uniqueness of $\hat{\beta}$. Since the matrix $(I_n - P)$ is a projection matrix, too, and therefore idempotent, we have

$$\hat{\varepsilon}^{\mathsf{T}}\hat{\varepsilon} = \mathbf{y}^{\mathsf{T}}(I_n - P)\mathbf{y}.$$

From here it follows that

$$
\begin{aligned}
\mathsf{E}_\theta \hat{\varepsilon}^{\mathsf{T}}\hat{\varepsilon} &= \mathsf{E}_\theta (X\beta + \varepsilon)^{\mathsf{T}}(I_n - P)(X\beta + \varepsilon) \\
&= \mathsf{E}_\theta \varepsilon^{\mathsf{T}}(I_n - P)\varepsilon = \mathsf{E}_\theta \mathrm{trace}(I_n - P)\varepsilon\varepsilon^{\mathsf{T}} \\
&= \sigma^2 \mathrm{trace}(I_n - P).
\end{aligned}
$$

Moreover, we have $\mathrm{trace}(I_n - P) = n - q$, where q is the rank of X. Thus, an unbiased estimator for σ^2 is given by

$$
\begin{aligned}
\hat{\sigma}^2 &= \frac{1}{n-q}\mathbf{Y}^{\mathsf{T}}(I_n - P)\mathbf{Y} = \frac{1}{n-q}(\mathbf{Y} - X\hat{\beta})^{\mathsf{T}}(\mathbf{Y} - X\hat{\beta}) \\
&= \frac{1}{n-q}\sum_{i=1}^{n}(Y_i - x_i^{\mathsf{T}}\hat{\beta})^2.
\end{aligned}
\tag{6.25}
$$

Let us consider two special cases:

Special case 6.4 (One-sample problem) Let $p = 1$ and $x_{i1} = 1$, then the problem is not really a regression problem, but simply the problem of estimating the mean $\beta \in \mathbb{R}$ by a sample of i.i.d. r.v.'s. With $X = (1, \ldots, 1)^{\mathsf{T}}$ we get $X^{\mathsf{T}}X = n$ and $X^{\mathsf{T}}\mathbf{y} = \sum_{i=1}^{n} y_i$, thus the LSE for β is simply the sample mean $\bar{y}$. Moreover, $q = 1$ and the estimate for the variance is the sample variance

$$\hat{\sigma}(\mathbf{y}) = s^2 = \frac{1}{n-1}\sum_{i=1}^{n}(y_i - \bar{y})^2.$$

□

Special case 6.5 (Two-sample problem) Let $p = 2$ and set $x_{i1} = 1$ for $i = 1, \ldots n_1$, $x_{i1} = 0$ for $i = n_1 + 1, \ldots, n_1 + n_2$ and $x_{i2} = 0$ for $i = 1, \ldots n_1$,

$x_{i2} = 1$ for $i = n_1 + 1, \ldots, n_1 + n_2$. Then the model is simply the two-sample problem. The first sample is a sample of n_1 i.i.d. r.v.'s with mean β_1. The second sample consists of n_2 i.i.d. r.v.'s with mean β_2. The variance is supposed to be the same for both samples. The design matrix is

$$X^\mathsf{T} = \begin{pmatrix} 1 & 1 & \cdots & 1 & 0 & 0 & \cdots & 0 \\ 0 & 0 & \cdots & 0 & 1 & 1 & \cdots & 1 \end{pmatrix}$$

and the resulting estimates are

$$\hat{\beta}_1 = \frac{1}{n_1} \sum_{i=1}^{n_1} y_i \quad \text{and} \quad \hat{\beta}_2 = \frac{1}{n_2} \sum_{i=n_1+1}^{n_1+n_2} y_i$$

and

$$\hat{\sigma}^2(\mathbf{y}) = \frac{\sum_{i=1}^{n_1}(y_i - \hat{\beta}_1)^2 + \sum_{i=n_1+1}^{n_1+n_2}(y_i - \hat{\beta}_2)^2}{n_1 + n_2 - 2}.$$

$\square$

We conclude this section with the definition of the variance estimator in the model with $\mathsf{Cov}_\theta \varepsilon = \sigma^2 \Sigma$, where Σ is known. An unbiased estimator for the variance σ^2 is given by

$$\begin{aligned} \hat{\sigma}^2 &= \frac{1}{n-q} \tilde{\mathbf{Y}}^\mathsf{T} (I_n - \tilde{P}) \tilde{\mathbf{Y}} \\ &= \frac{1}{n-q} (\mathbf{Y} - X\hat{\beta}_{\mathrm{GLSE}})^\mathsf{T} \Sigma^{-1} (\mathbf{Y} - X\hat{\beta}_{\mathrm{GLSE}}). \end{aligned}$$

6.4   The Normal Linear Model

6.4.1   Estimation in the Normal Linear Model

Now let the errors be normally distributed. We consider only the case where the covariance matrix is $\sigma^2 I_n$, i.e., the observations are independent with common variance σ^2. Corresponding results with covariance matrix $\sigma^2 \Sigma$, where Σ is known, follow by transformation.

Since we have specified the distribution, we can construct the maximum likelihood estimates. For simplicity we assume that X is of full rank p. The log-likelihood function is given by

$$l(\beta, \sigma^2; \mathbf{y}) = -\frac{1}{2} \left(n \log \sigma^2 + \frac{1}{\sigma^2} (\mathbf{y} - X\beta)^\mathsf{T} (\mathbf{y} - X\beta) \right).$$

Maximizing the log-likelihood function with respect to β is equivalent to minimizing the quadratic form $(\mathbf{y} - X\beta)^\mathsf{T}(\mathbf{y} - X\beta)$. But this is just the problem of determining the least squares estimate considered in (6.12). Hence the MLE for β is the LSE

$$\hat{\beta} = (X^\mathsf{T} X)^{-1} X^\mathsf{T} \mathbf{y}.$$

The MLE for σ^2 we obtain from

$$\frac{\partial}{\partial \sigma^2} l(\beta, \sigma^2; \mathbf{y}) = -\frac{1}{2}\left(\frac{n}{\sigma^2} - \frac{(\mathbf{y} - X\beta)^\mathsf{T}(\mathbf{y} - X\beta)}{\sigma^4} \right) = 0.$$

Thus

$$\hat{\sigma}^2_{\mathrm{MLE}} = \frac{1}{n}(\mathbf{y} - X\hat{\beta})^\mathsf{T}(\mathbf{y} - X\hat{\beta}).$$

As in the case of a sample of independent and $\mathsf{N}(\mu, \sigma^2)$-distributed r.v.'s this estimator is not unbiased. The expectation is

$$\mathsf{E}_{\beta, \sigma^2} \hat{\sigma}^2_{\mathrm{MLE}} = \frac{(n-p)}{n} \sigma^2.$$

Hence, multiplying this estimator by $n/(n-p)$ we get the unbiased estimator

$$\hat{\sigma}^2 = \frac{1}{n-p}(\mathbf{Y} - X\hat{\beta})^\mathsf{T}(\mathbf{Y} - X\hat{\beta}). \tag{6.26}$$

Again, this estimator is already known. It is the estimator (6.25), which we constructed without using the assumption of normality, but on the basis of the residuals as estimators for the unobservable errors.

The estimators in the semiparametric model and in the normal linear regression model coincide. The difference is the following:

In the normal model we can derive the distribution of our estimators, which is important for the further statistical analysis—namely for the construction of confidence regions and for the derivation of test procedures. Furthermore, optimality properties of the estimators and tests can be verified.

Since $\mathbf{Y} \sim \mathsf{N}(X\beta, \sigma^2 I_n)$, it follows immediately that $\hat{\beta}$, which is linear in $\mathbf{Y}$, is normally distributed with expectation β and covariance matrix $\sigma^2(X^\mathsf{T} X)^{-1}$. Moreover, the MLE for a linear parameter $\gamma = F\beta$ is given by $\hat{\gamma} = F\hat{\beta}$, and we have

$$F\hat{\beta} \sim \mathsf{N}(F\beta, \sigma^2 F(X^\mathsf{T} X)^{-1} F^\mathsf{T}).$$

To derive optimality properties of the proposed estimators we apply the theory described in Chapter 4: The density of the vector $\mathbf{Y}$ is given by

$$\frac{1}{(2\pi\sigma^2)^{\frac{n}{2}}} \exp\left(-\frac{1}{2\sigma^2}(\mathbf{y} - X\beta)^\mathsf{T}(\mathbf{y} - X\beta) \right).$$

Thus, the factorization criterion formulated in Theorem 3.7 on 54 implies that the statistic defined by $T(\mathbf{y}) = (\mathbf{y}^\mathsf{T}\mathbf{y}, X^\mathsf{T}\mathbf{y})$ is sufficient. Moreover, since the distribution of $\mathbf{Y}$ belongs to an exponential family, by Theorem 4.6 the statistic T is complete. The estimators $\hat{\beta}$ and $\hat{\sigma}^2$ are unbiased and depend only on this sufficient and complete statistic. Therefore, the Lehmann–Scheffé theorem implies the following statement:

Theorem 6.5 (Best unbiased estimator) *The estimators $\hat{\beta}$ and $\hat{\sigma}^2$ are best unbiased estimators (BUE), that is, they have the smallest covariance matrix in the class of all unbiased estimators.*

OBS! Note that without assuming normality, the Gauss–Markov theorem states only that $\hat{\beta}$ is the best estimator in the class of all *linear* unbiased estimators.

6.4.2   Testing Hypotheses in the Normal Linear Model

We consider the normal linear model

$$\mathbf{Y} = X\beta + \varepsilon \qquad \text{with } \varepsilon \sim \mathsf{N}(0, \sigma^2 I_n).$$

In many statistical applications one has to check hypotheses about the coefficients β_j. For example:

- In a simple linear regression it can be interesting to find out whether the slope of the regression line is equal to a certain value.

- Comparing two regression lines we would like to know whether they can be assumed to be parallel.

- If we fit a polynomial we have to choose the degree of this polynomial. Thus, the question can be: Should we take a polynomial of degree three or four?

- In a regression set–up with three different covariates one has to check whether it is necessary to include interaction terms between these covariates.

- In the variance analysis models one wishes to check whether the factors have an effect on the response.

All these test problems can be written in the form

$$H_0\colon \ G\beta = g \qquad \text{versus} \qquad H_1\colon \ G\beta \neq g, \tag{6.27}$$

where G is a suitable $m \times p$ matrix and g is a $m \times 1$ vector. The null hypothesis says that the vector β lies in a subspace of $\mathbb{R}^p$, and this subspace is defined by the equality $G\beta = g$. Let us specify G and g for the examples stated above:

Special case 6.6 (Simple linear regression) For the simple linear model

$$Y_i = \beta_1 + \beta_2 x_i + \varepsilon_i, \qquad i = 1, \ldots, n$$

we have $\beta = (\beta_1, \beta_2)^{\mathsf{T}}$, $p = 2$. The hypothesis concerning the slope β_2 is

$$H_0\colon \beta_2 = \beta_{20} \qquad \text{versus} \qquad H_1\colon \beta_2 \neq \beta_{20},$$

where β_{20} is the hypothetical slope, and with $G = (0, 1)$ and $g = \beta_{20}$ $(m = 1)$, we get the form (6.27).

Special case 6.7 (Comparing two regression lines) Let (y_{1l}, z_{1l}), $l = 1, \ldots, n_1$ be the observations for the first regression line and (y_{2r}, z_{2r}), $r = 1, \ldots, n_2$ those for the second, i.e., the observations satisfy

$$y_{1l} = \beta_{11} + \beta_{12}z_{1l} + \varepsilon_{1l}$$

$$y_{2r} = \beta_{21} + \beta_{22}z_{2r} + \varepsilon_{2r}.$$

We set

$$y_i = \begin{cases} y_{1i} & i = 1, \ldots, n_1 \\ y_{2\,i-n_1} & i = n_1 + 1, \ldots, n \end{cases} , \quad \varepsilon_i = \begin{cases} \varepsilon_{1i} & i = 1, \ldots, n_1 \\ \varepsilon_{2\,i-n_1} & i = n_1 + 1, \ldots, n \end{cases} ,$$

$$x_{i1} = \begin{cases} 1 & i = 1, \ldots, n_1 \\ 0 & i = n_1 + 1, \ldots, n \end{cases} , \quad x_{i2} = \begin{cases} z_{1i} & i = 1, \ldots, n_1 \\ 0 & i = n_1 + 1, \ldots, n \end{cases} ,$$

$$x_{i3} = \begin{cases} 0 & i = 1, \ldots, n_1 \\ 1 & i = n_1 + 1, \ldots, n \end{cases} , \quad x_{i4} = \begin{cases} 0 & i = 1, \ldots, n_1 \\ z_{2\,i-n_1} & i = n_1 + 1, \ldots, n \end{cases}$$

where $n = n_1 + n_2$.
These "new" responses satisfy the model $\mathbf{Y} = X\beta + \varepsilon$ with

$$\beta = (\beta_{11}, \beta_{12}, \beta_{21}, \beta_{22})^{\mathsf{T}} \quad \text{and} \quad \varepsilon = (\varepsilon_{11}, \ldots, \varepsilon_{1n_1}, \varepsilon_{21}, \ldots, \varepsilon_{2n_2})^{\mathsf{T}}.$$

Thus we have $p = 4$ and the hypothesis is

$$H_0 \colon \beta_{12} = \beta_{22} \quad \text{versus} \quad H_1 \colon \beta_{12} \neq \beta_{22},$$

and with the 1×4 vector $G = (0, 1, 0, -1)$ and $g = 0$ we get form (6.27).

Special case 6.8 (Polynomial regression) The design matrix X for the polynomial regression of degree 4 (i.e., the regression function is $m(z) = \beta_1 + \sum_{j=1}^{4} \beta_{j+1}z^j$) has the elements

$$x_{i1} = 1, \quad x_{ij} = z_i^{j-1}, \quad j = 2, \ldots, 5, \quad i = 1, \ldots, n,$$

and the parameter is $\beta = (\beta_1, \ldots, \beta_5)^{\mathsf{T}}$. We wish to check whether a polynomial of degree three is sufficient for fitting the data. Thus we have to test

$$H_0 \colon \beta_5 = 0 \quad \text{versus} \quad H_1 \colon \beta_5 \neq 0,$$

which can be written in the matrix form (6.27) with $G = (0, 0, 0, 0, 1)$ and $g = 0$.

Special case 6.9 (Testing interactions) Consider a model with three different covariates, say z_1, z_2 and z_3. In the starting model we assume pairwise interactions between these covariates; thus the regression function is

$$m(z_1, z_2, z_3) = \beta_1 + \beta_2 z_1 + \beta_3 z_2 + \beta_4 z_3 + \beta_5 z_1 z_2 + \beta_6 z_1 z_3 + \beta_7 z_2 z_3.$$

The matrix X for this model is a $n \times 7$ matrix with elements $x_{i1} = 1$

$$x_{i2} = z_{i1}, \quad x_{i3} = z_{i2}, \quad x_{i4} = z_{i3}, \quad x_{i5} = z_{i1} z_{i2}, \quad x_{i6} = z_{i1} z_{i3}, \quad x_{i7} = z_{i2} z_{i3}.$$

The parameter β is 7-dimensional. With the 3×7 matrix

$$G = \begin{pmatrix} 0 & 0 & 0 & 0 & 1 & 0 & 0 \\ 0 & 0 & 0 & 0 & 0 & 1 & 0 \\ 0 & 0 & 0 & 0 & 0 & 0 & 1 \end{pmatrix} \quad \text{and} \quad g = (0,0,0)^{\mathsf{T}}$$

we can formulate the hypothesis that the interaction terms can be deleted from the model.

Example 6.10 Consider a model with two covariates z_1 and z_2 and regression function $m(z_1, z_2) = \beta_1 + \beta_2 z_1 + \beta_3 z_2 + \beta_4 z_1 z_2$. The parameter β is 4-dimensional, and the corresponding linear model has the design matrix X with

$$x_{i1} = 1, \quad x_{i2} = z_{i1}, \quad x_{i3} = z_{i2}, \quad x_{i4} = z_{i1} z_{i2}.$$

Consider the hypothesis H_0 with

$$G = \begin{pmatrix} 0 & 0 & 0 & 1 \\ 0 & 1 & -1 & 0 \end{pmatrix} \quad \text{and} \quad g = \begin{pmatrix} 0 \\ 0 \end{pmatrix}.$$

Since

$$G\beta = \begin{pmatrix} \beta_4 \\ \beta_2 - \beta_3 \end{pmatrix} = \begin{pmatrix} 0 \\ 0 \end{pmatrix} = g$$

the null hypothesis is $H_0 : \beta_2 = \beta_3$ and $\beta_4 = 0$, and the hypothetical model is

$$Y_i = \beta_1 + \beta_2 (z_{i1} + z_{i2}) + \varepsilon_i.$$

Special case 6.10 (Test of effects) In Special case 6.2 the one-way ANOVA model was introduced. Testing that there is no effect leads to the hypothesis

$$H_0 : \alpha_1 = \cdots = \alpha_{I-1} = 0 \quad \text{versus} \quad H_1 : \alpha_i \neq 0 \text{ for at least one } i.$$

Setting $\beta = (\mu, \alpha_1, \ldots, \alpha_{I-1})^{\mathsf{T}}, p = I$, this test problem is equivalent to testing

$$H_0 : G\beta = g \quad \text{versus} \quad H_1 : G\beta \neq g$$

with

$$G = \begin{pmatrix} 0 & 1 & 0 & \cdots & 0 \\ \vdots & 0 & 1 & \cdots & 0 \\ \vdots & 0 & 0 & \ddots & 0 \\ 0 & 0 & 0 & \cdots & 1 \end{pmatrix}, \qquad g = \begin{pmatrix} 0 \\ \vdots \\ \vdots \\ 0 \end{pmatrix}, \qquad m = p - 1. \qquad (6.28)$$

□

Special case 6.11 (Automatic tests in software packages) Note that software packages for regression analysis carry out automatically the following tests:

1. The so-called F-test testing the null hypothesis

$$H_0: \ \beta_2 = \beta_3 = \cdots = \beta_p = 0.$$

The alternative is that there exists at least one $\beta_j \neq 0$ for $j = 2, \ldots, p$. Written in matrix form this test problem is described by G and g defined in (6.28).

2. The t-tests testing the p single hypotheses—these problems are defined by:

$$H_{j0}: \ \beta_j = 0 \qquad \text{versus} \qquad H_{j1}: \ \beta_j \neq 0, \qquad j = 1, \ldots, p.$$

Here the matrices G_j are the vectors

$$G_j = (0, \ldots, \underbrace{1}_{j}, \ldots, 0), \quad \text{and} \quad g = 0 \qquad (m = 1),$$

and the hypothetical model is $\ Y_i = \sum_{\substack{r=1 \\ r \neq j}}^{p} \beta_r \, x_{ir} + \varepsilon_i.$

Special case 6.12 Finally let us remark that we can write the hypothesis that $\beta \in \mathbb{R}^p$ is equal to some $\beta_0 \in \mathbb{R}^p$ by choosing $G = I_p$ and $g = \beta_0$, that is $m = p$.

We start our considerations with the case $g = 0$. We allow that the rank q of the matrix X is smaller than p. Similar to the estimation problem it will turn out that we can derive tests only for those parameters $G\beta$ which are estimable. Having in mind the condition for estimability (6.20) we consider only such hypotheses $G\beta$ for which there exists a $n \times m$ matrix L such that $L^{\mathsf{T}}X = G$. (This is always satisfied if X is of full rank.) This equation leads to the following approach to our test problem: We have

$$G\beta = L^{\mathsf{T}}X\beta = L^{\mathsf{T}}\mu,$$

and instead of the test problem (6.27) with $g = 0$ we consider

$$H_0: \ L^{\mathsf{T}} \mu = 0 \qquad \text{versus} \qquad H_1: \ L^{\mathsf{T}} \mu \neq 0.$$

Geometrically speaking, the hypothesis H_0 claims that the $n \times 1$ vector μ lies in a subspace of $\mathcal{R}\,[X]$, say

$$\mathcal{M}_0 = \{ \mu : L^{\mathsf{T}} \mu = 0, \ \mu \in \mathcal{M} \} \qquad \text{with} \quad \dim \mathcal{M}_0 = r,$$

where $\mathcal{M} = \mathcal{R}\,[X]$, that is

$$H_0: \ \mu \in \mathcal{M}_0 \qquad \text{versus} \qquad H_1: \ \mu \in \mathcal{M} \setminus \mathcal{M}_0.$$

The idea for testing this hypothesis is to compare a good estimator for μ under the condition that μ lies in $\mathcal{M}_0$ with a good estimator without assuming this restriction. In the previous section it was shown that the projection method is useful for estimating. So we also apply this method here. As an estimate for $\mu \in \mathcal{M}$ we choose the projection $\hat{\mu}(\mathbf{y}) = P\mathbf{y}$ already introduced, and for estimating μ under H_0 we take the projection of $\mathbf{y}$ into $\mathcal{M}_0$ denoted by $\hat{\mu}_0(\mathbf{y}) = P_0\mathbf{y}$. The squared distance between both is

$$\varrho^2(\mathbf{y}) = ||\hat{\mu}(\mathbf{y}) - \hat{\mu}_0(\mathbf{y})||^2 = (P\mathbf{y} - P_0\mathbf{y})^{\mathsf{T}}(P\mathbf{y} - P_0\mathbf{y}) = \mathbf{y}^{\mathsf{T}}(P - P_0)\mathbf{y}.$$

Heuristically it is clear: If $\varrho^2(\mathbf{y})$ is large, we will reject H_0. If it is small, the data $\mathbf{y}$ do not contradict H_0. To quantify what is "large" and what is "small," we have to determine the distribution of $\varrho^2(\mathbf{Y})$. To do this we need results about the distribution of quadratic forms of normally distributed random vectors. The following lemma is proved in Rao (1973):

Lemma 6.1 (Quadratic forms) *Suppose that* $\mathbf{Y} \sim \mathrm{N}_n(\eta, \sigma^2 I_n)$, *then* $\mathbf{Y}^{\mathsf{T}} A \mathbf{Y}/\sigma^2$ *is distributed according to the noncentral* χ^2-*distribution with* m *degrees of freedom (df) and noncentrality parameter* $\lambda = \eta^{\mathsf{T}} A \eta$, *iff* A *is a symmetric idempotent matrix of rank* m.
Two quadratic forms $\mathbf{Y}^{\mathsf{T}} A \mathbf{Y}$ *and* $\mathbf{Y}^{\mathsf{T}} B \mathbf{Y}$ *with symmetric matrices* A *and* B *are stochastically independent, iff* $AB = 0$.
Let Z *be distributed according to a* χ^2-*distribution with* m *df and noncentrality parameter* λ *and* D *be central* χ^2-*distributed with* k *df. If* Z *and* D *are independent, then the distribution of the ratio*

$$\frac{k\,Z}{m\,D}$$

is the F-*distribution with* m *and* k *df and noncentrality parameter* λ.

Under the assumption of the error ε being normally distributed with mean zero and covariance $\sigma^2 I_n$ the quantity $\mathbf{Y}^{\mathsf{T}}(P - P_0)\mathbf{Y}/\sigma^2$ is a quadratic form of independent normally distributed r.v.'s with variance one and mean μ_i.

The rank of the idempotent matrix $P - P_0$ is $q - r$, where $q = \dim \mathcal{M}$ and $r = \dim \mathcal{M}_0$. Applying Lemma 6.1 we obtain that

$$Z := \frac{\mathbf{Y}^\top (P - P_0) \mathbf{Y}}{\sigma^2} \sim \chi^2_{q-r}(\lambda),$$

where the parameter of noncentrality λ is equal to $\mu^\top (P - P_0)\mu$. If H_0 holds, then $\lambda = 0$. That is, we have a central χ^2- distribution with $q - r$ df.

But, of course, the variance σ^2 cannot be assumed to be known, thus we have to plug in an estimator. From the previous section we know that $\hat{\sigma}^2$ is the best unbiased estimator. From (6.25) we obtain

$$D := \frac{(n-q)\hat{\sigma}^2}{\sigma^2} = \frac{\mathbf{Y}^\top (I_n - P)\mathbf{Y}}{\sigma^2} \sim \chi^2_{n-q}.$$

Furthermore, since

$$(P - P_0)(I_n - P) = PI_n - P^2 - P_0 I_n + P_0 P = P - P - P_0 + P_0 = 0,$$

Z and D are stochastically independent. Hence, by Lemma 6.1, under H_0 the statistic

$$F := F(\mathbf{Y}) = \frac{\frac{1}{q-r} Z}{\frac{1}{n-q} D} = \frac{\frac{1}{q-r} \varrho^2(\mathbf{Y})}{\hat{\sigma}^2(\mathbf{Y})} = \frac{n-q}{q-r} \frac{\mathbf{Y}^\top (P - P_0)\mathbf{Y}}{\mathbf{Y}^\top (I_n - P)\mathbf{Y}} \qquad (6.29)$$

is a ratio of independent χ^2-distributed r.v.'s. Therefore it is F-distributed with $q - r$ and $n - q$ df.

We add $F(\mathbf{Y}) = \infty$ for $\mathbf{Y}^\top (I_n - P)\mathbf{Y} = 0$ and $\mathbf{Y}^\top (P - P_0)\mathbf{Y} \neq 0$, and $F(\mathbf{Y}) = -1$ if $\mathbf{Y}^\top (I_n - P)\mathbf{Y} = 0$ and $\mathbf{Y}^\top (P - P_0)\mathbf{Y} = 0$.

Summarizing, we formulate the following theorem:

Theorem 6.6 (F-Test) *Assume the model* $\mathbf{Y} \sim \mathsf{N}(X\beta, \sigma I_n)$. *An α-test for testing the hypothesis* $H : G\beta = 0$ *versus* $K : G\beta \neq 0$ *for a $m \times p$ matrix G (satisfying $L^\top X = G$ for some $n \times m$ matrix L) is given by*

$$\varphi_F(\mathbf{y}) = \begin{cases} 1 & \text{for } F(\mathbf{y}) \geq F_{q-r,n-q;1-\alpha} \\ 0 & \text{for } F(\mathbf{y}) < F_{q-r,n-q;1-\alpha} \end{cases},$$

*where $F_{q-r,n-q;1-\alpha}$ is the $(1-\alpha)$-quantile of the $\mathsf{F}_{q-r,n-q}$-distribution. The test φ_F is called an **F-test**.*

Let us give an equivalent expression for (6.29): With

$$\|\mathbf{Y} - \hat{\mu}_0(\mathbf{Y})\|^2 = \mathbf{Y}^\top (I_n - P_0)\mathbf{Y} \qquad \text{and} \qquad \|\mathbf{Y} - \hat{\mu}(\mathbf{Y})\|^2 = \mathbf{Y}^\top (I_n - P)\mathbf{Y}$$

we get

$$\|\mathbf{Y} - \hat{\mu}_0(\mathbf{Y})\|^2 - \|\mathbf{Y} - \hat{\mu}(\mathbf{Y})\|^2 = \mathbf{Y}^\top (P - P_0)\mathbf{Y}$$

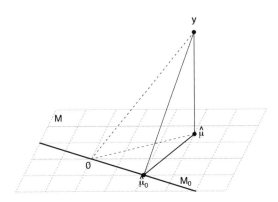

Figure 6.7: Here the hypothetical set of μ's is denoted by M_0. The estimators $\hat{\mu}$ and $\hat{\mu}_0$ are the projections of $\mathbf{y}$ on M and M_0. The denominator and numerator of the F-statistic, except of the factors, are the squares of the distances between $\mathbf{y}$ and $\hat{\mu}$ and between $\hat{\mu}$ and $\hat{\mu}_0$, respectively.

and therefore we can write

$$F = \frac{n-q}{q-r} \frac{||\mathbf{Y} - \hat{\mu}_0(\mathbf{Y})||^2 - ||\mathbf{Y} - \hat{\mu}(\mathbf{Y})||^2}{||\mathbf{Y} - \hat{\mu}(\mathbf{Y})||^2}. \tag{6.30}$$

The estimator $\hat{\mu}(\mathbf{Y})$ is the least squares estimator for $X\beta$ in the underlying model, that is $\hat{\mu}(\mathbf{Y}) = X\hat{\beta}(\mathbf{Y})$ with

$$\hat{\beta} = \arg\min_{\beta \in \mathbb{R}^p} ||\mathbf{Y} - X\beta||^2$$

and, as shown in the previous section, $\hat{\beta} = (X^\mathsf{T} X)^- X^\mathsf{T} \mathbf{Y}$.
The estimator $\hat{\mu}_0(\mathbf{Y})$ is the least squares estimator for $X\beta$ in the hypothetical model, i.e., $\hat{\mu}_0(\mathbf{Y}) = X\hat{\beta}_0$ with

$$\hat{\beta}_0 = \arg\min_{\substack{\beta:G\beta=0 \\ \beta \in \mathbb{R}^p}} ||\mathbf{Y} - X\beta||^2. \tag{6.31}$$

Hence an equivalent form of (6.30) is

$$F = \frac{n-q}{m} \frac{||\mathbf{Y} - X\hat{\beta}_0||^2 - ||\mathbf{Y} - X\hat{\beta}||^2}{||\mathbf{Y} - X\hat{\beta}||^2}. \tag{6.32}$$

This form of the F-statistic can be interpreted as follows: Suppose that the hypothetical model is true, i.e., we have $E_\theta Y = X\beta_0 \in \mathcal{M}_0$. Then $||Y - X\hat{\beta}_0||^2$ is a measure of data variability, which will not be very different from the variability obtained by fitting the "larger model" $\mathcal{M}$. But if the hypothetical model is not true, then $||Y - X\hat{\beta}_0||^2$ contains also a systematic error part, and this is larger than $||Y - X\hat{\beta}||^2$, so that F becomes large and this leads to a rejection of H_0.

Let us derive a formula for $\hat{\beta}_0$. This minimization problem (6.31) can be solved by the method of Lagrange multipliers: We set

$$\eta(\beta, \lambda) := ||Y - X\beta||^2 + \lambda^\mathsf{T} G\beta,$$

where λ is a $m \times 1$ vector. Differentiation of η with respect to β and to λ leads to the following system of equations

$$2X^\mathsf{T} X\beta - 2X^\mathsf{T} Y + G^\mathsf{T}\lambda = 0 \qquad (6.33)$$

$$G\beta = 0. \qquad (6.34)$$

Multiplying (6.33) by $G(X^\mathsf{T} X)^-$ finally leads to the following solution

$$\begin{aligned}
\hat{\beta}_0 &= \left(I_n - (X^\mathsf{T} X)^- G^\mathsf{T} \left[G(X^\mathsf{T} X)^- G^\mathsf{T}\right]^- G\right)(X^\mathsf{T} X)^- X^\mathsf{T} Y \\
&= \hat{\beta} - (X^\mathsf{T} X)^- G^\mathsf{T} \left[G(X^\mathsf{T} X)^- G^\mathsf{T}\right]^- G\hat{\beta}. \qquad (6.35)
\end{aligned}$$

The estimability of $G\beta$ and the assumption that G is of full rank m imply that the dimension of $\mathcal{M}_0$ is equal to $r = q - m$. Now, substituting (6.35) in (6.29) we get

$$\begin{aligned}
F &= \frac{n - q}{m} \frac{\hat{\beta}^\mathsf{T} G^\mathsf{T} [G(X^\mathsf{T} X)^- G^\mathsf{T}]^- G\hat{\beta}}{Y^\mathsf{T}(I_n - P)Y} \\
&= \frac{\hat{\beta}^\mathsf{T} G^\mathsf{T} [G(X^\mathsf{T} X)^- G^\mathsf{T}]^- G\hat{\beta}}{m\hat{\sigma}^2}. \qquad (6.36)
\end{aligned}$$

This form yields a further interpretation of the F-statistic. The difference between the parameter $G\beta$ and its hypothetical value 0 is estimated (in the underlying model $E_\theta Y = X\beta$) by $G\hat{\beta} - 0$. The covariance of this estimator is

$$\mathsf{Cov}_\theta G\hat{\beta} = \sigma^2 G(X^\mathsf{T} X)^- G^\mathsf{T} =: \sigma^2 GVG^\mathsf{T}.$$

An estimator of this covariance is $\hat{\sigma}^2 GVG^\mathsf{T}$. Roughly speaking (assuming that GVG^T is of full rank) we can consider F given in (6.36) as "squared difference between estimated parameter and hypothetical parameter, divided by m times estimated variance," i.e.

$$(G\hat{\beta} - 0)^\mathsf{T}(\hat{\sigma}^2 GVG^\mathsf{T})^{-1}(G\hat{\beta} - 0)/m.$$

And we reject the null hypothesis if this normalized squared difference between hypothetical value and its estimator is large. This form of the test statistic leads us, without going into further details, to the test statistic for testing the hypothesis $H_0 : G\beta = g$ for general g. It is given by

$$F = \frac{(G\hat{\beta} - g)^{\mathsf{T}}[G(X^{\mathsf{T}}X)^{-}G^{\mathsf{T}}]^{-}(G\hat{\beta} - g)}{m\,\hat{\sigma}^2}. \tag{6.37}$$

Now let us return to our examples:

Special case 6.13 (Simple linear regression) In Special case 6.6 on page 208 we formulated $H : \beta_2 = \beta_{20}$. Using the formulas given in Special case 6.3 the numerator of the F-statistic is

$$G(\hat{\beta} - g)^{\mathsf{T}}[G(X^{\mathsf{T}}X)^{-1}G^{\mathsf{T}}]^{-}(G\hat{\beta} - g) = (\hat{\beta}_2 - \beta_{20})^2 \sum_{i=1}^{n}(x_i - \bar{x})^2$$

and the estimate for σ^2 is given by

$$\frac{1}{n-2}\sum_{i=1}^{n}(y_i - \hat{\beta}_1 - \hat{\beta}_2 x_i)^2,$$

where $\hat{\beta}_1$ and $\hat{\beta}_2$ are given in (6.17). The resulting test statistic is $F_{1,(n-2)}$-distributed. The square root of this test statistic has a t-distribution with $n-2$ degrees of freedom. □

Example 6.11 (Soybeans) (Continuation of Example 6.6 on page 198.) The F-statistic for testing the hypothesis $\beta_1 = 0$ is F-distributed with 1 and 5 degrees of freedom. Its value is 45.55. Note that $\sqrt{45.55} = 6.749$. This value is given in the fourth column of the table in Example 6.6. The p-value is 0.001; thus the hypothesis is rejected. The relationship between weeks and height of the plants is highly significant. Compare Figure 6.4. □

Special case 6.14 (Polynomial regression) Consider Special case 6.8 on page 209. The test statistic for testing whether a polynomial of degree three is appropriate has the form

$$\frac{\hat{\beta}_5^2}{\hat{\sigma}^2 V_{55}},$$

where $\hat{\beta}_5$ is the 5-th component of the LSE $\hat{\beta}$, V_{55} is the 5-th diagonal component of the 5×5 matrix $V = (X^{\mathsf{T}}X)^{-1}$ and

$$\hat{\sigma}^2 = \frac{1}{n-5}\sum_{i=1}^{n}(Y_i - \hat{\beta}_1 - \hat{\beta}_2 z_i - \hat{\beta}_3 z_i^2 - \hat{\beta}_4 z_i^3 - \hat{\beta}_5 z_i^4)^2.$$

The test statistic is $F_{1,(n-5)}$-distributed. It is the square of a t-distributed r.v. with $n - 5$ degrees of freedom. $\qquad\square$

Special case 6.15 (Testing interactions) In Special case 6.9 on page 210 the resulting test statistic has the F-distribution with $n - 7$ and 3 degrees of freedom. $\qquad\square$

Example 6.12 (Effects of brands of wheat) Let us continue Example 6.7. We suppose that the crop yields of wheat are values of normally distributed random variables, i.e., the errors in model 6.6 satisfy $\varepsilon_{ij} \sim N(0, \sigma^2)$. Here it is not useful to apply the general formula for computing the LSE to derive estimates for μ and the α_i's. It is better to decompose the observations into orthogonal parts: Using the condition $\sum_{i=1}^{I}\alpha_i = 0$ we get with

$$\overline{y}_{..} = \frac{1}{n}\sum_{i=1}^{I}\sum_{j=1}^{N}y_{ij} \quad \text{and} \quad \overline{y}_{i.} = \frac{1}{N}\sum_{j=1}^{N}y_{ij}:$$

$$\sum_{i=1}^{I}\sum_{j=1}^{N}(y_{ij} - \mu - \alpha_i)^2$$

$$= \sum_{i=1}^{I}\sum_{j=1}^{N}((y_{ij} - \overline{y}_{i.}) - (\mu - \overline{y}_{..}) - (\alpha_i - \overline{y}_{i.} + \overline{y}_{..}))^2$$

$$= \sum_{i=1}^{I}\sum_{j=1}^{N}(y_{ij} - \overline{y}_{i.})^2 + IN(\mu - \overline{y}_{..})^2 + N\sum_{i=1}^{I}(\alpha_i - \overline{y}_{i.} + \overline{y}_{..})^2.$$

Minimizing this sum w.r.t. α_i and μ gives

$$\hat{\alpha}_i = \overline{y}_{i.} - \overline{y}_{..} \quad \text{and} \quad \hat{\mu} = \overline{y}_{..}.$$

The variance σ^2 is estimated by formula (6.26) (with $I = 4, N = 5, n = 20$)

$$\hat{\sigma}^2 = \frac{1}{n-I}(\mathbf{y} - X\hat{\beta})^{\mathsf{T}}(\mathbf{y} - X\hat{\beta}) = \frac{1}{n-I}\sum_{i=1}^{I}\sum_{j=1}^{N}(y_{ij} - \overline{y}_{i.})^2 =: \frac{\text{SSRes}}{n-I}.$$

The difference $||\mathbf{y} - X\hat{\beta}_0||^2 - ||\mathbf{y} - X\hat{\beta}||^2$ has the form

$$\text{SSA} := N \sum_{i=1}^{I} (\bar{y}_{i.} - \bar{y}_{..})^2.$$

From (6.28) we know that $m = I - 1$. Thus, the test statistic is

$$F = \frac{n-I}{I-1} \frac{N \sum_{i=1}^{I}(\bar{y}_{i.} - \bar{y}_{..})^2}{\sum_{i=1}^{I} \sum_{j=1}^{N}(y_{ij} - \bar{y}_{i.})^2} = \frac{n-I}{I-1} \frac{\text{SSA}}{\text{SSRes}}.$$

The results for the data considered in the example are:

	Df	SumSq	MeanSq	F value	Pr(>F)
Factor	3	1646.00	548.67	12.849	0.0001573
Residuals	16	683.20	42.70		

Thus the value of the test statistic is

$$F(\mathbf{y}) = \frac{16}{3} \frac{1646.0}{683.2} = \frac{548.67}{42.70} = 12.849.$$

The p-value is 0.00016; thus H_0 is rejected. The factor brand has an effect on the yield of wheat. □

Let us apply F-tests in a more complicated model of the variance analysis:

Example 6.13 (Two-way model with interaction) Consider the nice example Tooth growth from the R-package *Using R* written by Verzani (2005) about the effect of vitamin C on tooth growth in guinea pigs. The response is the length of odontoblasts (teeth) in each of 10 guinea pigs at each of three dose levels of vitamin C, namely 0.5, 1 and 2 mg and with each of two delivery methods—orange juice or ascorbic acid. The data are given in Table 6.4. Using here the standard notation for ANOVA-models we can write the following model:

$$Y_{ijk} = \mu + \alpha_i + \beta_j + \gamma_{ij} + \varepsilon_{ijk}, \qquad i = 1, 2, 3, \;\; j = 1, 2 \;\; \text{and} \;\; k = 1, \ldots, 10.$$

Here y_{ijk} is the measurement at pig k from the group which got dose level i with method j. Further, α_i denotes the effect of the level i of the factor "dose level," the β_j's are the effects of the factor "delivery method" and γ_{ij} stands for a possible interaction between both factors. Assuming

$$\sum_{i=1}^{3} \alpha_i = 0, \quad \sum_{j=1}^{2} \beta_j = 0, \quad \sum_{i=1}^{3} \gamma_{ij} = 0 \quad \text{and} \quad \sum_{j=1}^{2} \gamma_{ij} = 0,$$

we have a linear model with an unknown parameter of dimension $p = 6$:

Length of teeth

Method	Dose levels		
	0.5mg	1mg	2mg
Ascorbic acid	4.2	16.5	23.6
	11.5	16.5	18.5
	7.3	15.2	33.9
	5.8	17.3	25.5
	6.4	22.5	26.4
	10.0	17.3	32.5
	11.2	13.6	26.7
	11.2	14.5	21.5
	5.2	18.8	23.3
	7.0	15.5	29.5
Orange juice	15.2	19.7	25.5
	21.5	23.3	26.4
	17.6	23.6	22.4
	9.7	26.4	24.5
	14.5	20.0	24.8
	10.0	25.2	30.9
	8.2	25.8	26.4
	9.4	21.2	27.3
	16.5	14.5	29.4
	9.7	27.3	23.0

Table 6.4: Length of teeth of guinea pigs depending on vitamin C

$\boldsymbol{\beta} = (\mu, \alpha_1, \alpha_2, \beta_1, \gamma_{12}, \gamma_{21})$. (To avoid confusions with the betas we use the bold beta to denote the complete parameter vector.) The 60×6 matrix X has the form

$$
\begin{pmatrix}
\mathbb{1}_{10} & \mathbb{1}_{10} & \mathbb{0}_{10} & \mathbb{1}_{10} & \mathbb{1}_{10} & \mathbb{0}_{10} \\
\mathbb{1}_{10} & \mathbb{0}_{10} & \mathbb{1}_{10} & \mathbb{1}_{10} & \mathbb{0}_{10} & \mathbb{1}_{10} \\
\mathbb{1}_{10} & -\mathbb{1}_{10} & -\mathbb{1}_{10} & \mathbb{1}_{10} & -\mathbb{1}_{10} & -\mathbb{1}_{10} \\
\mathbb{1}_{10} & \mathbb{1}_{10} & \mathbb{0}_{10} & -\mathbb{1}_{10} & -\mathbb{1}_{10} & \mathbb{0}_{10} \\
\mathbb{1}_{10} & \mathbb{0}_{10} & \mathbb{1}_{10} & -\mathbb{1}_{10} & \mathbb{0}_{10} & -\mathbb{1}_{10} \\
\mathbb{1}_{10} & -\mathbb{1}_{10} & -\mathbb{1}_{10} & -\mathbb{1}_{10} & \mathbb{1}_{10} & \mathbb{1}_{10}
\end{pmatrix} .
$$

As estimates for the effects we obtain

$$\hat{\mu} = \frac{1}{n} \sum_{i=1}^{I} \sum_{j=1}^{J} \sum_{k=1}^{K} y_{ijk} = \bar{y}_{...},$$

$$\hat{\alpha}_i = \bar{y}_{i..} - \bar{y}_{...}, \quad \bar{y}_{i..} = \frac{1}{JK} \sum_{j=1}^{J} \sum_{k=1}^{K} y_{ijk}, \qquad i = 1, \ldots, I,$$

$$\hat{\beta}_j = \bar{y}_{.j.} - \bar{y}_{...}, \quad \bar{y}_{.j.} = \frac{1}{IK} \sum_{i=1}^{I} \sum_{k=1}^{K} y_{ijk}, \qquad j = 1, \ldots, J,$$

$$\hat{\gamma}_{ij} = \bar{y}_{ij.} - \bar{y}_{i..} - \bar{y}_{.j.} + \bar{y}_{...}, \quad \bar{y}_{ij.} = \frac{1}{K} \sum_{k=1}^{K} y_{ijk}.$$

We consider the following three hypotheses:

$H_{A0}: \alpha_1 = \alpha_2 = 0$ versus $H_{A1}: \alpha_1 \neq 0$ or $\alpha_2 \neq 0$.

$H_{B0}: \beta_1 = 0$ versus $H_{B1}: \beta_1 \neq 0$.

$H_{C0}: \gamma_{11} = \gamma_{12} = 0$ versus $H_{C1}: \gamma_{11} \neq 0$ or $\gamma_{12} \neq 0$.

As an estimate for the variance and therefore the denominator of the test statistics we get $\text{SSRes}/(60 - 6)$ with

$$\text{SSRes} = ||\mathbf{y} - X\hat{\boldsymbol{\beta}}||^2 = \sum_{i=1}^{I} \sum_{j=1}^{J} \sum_{k=1}^{K} (y_{ijk} - \bar{y}_{ij.})^2.$$

The difference $||\mathbf{y} - X\hat{\boldsymbol{\beta}}_A||^2 - ||\mathbf{y} - X\hat{\boldsymbol{\beta}}||^2$, where $\hat{\boldsymbol{\beta}}_A$ is the estimator of the parameter in the model under H_{A0}, is given by

$$\text{SSA} = KJ \sum_{i=1}^{I} (\bar{y}_{i..} - \bar{y}_{...})^2.$$

The corresponding m is equal to 2. The difference $||\mathbf{y} - X\hat{\boldsymbol{\beta}}_B||^2 - ||\mathbf{y} - X\hat{\boldsymbol{\beta}}||^2$, where $\hat{\boldsymbol{\beta}}_B$ is the estimator in the model under H_{B0}, is given by

$$\text{SSB} = KI \sum_{j=1}^{J} (\bar{y}_{.j.} - \bar{y}_{...})^2 \qquad \text{and} \quad m = 1.$$

We use the R-code to compute the corresponding test statistics: The output is given by

	Df	Sum Sq	Mean Sq	F value	Pr(>F)
dose	2	2426.43	1213.22	92.000	< 2.2e-16
method	1	205.35	205.35	15.572	0.0002312
dose:method	2	108.32	54.16	4.107	0.0218603
Residuals	54	712.11	13.19		

We see SSA $= 2426.43$ and SSRes $= 712.11$. The value of the F-statistic is

$$F = \frac{2426.43}{2} \frac{54}{712} = \frac{1213.22}{13.19} = 92.$$

This is a value of an F-statistic with 2 and 54 degrees of freedom. The p-value is very small, so H_{A0} is rejected. The effect of the dose is highly significant. For the test of H_{B0} we obtain

$$F = \frac{205.35}{1} \frac{54}{712} = \frac{205.35}{13.19} = 15.57.$$

Since the p-value is small, too, H_{B0} is rejected. Also the method of delivery has a significant effect. □

Exercise 6.5 Give the general form of the test statistic for testing the interaction term in Example 6.13. Carry out the corresponding test.

Remarks:

1. If $m = 1$ the resulting test statistic is $F_{1,n-q}$-distributed. In this case the test statistic is the square of a statistic which is distributed according to a t-distribution with $n - q$ degrees of freedom.

2. In many statistical software packages regression procedures carry out the "F-test" automatically. What they do is: to check the hypothesis

$$H_0: \beta_2 = \beta_3 = \cdots = \beta_p = 0 \qquad \text{versus} \qquad H_1: \beta_j \neq 0 \quad \text{for some} \ j > 1.$$

In this case (if X is of full rank) the F-statistic has a F-distribution with $p - 1$ and $n - p$ degrees of freedom.

3. The combination of F-tests in the so-called backward elimination and forward selection procedures is applied for an automatic model choice.

6.4.3 Confidence Regions

The estimators for the parameter β are point estimators. Sometimes one is not only interested in a plausible value for the unknown parameter, but in a *set* of plausible values. Thus, based on the data we construct a set $\mathcal{C}$ of values such that with a high probability the unknown parameter is covered by $\mathcal{C}$. Since the problem of deriving such a confidence region is strongly connected with the construction of a significance test, the coverage probability will be denoted by $1 - \alpha$, where α is a small value.

There are different possibilities for confidence regions in the context of linear regression: We can construct

- confidence intervals for the single (real) parameter β_j, $j = 1, \ldots, p$,
- confidence intervals for the value of the regression function at a fixed point $\tilde{x}$, that is for the (real) value

$$\tilde{x}_1 \beta_1 + \cdots + \tilde{x}_p \beta_p,$$

- prediction intervals for a single (new) observation $Y_i = \tilde{x}_i^{\mathsf{T}} \beta + \varepsilon_i$ at $\tilde{x}$,
- confidence regions for a m-dimensional parameter $\gamma = G\beta$.

As an example for such confidence sets we will give here a confidence region for an estimable parameter $\gamma = G\beta$, where G is a $m \times p$ matrix of full rank. For simplicity we assume that X is of full rank p. From the previous section we know (see equation (6.37)) that for $\gamma = G\beta$ the statistic

$$\frac{(G\hat{\beta} - \gamma)^{\mathsf{T}}[G(X^{\mathsf{T}}X)^{-1}G^{\mathsf{T}}]^{-}(G\hat{\beta} - \gamma)}{m\hat{\sigma}^2}$$

is $\mathsf{F}_{m,n-p}$-distributed. Thus, we obtain for the set

$$\mathcal{C}(\mathbf{y}) = \{\gamma : (G\hat{\beta} - \gamma)^{\mathsf{T}}[G(X^{\mathsf{T}}X)^{-1}G^{\mathsf{T}}]^{-}(G\hat{\beta} - \gamma) \leq m\hat{\sigma}^2 F_{m,n-p;1-\alpha}\}$$

$$P_\theta(\mathcal{C}(\mathbf{Y})) = 1 - \alpha.$$

In other words, the set $\mathcal{C}(\mathbf{Y})$ forms a confidence region with coverage probability $1 - \alpha$. Note that this is an m-dimensional ellipsoid.
From the confidence ellipsoid we get a confidence interval for a single parameter β_j by choosing $G_{(j)} = (0, 0, \ldots, 0, 1, 0, \ldots, 0)$. Then we get that

$$\frac{(\hat{\beta}_j - \beta_j)^{\mathsf{T}}[G_{(j)}(X^{\mathsf{T}}X)^{-1}G_{(j)}^{\mathsf{T}}]^{-}(\hat{\beta}_j - \beta_j)}{\hat{\sigma}^2} = \frac{(\hat{\beta}_j - \beta_j)^2}{\hat{\sigma}^2 V_{jj}}$$

is $\mathsf{F}_{1,n-q}$-distributed, or equivalently the square root of this statistic

$$\frac{\hat{\beta}_j - \beta_j}{\hat{\sigma}\sqrt{V_{jj}}}$$

has a t-distribution with $n - p$ degrees of freedom, where V_{jj} is the j-th diagonal element of the matrix $(X^{\mathsf{T}}X)^{-1}$. Thus, a confidence interval for β_j is given by:

$$\mathcal{I}_j(\mathbf{y}) = \{\beta_j \mid \hat{\beta}_j - t_{n-p;1-\frac{\alpha}{2}}\hat{\sigma}\sqrt{V_{jj}} \leq \beta_j \leq \hat{\beta}_j + t_{n-p;1-\frac{\alpha}{2}}\hat{\sigma}\sqrt{V_{jj}}\}.$$

Note that:

1. Testing the single hypothesis

$$H_0 : \quad \beta_j = g \qquad \text{versus} \qquad H_1 : \quad \beta_j \neq g,$$

where g is a real number, is equivalent to checking whether $g \in \mathcal{I}_j(\mathbf{y})$.

2. Consider the hypothesis

$$H_0: \quad \beta_2 = \beta_3 = \cdots = \beta_p = 0 \qquad \text{versus} \qquad H_1: \quad \beta_j \neq 0 \text{ for some } j > 1.$$

In the regression model, where the first column of the X-matrix consists only of 1's, the test is equivalent to checking whether the $(p-1)$-dimensional null vector is covered by the ellipsoid

$$\mathcal{C}_0(\mathbf{y})$$
$$= \{ \gamma \mid (G\hat{\beta} - \gamma)^\mathsf{T} [G(X^\mathsf{T} X)^{-1} G^\mathsf{T}]^{-} (G\hat{\beta} - \gamma) \leq (p-1)\,\hat{\sigma}^2 F_{p-1,n-p;1-\alpha} \}$$

where G is the $((p-1) \times p)$-matrix

$$\begin{pmatrix} 0 & 1 & 0 & 0 & \cdots & 0 \\ 0 & 0 & 1 & 0 & \cdots & 0 \\ 0 & 0 & 0 & 1 & \cdots & 0 \\ \vdots & \vdots & \vdots & \vdots & \vdots & \vdots \\ 0 & 0 & 0 & 0 & \cdots & 1 \end{pmatrix}.$$

The following cases can occur:

- All intervals $\mathcal{I}_j(\mathbf{y})$, $j = 2, \ldots, p$ cover the number zero, but the $(p-1) \times 1$ vector of zeros is not covered by $\mathcal{C}_0(\mathbf{y})$.
- The $(p-1) \times 1$ vector of zeros lies in $\mathcal{C}_0(\mathbf{y})$, but there exists an interval $\mathcal{I}_j(\mathbf{y})$, which does not cover the number zero.
- Of course, it can happen that the vector of zeros is covered by the confidence ellipsoid and also all confidence intervals contain the number zero.

Example 6.14 For illustration consider the following data. The y_i's are assumed to be values of normally distributed r.v.'s with equal variance.

x_1	-1	1	-1	1	0	0	0	2	3	1
x_2	-1	-1	1	1	0	1	2	2	1	1
y	-0.93	4.30	2.09	6.95	1.00	5.36	6.74	13.85	12.33	8.31

We assume a model of the form

$$Y_i = \beta_1 + x_{i1}\beta_2 + x_{i2}\beta_3 + \varepsilon_i.$$

The point estimates for β are

$$\hat{\beta}_1 = 2.941, \qquad \hat{\beta}_2 = 2.736, \qquad \hat{\beta}_3 = 2.025.$$

The confidence ellipsoid for (β_2, β_3) is shown in Figure 6.8. $\square$

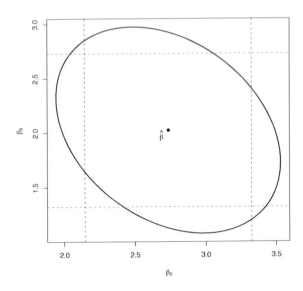

Figure 6.8: Confidence ellipsoid with coverage probability of 90%. The dashed lines show the corresponding single confidence intervals for β_2 and β_3.

6.4.4   Optimality of the F-Test*

To characterize the F-test as an optimal test we need the notation of **invariance**. Many statistical inferences are based on data which are measured in certain units. If, for instance, data **y** are recorded in Celsius degrees, one might obtain a conclusion $T(\mathbf{y})$ directly from the original data, or one might transform them to Fahrenheit degrees, giving $\tilde{y} = \pi(\mathbf{y})$ and conclusion $\tilde{T}(\tilde{\mathbf{y}})$. Of course, we expect from a good test procedure that the decision for one of the hypotheses does not depend on the choice of temperature unit—in other words, that our inference is invariant under the transformation from Celsius to Fahrenheit.

Let Q be a group of transformations π of the sample space $\mathcal{X}$. Then the transformed r.v. $\pi(\mathbf{Y})$ has the distribution P^π, and we denote the family of these distributions by $\mathcal{P}^\pi$.

Definition 6.5 (Invariance) A statistical model $\mathcal{P} = \{\mathsf{P}_\theta : \theta \in \Theta\}$ is **invariant** with respect to a group of transformations Q if $\mathcal{P} = \mathcal{P}^\pi$ for all $\pi \in Q$. A test problem $H_0 \colon \theta \in \Theta_0$ versus $H_1 \colon \theta \in \Theta_1$ is invariant w.r.t. Q if the families $\mathcal{P}_0 = \{\mathsf{P}_\theta : \theta \in \Theta_0\}$ and $\mathcal{P}_1 = \{\mathsf{P}_\theta : \theta \in \Theta_1\}$ are invariant.

Moreover, a statistic $T : \mathcal{X} \to \mathcal{T}$ is called invariant w.r.t. Q if for all $\pi \in Q$ and for all $\mathbf{y} \in \mathcal{X}$

$$T(\pi \mathbf{y}) = T(\mathbf{y}).$$

A **statistic** T is called **maximal invariant** w.r.t. Q if it is invariant and if for all $\mathbf{y}_1$ and $\mathbf{y}_2$ with $\mathbf{y}_1 \neq \mathbf{y}_2$ and $T(\mathbf{y}_1) = T(\mathbf{y}_2)$ there exists a transformation $\pi \in Q$, such that $\mathbf{y}_1 = \pi(\mathbf{y}_2)$.

Let us now consider the **invariance in the normal linear model**. We assume

$$\mathcal{P} = \{N(\mu, \sigma^2 I_n) \,|\, (\mu, \sigma^2) \in \mathcal{M} \times \mathbb{R}_+\},$$

where $\mathcal{M}$ is the q-dimensional linear subspace introduced above. We wish to test

$$H_0 : \mu \in \mathcal{M}_0 \quad \text{versus} \quad H_1 : \mu \in \mathcal{M} \backslash \mathcal{M}_0. \tag{6.38}$$

Let Q_0 be the group of the following affine transformations: $\pi(\mathbf{y}) = cA\mathbf{y} + d$, where $c \neq 0$ is a real number, the vector $d \in \mathcal{M}_0$ and the $n \times n$ matrix A is orthogonal with

$$A(\mathcal{M}_0) = \mathcal{M}_0 \quad \text{and} \quad A(\mathcal{M}) = \mathcal{M}. \tag{6.39}$$

We can prove the following statements:

Theorem 6.7 (Invariance of the F-test) *1. The test problem (6.38) is invariant with respect to Q_0.*

2. The test statistic F of the F-test defined in (6.29) is maximal invariant.

PROOF: If $\mathbf{Y} \sim N(\mu, \sigma^2 I_n)$, $\theta = (\mu, \sigma^2)$, then $\tilde{\mathbf{Y}} = \pi(\mathbf{Y}) = cA\mathbf{Y} + d$ is normally distributed with

$$E_\theta \tilde{\mathbf{Y}} = \tilde{\mu} = cA\mu + d \quad \text{and} \quad \text{Cov}_\theta \tilde{\mathbf{Y}} = c^2 \sigma^2 AA^\mathsf{T} = c^2 \sigma^2 I_n = \tilde{\sigma}^2 I_n,$$

where $\tilde{\mu} \in \mathcal{M}$ and $\tilde{\sigma}^2 \in \mathbb{R}_+$. That is, the distribution of the transformed vector belongs to the assumed family of distributions. Moreover, since the transformation π is invertible, we have also $\mathcal{P}^\pi \subseteq \mathcal{P}$, thus $\mathcal{P} = \mathcal{P}^\pi$. Now suppose that H_0 holds, i.e., $\mu \in \mathcal{M}_0$. Since $A(\mathcal{M}_0) = \mathcal{M}_0$ and $d \in \mathcal{M}_0$, we have $\tilde{\mu} \in \mathcal{M}_0$. If the alternative holds, then the transformed expectation lies also in $\mathcal{M} \backslash \mathcal{M}_0$. Thus, the first statement is proved. Consider now the projections of the transformed data vector:

$$\begin{aligned} (P - P_0)\tilde{\mathbf{y}} &= (P - P_0)(cA\mathbf{y} + d) = cA(P - P_0)\mathbf{y}, \\ (I_n - P)\tilde{\mathbf{y}} &= (I_n - P)(cA\mathbf{y} + d) = cA(I_n - P)\mathbf{y}. \end{aligned}$$

Because of the orthogonality of A we have

$$\begin{aligned} \tilde{\mathbf{y}}^\mathsf{T}(P - P_0)\tilde{\mathbf{y}} &= c^2 \mathbf{y}^\mathsf{T}(P - P_0)\mathbf{y}, \\ \tilde{\mathbf{y}}^\mathsf{T}(I_n - P)\tilde{\mathbf{y}} &= c^2 \mathbf{y}^\mathsf{T}(I_n - P)\mathbf{y}, \end{aligned}$$

and therefore

$$F(\pi(\mathbf{y})) = F(\mathbf{y}).$$

It remains to show the maximal invariance: Let $\mathbf{y}_1$ and $\mathbf{y}_2$ be two different points with $F(\mathbf{y}_1) = F(\mathbf{y}_2) \in [0, \infty)$. From

$$\frac{\mathbf{y}_1^\mathsf{T}(P - P_0)\mathbf{y}_1}{\mathbf{y}_1^\mathsf{T}(I_n - P)\mathbf{y}_1} = \frac{\mathbf{y}_2^\mathsf{T}(P - P_0)\mathbf{y}_2}{\mathbf{y}_2^\mathsf{T}(I_n - P)\mathbf{y}_2}$$

it follows that there exists a $c \neq 0$, such that

$$\mathbf{y}_1^\mathsf{T}(P - P_0)\mathbf{y}_1 = c^2 \mathbf{y}_2^\mathsf{T}(P - P_0)\mathbf{y}_2 \quad \text{and} \quad \mathbf{y}_1^\mathsf{T}(I_n - P)\mathbf{y}_1 = c^2 \mathbf{y}_2^\mathsf{T}(I_n - P)\mathbf{y}_2.$$

Moreover, there exists an orthogonal matrix A with (6.39) and $(P - P_0)\mathbf{y}_1 = cA(P - P_0)\mathbf{y}_2$ and $(I_n - P)\mathbf{y}_1 = cA(I_n - P)\mathbf{y}_2$. (Note that $(P - P_0)\mathbf{y}_1$ and $c(P - P_0)\mathbf{y}_2$ are vectors of the same length in the same subspace; the same holds for $(I_n - P)\mathbf{y}_1$ and $c(I_n - P)\mathbf{y}_2$.) With $d := P_0\mathbf{y}_1 - cAP_0\mathbf{y}_2 \in \mathcal{M}_0$ we get

$$\begin{aligned}
\mathbf{y}_1 &= P_0\mathbf{y}_1 + (P - P_0)\mathbf{y}_1 + (I_n - P)\mathbf{y}_1 \\
&= P_0\mathbf{y}_1 + cA(P - P_0)\mathbf{y}_2 + cA(I_n - P)\mathbf{y}_2 \\
&= cA\mathbf{y}_2 + d.
\end{aligned}$$

For $F(\mathbf{y}_1) = F(\mathbf{y}_2) = \infty$ and $F(\mathbf{y}_1) = F(\mathbf{y}_2) = -1$ the transformation π is similarly constructed.

$\square$

Given an invariant test problem it seems to be natural to consider only test procedures which are invariant. In this way we can reduce the problem of testing the parameter μ to the problem of testing a one-dimensional parameter λ. This parameter parameterized the class of distributions of the maximal invariant statistic F which is a class with monotone likelihood ratio. So we can apply methods for constructing optimal tests which were considered in Chapter 5.

By doing this we get the following

Theorem 6.8 *For testing the hypothesis (6.38) the F-test is the uniformly best test in the class of all α-tests which are invariant with respect to the group Q_0 of affine transformations.*

PROOF: We give only the main ideas of the proof:

(i) Each invariant test ϕ depends on $\mathbf{y}$ only via $F(\mathbf{y})$, i.e., $\phi(\mathbf{y}) = \psi(F(\mathbf{y}))$ for some suitable ψ.

Proof: Let ϕ be an invariant test, i.e., $\phi(\mathbf{y}) = \phi(\pi(\mathbf{y}))$ for all $\pi \in Q_0$ and all $\mathbf{y}$. Since F is maximal invariant it is constant on the orbits but for each orbit it takes on a different value. Thus we can define $\psi(t) = \phi(\mathbf{y})$ iff $F(\mathbf{y}) = t$ and we have $\phi(\mathbf{y}) = \psi(F(\mathbf{y}))$.

(ii) Using an orthogonal transformation one can prove that the statistic F is distributed according to the noncentral F-distribution with $q - r$ and $n - q$ degrees of freedom and noncentrality parameter

$$\lambda(\mu) = \frac{\mu^{\mathsf{T}}(P - P_0)\mu}{\sigma^2}.$$

(iii) Testing $H_0\colon \mu \in \mathcal{M}_0$ versus $H_1\colon \mu \in \mathcal{M} \setminus \mathcal{M}_0$ is equivalent to testing

$$H_0\colon \lambda(\mu) = 0 \quad \text{versus} \quad H_1\colon \lambda(\mu) > 0. \tag{6.40}$$

Proof: From (ii) it follows, that for different values μ and μ' with $\lambda(\mu) = \lambda(\mu')$, the distribution of the test statistic F is the same. In other words, the distribution of the test statistic of an invariant test depends on μ only via $\lambda(\mu)$. Further, for all $\mu \in \mathcal{M}_0$ we have $\lambda(\mu) = 0$ and for all $\mu \in \mathcal{M} \setminus \mathcal{M}_0$ $\lambda(\mu) > 0$.

The class of noncentral F-distributions has a monotone likelihood ratio. See Special case 5.4 on page 162. Hence, for the test problem (6.40) we have a uniformly best α-test, namely φ_F.

$\square$

6.5 List of Problems

1. Consider the data $(0, 2), (1, 1), (4, 3), (5, 2)$. Find the least squares line for these points. Plot the points, the line and mark the residuals.

2. Let $Y_1, \ldots, Y_n$ be uncorrelated r.v.'s with expectation m and variance σ_i^2 with $\sigma_i^2 \neq \sigma_j^2$ for $i \neq j$, $i, j = 1, \ldots, n$. Suppose the variances are known. Derive the best linear unbiased estimator for m.

3. Derive the normal equations for the problem of fitting a quadratic regression function $\beta_1 + \beta_2 x + \beta_3 x^2$, and solve them for the data points $(-1, 1), (0, 2), (1, 1), (2, -2)$.

4. Let $\mathbf{Y}$ be a vector of independent r.v.'s which are distributed according to $N(\alpha + \beta x_i, \sigma^2)$. The x_i's are known. Find the minimal sufficient statistic for the unknown parameter $\theta = (\alpha, \beta, \sigma^2)$.

5. Consider the regression model $Y_i = \beta_1 + \beta_2 x_i + \beta_3 x_i^2 + \varepsilon_i$, $i = 1, \ldots, n$ with $\mathsf{E}_\theta \varepsilon_i = 0$, $\mathsf{E}_\theta \varepsilon_i^2 = \sigma^2$ and $\mathsf{E}_\theta \varepsilon_i \varepsilon_j = 0$ for $i \neq j$.
a) Measurements can be made at points from the set $\{0, 1, 1.5\}$. How many measurements at which points have to been taken at least such that the 3-dimensional parameter $\gamma = (\beta_1 - \beta_2, \beta_2 - \beta_3, \beta_3 - \beta_1)^{\mathsf{T}}$ is estimable?
b) Suppose we have the following observations:

x_i	0	0	0	1	1	1	1.5	1.5	1.5	1.5
y_i	3.1	3.0	3.1	4.5	4.4	4.5	5.8	5.4	5.7	5.7

Compute the value of the best linear unbiased estimator for γ.

6. Consider the simple normal linear regression model in the form $y_i = \alpha + \beta (x_i - \bar{x}) + \varepsilon_i$ with $\varepsilon_i \sim N(0, \sigma^2)$.
   a) Determine the least squares estimators for α and β, say $\hat{\alpha}$ and $\hat{\beta}$.
   b) Under which conditions are the least squares estimators consistent?
   c) Determine the joint distribution of $(\hat{\alpha}, \hat{\beta})$.
   d) Consider the residuals: $\hat{\varepsilon}_i = y_i - \hat{y}_i$ with $\hat{y}_i = \hat{\alpha} + \hat{\beta}(x_i - \bar{x})$. Are they independent?
   e) Determine the distribution of $\hat{\alpha} + \hat{\beta}(x - \bar{x})$ for all x.
   f) Determine a confidence interval for $m(x) = \alpha + \beta(x - \bar{x})$. (Confidence interval $(l(x), u(x))$: $P_\theta(m(x) \in (l(x), u(x))) = 0.95$).
   g) Determine a prediction interval for a new observation $Y = m(x) + \varepsilon$.
   h) Compare f) and g).

7. Construct the F-statistic for testing $H_0 : \beta = 0$ versus $H_1 : \beta \neq 0$ in the simple linear regression model $y_i = \alpha + \beta(x_i - \bar{x}) + \varepsilon_i$ with $\varepsilon_i \sim N(0, \sigma^2)$. Discuss the relationship to the corresponding t-test.

8. Derive the MLE estimators in a two-way ANOVA model $y_{ij} = \mu + \alpha_i + \beta_j + \varepsilon_{ij}$, $i = 1, \ldots, I$, $j = 1, \ldots, J$.

9. Derive the F-Test for $H_0 : \alpha_1 = \cdots = \alpha_I = 0$ in the two-way ANOVA model $y_{ij} = \mu + \alpha_i + \beta_j + \varepsilon_{ij}$).

6.6 Further Reading

Applying statistical methods in practice is always connected with the problem of finding an appropriate statistical model—or, in other words, there is, as Freedman (2009) writes, "... a lack of connection between ... models and the real phenomena." In *Statistical Models* Freedman explains the main statistical techniques used in modeling causality.

More details of the investigation of linear model are given in Draper and Smith (1966). In the book of Rao and Toutenburg (1995) we can find more than the least squares approach. It includes for instance missing observations and robustness properties. Robustness is also studied in the textbook of Atkinson (2000).

A theory of statistical inference in linear models is given in Bunke and Bunke (1986). This book is written on a high theoretic level. It contains a decision-theoretic approach to linear models, the estimation of parameters under additional restrictions in form of inequalities, Bayesian inference and methods of experimental design.

A more applied treatment of linear models under additional restrictions you can find in Hastie et al. (2001).

In Section 6.2 nonlinear regression models are mentioned. In Bates and Watts

(1988) methods for the investigation of nonlinear relationships are presented. Employing real data sets the authors give examples for various model types. They provide background material for linear and nonlinear least squares estimation, including geometrical presentations.

For a more advanced study of the asymptotic properties of least squares estimators in nonlinear regression models, including higher order results, we refer to the book of Ivanov (1997).

We have only considered one-dimensional responses Y. Multidimensional outputs are the subject of multivariate analysis.

The following textbooks provide important statistical tools to treat multivariate models. They are not only useful to handle multivariate outputs in linear models but also to consider multivariate distributions, to analyze classification models and to describe principal components and factor analysis: Mardia et al. (1979), Giri (2004) and Kollo and von Rosen (2005).

German readers are referred to the book of Läuter and Pincus (1989); here methods for multivariate linear models, including classification models and multiple test problems, are considered. Moreover approaches for the analysis of nonlinear models are presented.

Methods derived for normal linear models where the responses are normally distributed can be extended to the treatment of models where the distribution of Y belongs to another parametric family, for example where Y given the covariates is distributed according to a Bernoulli, binomial or Poisson distribution. For example, in the case of the Bernoulli distribution: If Y takes only the value one or zero, then the expected value of Y given x is the probability of success depending on the covariates. It is useless to model this by a linear model; however using a link function this expectation can be expressed as a function of a linear combination of the covariates. This approach leads to nonlinear models, but because of its special structure it allows to apply inference methods coming from the linear model. For a detailed discussion of these so-called generalized linear models see McCullagh and Nelder (1990).

Nonparametric methods for the estimation of the relationship between input and output variables are based on smoothing. Here no parametric form of the regression function is assumed. For an introduction in this field we recommend Wand and Jones (1995) and Härdle (1990).

Chapter 7

Solutions

7.1 Solutions for Chapter 2: Statistical Model

1. The most general model is a model which takes into account different probabilities for defective parts at each day of the week. Such a model is given by a sample $\mathbf{X} = (X_1, \ldots, X_7)$ of independent r.v.'s with $X_i \sim \mathrm{Bin}(n, p_i)$, where $p_i \in (0, 1)$ and, since X_6 and X_7 describe the number of defective items at the weekends, $p_j > p_i$ for $j = 6, 7$ and $i < 6$. If one can assume that during the working days the production is of the same quality, and the probability of producing defective parts at the weekends is 50% higher, then we can assume the following model:

$$\mathcal{P} = \{\mathrm{Bin}(n, \theta)^{\otimes 5} \otimes \mathrm{Bin}(n, 1.5\theta)^{\otimes 2} : \theta \in (0, \tfrac{2}{3})\}.$$

(*Here also other answers are possible.*)

2. Let X_{it} be the color of an arbitrary flower collected in region $i = 1, \ldots, 5$ on day $t = 1, 2$. Define

$$\mathrm{P}(X_{it} = \text{violet}) = p_{it,v}, \quad \mathrm{P}(X_{it} = \text{white}) = p_{it,w}, \quad \mathrm{P}(X_{it} = \text{pink}) = p_{it,p},$$

with $p_{it,p} = 1 - p_{it,v} - p_{it,w}$. Note that we have for each pair (i, t) a three-point distribution.

3. Let X be a r.v. drawn from the described mixture of normal distributions. Denote the event, that X is drawn from the first part, by F. Then

$$
\begin{aligned}
\mathrm{P}_\theta(X \leq x) &= \mathrm{P}_\theta(X \leq x, F) + \mathrm{P}_\theta(X \leq x, \overline{F}) \\
&= \mathrm{P}_\theta(X \leq x|F)\mathrm{P}_\theta(F) + \mathrm{P}_\theta(X \leq x|\overline{F})\mathrm{P}_\theta(\overline{F}) \\
&= \Phi\left(\frac{x - \mu_1}{\sigma_1}\right)\pi + \Phi\left(\frac{x - \mu_2}{\sigma_2}\right)(1 - \pi),
\end{aligned}
$$

where π is the proportion of the first part in the population. Thus, the statistical model $\mathcal{P}$ for X consists of all distributions with distribution functions

$$\left\{\Phi\left(\frac{\cdot\; - \mu_1}{\sigma_1}\right)\pi + \Phi\left(\frac{\cdot\; - \mu_2}{\sigma_2}\right)(1 - \pi):\right.$$

$$\left. \mu_j \in \mathbb{R}, \sigma_j^2 \in \mathbb{R}_+, j = 1, 2, \pi \in (0,1)\right\}.$$

4. With the same argumentation as in Example 2.18 we can show that uniform distributions do not form an exponential family. Uniform distributions have the density $f(x; \theta) = \mathbb{1}_{[0,\theta]}(x)$. Take $\theta_1 < \theta_2$. The set $N = (\theta_1, \theta_2)$ is a P_{θ_1}-null set, but $P_{\theta_2}(N) > 0$. Thus, the measures are not equivalent, which is a contradiction to being an exponential family.

5. a) $N(0, \sigma^2)$ $\theta = \sigma^2$. Since $f(x; \theta) = \frac{1}{\sqrt{2\pi}}\frac{1}{\sqrt{\theta}}\exp\left(-\frac{1}{2\theta}x^2\right)$ this family forms a 1-parameter exponential family with

$$A(\theta) = (2\pi\theta)^{-\frac{1}{2}}, \quad h(x) = 1, \quad \zeta(\theta) = -\frac{1}{2\theta} \text{ and } T(x) = x^2.$$

b) $N(1, \sigma^2)$, $\theta = \sigma^2$. Since $f(x; \theta) = \frac{1}{\sqrt{2\pi}}\frac{1}{\sqrt{\theta}}\exp\left(-\frac{1}{2\theta}(x - 1)^2\right)$ this family forms a 1-parameter exponential family with

$$A(\theta) = (2\pi\theta)^{-\frac{1}{2}}, \quad h(x) = 1, \quad \zeta(\theta) = -\frac{1}{2\theta} \text{ and } T(x) = (x - 1)^2,$$

alternatively

$$\begin{aligned} f(x; \theta) &= \frac{1}{\sqrt{2\pi\theta}}\exp\left(-\frac{1}{2\theta}(x^2 - 2x + 1)\right) \\ &= \frac{1}{\sqrt{2\pi\theta}}\exp(-\frac{1}{2\theta})\exp\left(-\frac{1}{2\theta}(x^2 - 2x)\right) \end{aligned}$$

thus

$$A(\theta) = (2\pi\theta)^{-\frac{1}{2}}\exp(-\frac{1}{2\theta}), \quad h(x) = 1, \quad \zeta(\theta) = -\frac{1}{2\theta} \text{ and } T(x) = x^2 - 2x.$$

c) $N(\mu, \sigma^2)$, $\theta = (\mu, \sigma^2)$. Since

$$f(x; \theta) = \frac{1}{\sqrt{2\pi}}\frac{1}{\sqrt{\theta_1}}\exp\left(-\frac{1}{2\theta_1}(x - \theta_2)^2\right) = A(\theta)\exp\left(-\frac{1}{2\theta_1}x^2 + \frac{\theta_2}{\theta_1}x\right)$$

with

$$A(\theta) = (2\pi\theta)^{-\frac{1}{2}}\exp(-\frac{\theta_2^2}{2\theta_1}), \quad h(x) = 1,$$

$$\zeta_1(\theta) = -\frac{1}{2\theta_1}, \quad T_1(x) = x^2, \quad \zeta_2(\theta) = \frac{\theta_2}{\theta_1}, \quad T_2(x) = x$$

the normal distributions form a 2-parameter exponential family.

d) $N(\mu, \mu)$, $\theta = \mu$. Since

$$f(x; \theta) = \frac{1}{\sqrt{2\pi\theta}} \exp\left(-\frac{1}{2\theta}(x - \theta)^2\right) = \frac{1}{\sqrt{2\pi\theta}} \exp\left(-\frac{1}{2\theta}x^2 + x - \frac{\theta}{2}\right)$$

these distributions form a 1-parameter exponential family with

$$A(\theta) = (2\pi\theta)^{-\frac{1}{2}} \exp(-\frac{\theta}{2}), \quad h(x) = \exp(x), \quad \zeta_1(\theta) = -\frac{1}{2\theta}, \quad T_1(x) = x^2.$$

6. a) Poisson: Since

$$p(x; \lambda) = \exp(-\lambda)\frac{\lambda^x}{x!} = \exp(-\lambda)\exp(x \ln \lambda)\frac{1}{x!} \qquad \text{for } x = 0, 1, \ldots$$

we obtain that Poisson distributions form a 1-parameter exponential family with

$$A(\lambda) = \exp(-\lambda), \ h(x) = \frac{1}{x!}, \quad \zeta(\lambda) = \ln \lambda, \quad \text{and } T(x) = x.$$

b) $Geo(p)$: Since

$$p(x) = p(1 - p)^x = p\exp(x \ln(1 - p)) \qquad \text{for } x = 0, 1, \ldots$$

geometric distributions form a 1-parameter exponential family with

$$A(p) = p, \ h(x) = 1, \quad \zeta(p) = \ln(1 - p), \quad T(x) = x.$$

c) The Rayleigh distribution, defined by the density

$$f(x; \alpha) = \frac{2}{\alpha}x \exp(-\frac{x^2}{\alpha}) \, \mathbb{1}_{[0,\infty)}(x) \qquad \alpha > 0,$$

belongs to a 1-parameter exponential family with

$$A(\alpha) = \frac{2}{\alpha}, \ h(x) = x \, \mathbb{1}_{[0,\infty)}(x), \quad \zeta(\alpha) = -\frac{1}{\alpha} \ \text{and } T(x) = x^2.$$

7. As an example consider the triangle distribution: $Tri(0, \theta)$, defined by

$$f(x; \theta) = \frac{2}{\theta}(1 - \frac{2}{\theta}|x - \frac{\theta}{2}|) \, \mathbb{1}_{[0,\theta]}(x).$$

As in Example 2.18 or Problem 4, we show that measures from this family are not pairwise equivalent. Consider the set $N = [\frac{1}{2}, 1]$. For the parameter $\theta = 1$ we obtain $P_1(N) = \frac{1}{2}$, and for $\theta = \frac{1}{2}$ $P_{\frac{1}{2}}(N) = 0$. Thus N is not a null-set for all parameters—this is a contradiction to the assumption that all measures in an exponential family are pairwise equivalent.
(*This is only one of the possible answers.*)

8. The joint distribution of both samples is given by

$$
\begin{aligned}
f(\mathbf{x}; \lambda_1, \lambda_2) &= \prod_{i=1}^{n_1} \lambda_1 \exp(-\lambda_1 x_{1i}) \prod_{j=1}^{n_2} \lambda_2 \exp(-\lambda_2 x_{2j}) \\
&= \lambda_1^{n_1} \exp(-\lambda_1 \sum_{i=1}^{n_1} x_{1i}) \lambda_2^{n_2} \exp(-\lambda_2 \sum_{j=1}^{n_2} x_{2j}) \\
&= \lambda_1^{n_1} \lambda_2^{n_2} \exp(-\lambda_1 \sum_{i=1}^{n_1} x_{1i} - \lambda_2 \sum_{j=1}^{n_2} x_{2j}).
\end{aligned}
$$

We obtain for $\theta = (\lambda_1, \lambda_2)$: $\qquad A(\theta) = \lambda_1^{n_1} \lambda_2^{n_2}$,

$$
\zeta_1(\theta) = -\lambda_1, \ \zeta_2(\theta) = -\lambda_2, \ T_1(\mathbf{x}) = \sum_{i=1}^{n_1} x_{1i}, \ T_2(\mathbf{x}) = \sum_{j=1}^{n_2} x_{2j}.
$$

So we have a two-parameter exponential family.

9. Multinomial distribution: The calculation is analogous to Example 2.19. We have

$$
p(x_1, \ldots, x_{m-1}; \pi) = \binom{n}{x_1, \ldots, x_m} \pi_1^{x_1} \cdots \pi_m^{x_m}
$$

with $\pi_m = \sum_{i=1}^{m-1} \pi_i$, $x_m = n - \sum_{i=1}^{m-1} x_i$. Set $h(\mathbf{x}) = \binom{n}{x_1, \ldots, x_m}$, then

$$
\begin{aligned}
p(x_1, \ldots, x_{m-1}; \pi) &= \\
&= h(\mathbf{x}) \exp\left(\ln(\pi_1^{x_1} \ldots \pi_m^{x_m}) \right) \\
&= h(\mathbf{x}) \exp\left(\sum_{i=1}^{m} x_i \ln(\pi_i) \right) \\
&= h(\mathbf{x}) \exp\left(\sum_{i=1}^{m-1} x_i \ln(\pi_i) + (n - \sum_{i=1}^{m-1} x_i) \ln(1 - \sum_{i=1}^{m-1} \pi_i) \right) \\
&= h(\mathbf{x}) \exp\left(\sum_{i=1}^{m-1} x_i(\ln(\pi_i) - \ln(1 - \sum_{i=1}^{m-1} \pi_i)) + n \ln(1 - \sum_{i=1}^{m-1} \pi_i) \right) \\
&= h(\mathbf{x})(1 - \sum_{i=1}^{m-1} \pi_i)^n \exp\left(\sum_{i=1}^{m-1} x_i \ln(\frac{\pi_i}{1 - \sum_{j=1}^{m-1} \pi_j}) \right).
\end{aligned}
$$

The multinomial distribution (with m outcomes) forms a $(m-1)$-parameter exponential family with $A(\theta) = (1 - \sum_{i=1}^{m-1} \pi_i)^n$, and

$$
\zeta_i(\pi) = \ln(\frac{\pi_i}{1 - \sum_{j=1}^{m-1} \pi_j}), \ T_i(\mathbf{x}) = x_i, \ i = 1, \ldots, m-1.
$$

7.2 Solutions for Chapter 3: Inference Principles

1. a) $\theta = \sigma^2$, $f(x; \theta) = \frac{1}{\sqrt{2\pi\theta}} \exp\left(-\frac{1}{2\theta}x^2\right)$

$$L(\theta; \mathbf{x}) = \prod_{i=1}^{n} f(x_i; \theta) \propto \theta^{-\frac{n}{2}} \exp\left(-\frac{1}{2\theta}\sum_{i=1}^{n} x_i^2\right).$$

b) $\theta = \sigma^2$, $f(x; \theta) = \frac{1}{\sqrt{2\pi\theta}} \exp\left(-\frac{1}{2\theta}(x-1)^2\right)$

$$L(\theta; \mathbf{x}) \propto \theta^{-\frac{n}{2}} \exp\left(-\frac{1}{2\theta}\sum_{i=1}^{n}(x_i-1)^2\right).$$

c) $\theta = (\mu, \sigma^2)$, $f(x; \theta) = \frac{1}{\sqrt{2\pi\theta_1}} \exp\left(-\frac{1}{2\theta_1}(x-\theta_2)^2\right)$

$$L(\theta; \mathbf{x}) \propto \theta_1^{-\frac{n}{2}} \exp\left(-\frac{1}{2\theta_1}\sum_{i=1}^{n}(x_i-\theta_2)^2\right).$$

d) $\theta = \mu$, $f(x; \theta) = \frac{1}{\sqrt{2\pi\theta}} \exp\left(-\frac{1}{2\theta}(x-\theta)^2\right)$

$$L(\theta; \mathbf{x}) \propto \theta^{-\frac{n}{2}} \exp\left(-\frac{1}{2\theta}\sum_{i=1}^{n}(x_i-\theta)^2\right).$$

2. a) $f(x; \lambda) = \lambda \exp(-\lambda x)\, \mathbb{1}_{(0,\infty)}(x)$

$$L(\lambda; \mathbf{x}) = \prod_{i=1}^{n} \lambda \exp(-\lambda x_i) = \lambda^n \exp\left(-\lambda \sum_{i=1}^{n} x_i\right).$$

b) $p(x; \lambda) = \frac{\lambda^x}{x!} \exp(-\lambda)$

$$L(\lambda; \mathbf{x}) = \prod_{i=1}^{n} \frac{\lambda^{x_i}}{x_i!} \exp(-\lambda) \propto \lambda^{\sum_{i=1}^{n} x_i} \exp(-n\lambda).$$

c) $p(x; \theta) = \theta(1-\theta)^x$

$$L(\theta; \mathbf{x}) = \prod_{i=1}^{n} \theta(1-\theta)^{x_i} = \theta^n (1-\theta)^{\sum_{i=1}^{n} x_i}.$$

3. a)

	1	2	3
$L(\mathsf{P}_j; \mathrm{a})$	0.1	0	0.8

P_3 maximizes the likelihood function.

b)

	1	2	3
$L(\mathsf{P}_j; \mathrm{b})$	0.4	0.2	0.1
$L(\mathsf{P}_j; \mathrm{d})$	0.2	0.1	0.05

The likelihood functions are proportional to each other: $L(\mathsf{P}_j; \mathrm{b}) \propto L(\mathsf{P}_j; \mathrm{d})$.

4. $f(x; \theta) = \mathbb{1}_{[\theta - \frac{1}{2}, \theta + \frac{1}{2}]}(x)$

$$
\begin{aligned}
L(\theta; \mathbf{x}) &= \prod_{i=1}^{n} f(x_i; \theta) \\
&= \prod_{i=1}^{n} \mathbb{1}_{[\theta - \frac{1}{2}, \theta + \frac{1}{2}]}(x_i) = \begin{cases} 1 & \text{iff for all } i \quad \theta - \frac{1}{2} \le x_i \le \theta + \frac{1}{2} \\ 0 & \text{else} \end{cases} \\
&= \begin{cases} 1 & \text{iff} \quad \theta - \frac{1}{2} \le \min x_i, \ \max x_i \le \theta + \frac{1}{2} \\ 0 & \text{else} \end{cases} \\
&= \begin{cases} 1 & \text{iff} \quad \max x_i - \frac{1}{2} \le \theta \le \min x_i + \frac{1}{2} \\ 0 & \text{else} \end{cases} \\
&= \mathbb{1}_{[\max_i x_i - \frac{1}{2}, \min_i x_i + \frac{1}{2}]}(\theta).
\end{aligned}
$$

The maximum of the likelihood function is achieved at $\hat{\theta}_1 = \max_i x_i - \frac{1}{2}$ and $\hat{\theta}_2 = \min_i x_i + \frac{1}{2}$.

5. a) $T(\mathbf{x}) = \sum x_i^2$. b) $T(\mathbf{x}) = \sum (x_i - 1)^2$.
   c) $T(\mathbf{x}) = \sum x_i$. d) Compare case 3.13 $T(\mathbf{x}) = (\bar{x}, s^2)$.
   e) Compare Problem 4c of Chapter 3. This distribution belongs to an exponential family with $k = 1$, $A(\mu) = \mu^{-\frac{n}{2}} \exp(-\frac{n}{2}\mu^2)$, $h(\mathbf{x}) = (2\pi)^{-\frac{n}{2}} \exp(\sum_{i=1}^{n} x_i)$, $\zeta(\mu) = -\frac{1}{2\mu}$ and $T(\mathbf{x}) = \sum_{i=1}^{n} x_i^2$. Thus, a minimal sufficient statistic is given by $T(\mathbf{x})$.

6. Possible answers are:
   a) $T(\mathbf{x}) = x_1$. b) $T(\mathbf{x}) = \left(\sum_{i=1}^{n} x_i, \sum_{i=1}^{n} x_i^2, \sum_{i=1}^{n} x_i^3, \sum_{i=1}^{n} x_i^4 \right)$.

7. We have to show that under the regularity conditions assumed in Theorem 3.6 for the element (i, j) of the Fisher information matrix the following holds:

$$
I_{\mathbf{X}}(\theta)_{ij} = -\mathsf{E}_\theta \left(\frac{\partial^2}{\partial \theta_i \theta_j} \ln f(\mathbf{X}; \theta) \right).
$$

First observe that

$$
\begin{aligned}
\frac{\partial^2}{\partial \theta_i \theta_j} \ln f(\mathbf{X}, \theta) &= \frac{\partial}{\partial \theta_j} \frac{\frac{\partial}{\partial \theta_i} f(\mathbf{X}, \theta)}{f(\mathbf{X}, \theta)} \\
&= \frac{\left(\frac{\partial^2}{\partial \theta_i \theta_j} f(\mathbf{X}, \theta) \right) f(\mathbf{X}, \theta) - \frac{\partial}{\partial \theta_i} f(\mathbf{X}, \theta) \frac{\partial}{\partial \theta_j} f(\mathbf{X}, \theta)}{f(\mathbf{X}, \theta)^2} \\
&= \frac{\frac{\partial^2}{\partial \theta_i \theta_j} f(\mathbf{X}, \theta)}{f(\mathbf{X}, \theta)} - \frac{\frac{\partial}{\partial \theta_i} f(\mathbf{X}, \theta) \frac{\partial}{\partial \theta_j} f(\mathbf{X}, \theta)}{f(\mathbf{X}, \theta)^2}.
\end{aligned}
$$

Now let us take the expectation. Here we make use of the assumption that integration and differentiation can be reversed. We obtain

$$
\mathsf{E}_\theta \frac{\frac{\partial^2}{\partial \theta_i \theta_j} f(\mathbf{X}, \theta)}{f(\mathbf{X}, \theta)} = \int_A \frac{\partial^2}{\partial \theta_i \theta_j} f(\mathbf{x}, \theta)\, d\mathbf{x} = \frac{\partial^2}{\partial \theta_i \theta_j} \int_A f(\mathbf{x}, \theta)\, d\mathbf{x} = 0.
$$

Thus

$$-\mathsf{E}_\theta\left(\frac{\partial^2}{\partial\theta_i\partial\theta_j}\ln f(\mathbf{X};\theta)\right)=\mathsf{E}_\theta\left(\frac{\frac{\partial}{\partial\theta_i}f(\mathbf{X},\theta)\frac{\partial}{\partial\theta_j}f(\mathbf{X},\theta)}{f(\mathbf{X},\theta)^2}\right).$$

Recall the definition of the Fisher information matrix as $\mathsf{Cov}_\theta V(\theta;\mathbf{X})$, where V is the score vector. (Compare Definition 3.6.) Thus, the element (i,j) is given by

$$\mathsf{E}_\theta\left(\frac{\partial}{\partial\theta_i}\ln f(\mathbf{X};\theta)\frac{\partial}{\partial\theta_j}\ln f(\mathbf{X};\theta)\right),$$

which is equal to

$$\mathsf{E}_\theta\left(\frac{\frac{\partial}{\partial\theta_i}f(\mathbf{X};\theta)}{f(\mathbf{X};\theta)}\frac{\frac{\partial}{\partial\theta_j}f(\mathbf{X};\theta)}{f(\mathbf{X};\theta)}\right).$$

8. Compare Example 3.16.

9. a) The density of the sample $\mathbf{X}$ is given by

$$f(\mathbf{x};\theta)=\frac{1}{(2\theta)^n}\exp\left(-\frac{\sum_{i=1}^n|x_i|}{\theta}\right).$$

It follows immediately from the factorization theorem that $\sum_{i=1}^n|X_i|$ is sufficient.

b) From $\mathsf{P}_\theta(|X_i|\le z)=F(z;\theta)-F(-z;\theta)$ it follows that the density of $|X_i|$ is given by $f(z;\theta)+f(-z;\theta)$. Thus $|X_i|$ is exponentially distributed with parameter θ. Therefore $T=\sum_{i=1}^n|X_i|$ has a gamma distribution with parameters n and θ. For the conditional distribution of $\mathbf{X}$ given $T=t$ we obtain

$$\frac{\frac{1}{(2\theta)^n}\exp(-\frac{t}{\theta})}{\frac{1}{\Gamma(n)\theta^n}t^{n-1}\exp(-\frac{t}{\theta})}=\frac{\Gamma(n)}{2^n}t^{-(n-1)}.$$

10. The density of (X,Y) is defined by

$$f(x,y;\rho)=\frac{1}{2\pi\sqrt{1-\rho^2}}\exp\left(-\frac{1}{2(1-\rho^2)}(x^2-2\rho xy+y^2)\right).$$

Thus the log-likelihood function and its derivative are given by

$$l(\rho;x,y)=-\ln(2\pi)-\frac{1}{2}\ln(1-\rho^2)-\frac{x^2-2\rho xy+y^2}{2(1-\rho^2)},$$

$$l'(\rho;x,y)=\frac{\rho}{1-\rho^2}+xy\frac{1+\rho^2}{(1-\rho^2)^2}-x^2\frac{\rho}{(1-\rho^2)^2}-y^2\frac{\rho}{(1-\rho^2)^2}.$$

Thus

$$I_{X,Y}(\rho)=\left(\frac{1+\rho^2}{1-\rho^2}\right)^2\mathsf{Var}_\rho(XY)+2\frac{\rho^2}{(1-\rho^2)^4}\mathsf{Var}_\rho X^2$$

$$-4\frac{(1+\rho^2)\rho}{(1-\rho^2)^4}\mathsf{Cov}_\rho(XY,X^2)+2\frac{\rho^2}{(1-\rho^2)^4}\mathsf{Cov}_\rho(X^2,Y^2).$$

We have

$$\mathsf{Var}_\rho X^2 = \mathsf{E}_\rho X^4 - (\mathsf{E}_\rho X^2)^2 = 3 - 1 = 2.$$

$\mathsf{E}_\rho(XY)^2$

$$= \frac{1}{\sqrt{2\pi(1-\rho^2)}} \int x^2 \frac{1}{\sqrt{2\pi}} \int y^2 \exp\left(\frac{(x^2 - 2\rho xy + y^2)}{2(1-\rho^2)}\right) dy dx$$

$$= \frac{1}{\sqrt{2\pi}} \int x^2 \frac{1}{\sqrt{2\pi(1-\rho^2)}} \int y^2 \exp\left(\frac{(x^2 - \rho^2 x^2 + (y - \rho x)^2)}{2(1-\rho^2)}\right) dy dx$$

$$= \frac{1}{\sqrt{2\pi}} \int x^2 \exp\left(-\frac{x^2 - \rho^2 x^2}{2(1-\rho^2)}\right) \mathsf{E}_\rho Z_x^2 dx$$

$$= \frac{1}{\sqrt{2\pi}} \int x^2 \exp(-\frac{x^2}{2})(1 - \rho^2 + \rho^2 x^2) dx$$

$$= 1 - \rho^2 + \rho^2 \mathsf{E}_\rho X^4 = 1 - \rho^2 + 3\rho^2$$

$$= 1 + 2\rho^2.$$

Here Z_x is a normally distributed r.v. with expectation ρx and variance $1 - \rho^2$. Thus

$$\mathsf{Var}_\rho(XY) = \mathsf{E}_\rho(XY)^2 - (\mathsf{E}_\rho(XY))^2 = 1 + 2\rho^2 - \rho^2 = 1 + \rho^2.$$

Further,

$\mathsf{E}_\rho X^3 Y$

$$= \frac{1}{\sqrt{2\pi}} \int x^3 \frac{1}{\sqrt{2\pi(1-\rho^2)}} \int y \exp\left(-\frac{x^2 - 2\rho xy + y^2}{2(1-\rho^2)}\right) dy dx$$

$$= \frac{1}{\sqrt{2\pi}} \int x^3 \exp(-\frac{x^2}{2}) \mathsf{E}_\rho Z_x dx$$

$$= \frac{1}{\sqrt{2\pi}} \int x^3 \exp(-\frac{x^2}{2}) \rho x dx$$

$$= \rho \mathsf{E}_\rho X^4 = 3\rho.$$

Hence

$$\mathsf{Cov}_\rho(XY, X^2) = \mathsf{E}_\rho(X^3 Y) - \mathsf{E}_\rho(XY)\mathsf{E}_\rho X^2 = 3\rho - \rho \cdot 1 = 2\rho.$$

And finally,

$$\mathsf{Cov}_\rho(X^2, Y^2) = \mathsf{E}_\rho(X^2 Y^2) - \mathsf{E}_\rho X^2 \mathsf{E}_\rho Y^2 = 1 + 2\rho^2 - 1 = 2\rho^2.$$

Putting all together, we obtain

$$I_{X,Y}(\rho) = \frac{1 + \rho^2}{(1 - \rho^2)^2}.$$

11. For the first experiment we obtain with $X_i \sim \text{Bin}(1, \gamma)$

$$I_{\mathbf{X}}(\gamma) = \frac{n}{\gamma(1-\gamma)}.$$

For the computation of $I_{\mathbf{Y}}(\gamma)$ we can use the transformation formula or the direct way. Let us apply the transformation formula: The Fisher information for θ in the model $Y_i \sim \text{Bin}(1, \theta)$ is given by $I_{\mathbf{Y}}(\theta) = \frac{n}{\theta(1-\theta)}$. We have

$$\theta = h(\gamma) = \sqrt{\gamma} \text{ with } h'(\gamma) = \frac{1}{2}\gamma^{-\frac{1}{2}}.$$

Thus

$$I_{\mathbf{Y}}^*(\gamma) = I_{\mathbf{Y}}(h(\gamma))[h'(\gamma)]^2 = \frac{n\sqrt{\gamma}}{4\gamma^2(1-\sqrt{\gamma})}.$$

Let us now compare $I_{\mathbf{X}}(\gamma)$ and $I_{\mathbf{Y}}^*(\gamma)$: We have

$$\frac{n\sqrt{\gamma}}{4\gamma^2(1-\sqrt{\gamma})} \geq \frac{n}{\gamma(1-\gamma)} \quad \Leftrightarrow \quad \sqrt{\gamma} \geq 3\gamma.$$

That is, for $\gamma \in (\frac{1}{9}, 1)$ the first approach is more informative; for $\gamma \in (0, \frac{1}{9})$ the second is more informative. For $\gamma = \frac{1}{9}$ the information in both experiments is the same.

12. The log-likelihood function of $\text{Poi}(\mu)$ is $l(\mu; x) = x \ln \mu - \ln x! - \mu$. The first derivative with respect to μ is $l'(\mu; x) = x/\mu - 1$. Thus

$$I_X(\mu) = \text{Var}_\mu l'(\mu; X) = \frac{\text{Var}_\mu X}{\mu^2} = \frac{\mu}{\mu^2} = \frac{1}{\mu}.$$

The truncated Poisson distribution is defined by

$$P_\mu(Z = z) = P_\mu(X = z | X > 0) = \frac{P_\mu(X = z)}{P_\mu(X > 0)} = \frac{\mu^z}{z!} \frac{e^{-\mu}}{1 - e^{-\mu}}.$$

The log-likelihood function and its derivative are given by

$$l(\mu; z) = z \ln \mu - \ln z! - \mu - \ln(1 - e^{-\mu}), \quad l'(\mu; z) = \frac{z}{\mu} - 1 - \frac{e^{-\mu}}{1 - e^{-\mu}}.$$

Since $I_Z(\mu) = \text{Var}_\mu l'(\mu; Z) = \frac{\text{Var}_\mu Z}{\mu^2}$ and

$$E_\mu Z = \frac{\mu}{1 - e^{-\mu}} \quad \text{and} \quad E_\mu Z^2 = \frac{\mu^2 + \mu}{1 - e^{-\mu}}$$

we obtain

$$I_Z(\mu) = \frac{1}{\mu(1 - e^{-\mu})} - \frac{e^{-\mu}}{(1 - e^{-\mu})^2}.$$

Comparing $I_X(\mu)$ and $I_Z(\mu)$ we see there is a loss of information by truncation. For large μ, mathematically described by $\mu \to \infty$, this loss tends to zero. But for small μ the loss is large. In other words, if the expected number μ is small, it is essential to observe the event $\{X = 0\}$.

13. Here we can compute the Fisher information w.r.t. γ by using the transformation formula or directly. Let us take the direct way: The log-likelihood function is given by $l(\lambda; x) = \ln \lambda - \lambda x$. Since $\gamma = \mathsf{P}_\lambda(X > t_0) = \exp(-t_0\lambda)$ the log-likelihood function w.r.t. γ is

$$l(\gamma; x) = \ln(-\ln \gamma) - \ln t_0 + \frac{x}{t_0} \ln \gamma.$$

Since

$$l'(\gamma; x) = -\frac{1}{\gamma} \frac{1}{\ln \gamma} + \frac{x}{t_0 \gamma}$$

we obtain

$$I_X(\gamma) = \frac{\mathrm{Var}_\gamma X}{t_0^2 \gamma^2} = \frac{t_0^2}{(\ln \gamma)^2} \frac{1}{t_0^2 \gamma^2} = \frac{1}{(\gamma \ln \gamma)^2}.$$

14. For the binomial model we have

$$I_X(\theta) = \frac{n}{\theta(1 - \theta)}.$$

For the geometric distribution with probability function $p(y; \theta) = \theta(1 - \theta)^{y-1}$ for $y = 1, \ldots$ we obtain

$$l(\theta; y) = \ln \theta + (y - 1) \ln(1 - \theta).$$

The first and second derivative are given by

$$l'(\theta; y) = \frac{1}{\theta} - \frac{y - 1}{1 - \theta}, \qquad l''(\theta; y) = -\frac{1}{\theta^2} - \frac{y - 1}{(1 - \theta)^2}.$$

Since $\mathsf{E}_\theta Y = \frac{1}{\theta}$ the Fisher information of the sample is

$$I_Y(\theta) = n\left(\frac{1}{\theta^2} + \frac{\frac{1}{\theta} - 1}{(1 - \theta)^2}\right) = \frac{n}{\theta^2(1 - \theta)}.$$

The comparison of $I_X(\theta)$ and $I_Y(\theta)$ shows that

$$I_X(\theta) < I_Y(\theta) \quad \Leftrightarrow \quad \frac{1}{\theta(1 - \theta)} < \frac{1}{\theta^2(1 - \theta)}.$$

This is satisfied for all $\theta \in (0, 1)$. The approach based on the geometric distribution is more informative.

7.3   Solutions for Chapter 4: Estimation

1. The density is given by $f_\alpha(t) = \alpha t^{\alpha-1} \mathbb{1}_{[0,1]}(t)$ (Beta-distribution with parameter α and $\beta = 1$).

a) All moments exist and are equal to

$$m_k = \alpha \int_0^1 t^{k+\alpha-1} dt = \frac{\alpha}{k+\alpha}.$$

Thus, we obtain $\alpha(k) = k\, m_k/(1-m_k)$ and therefore for $k=1$

$$\hat{\alpha}_{\text{MME}} = \frac{\bar{x}}{1-\bar{x}}.$$

b) The likelihood function and the log-likelihood function are:

$$L(\alpha;\mathbf{x}) = \alpha^n \prod_{i=1}^n x_i^{\alpha-1} \quad \text{and} \quad l(\alpha;\mathbf{x}) = n\ln\alpha + (\alpha-1)\sum_{i=1}^n \ln x_i;$$

$l(\cdot;\mathbf{x})$ is differentiable, hence $\hat{\alpha}_{\text{MLE}}$ is the solution of

$$l'(\alpha;\mathbf{x}) = \frac{n}{\alpha} + \sum_{i=1}^n \ln x_i = 0.$$

Thus, $\hat{\alpha}_{\text{MLE}} = -\dfrac{1}{\frac{1}{n}\sum_{i=1}^n \ln x_i}$. Note that $l''(\hat{\alpha}_{\text{MLE}}) = -\dfrac{n}{\hat{\alpha}_{\text{MLE}}^2} < 0.$

2. The likelihood function is

$$L(\theta;\mathbf{x}) = \theta^{2n_1} [2\theta(1-\theta)]^{n_2} (1-\theta)^{2(n-n_1-n_2)}$$

where n_1 and n_2 are the numbers of elements in the sample with values aa and Aa, respectively. The log-likelihood is

$$l(\theta;\mathbf{x}) = 2n_1 \ln\theta + n_2 \ln[2\theta(1-\theta)] + 2(n-n_1-n_2)\ln(1-\theta).$$

This function is differentiable, and the solution of

$$l'(\theta;\mathbf{x}) = \frac{2n_1}{\theta} + \frac{n_2(1-2\theta)}{\theta(1-\theta)} - \frac{2(n-n_1-n_2)}{1-\theta}$$

is $\hat{\theta}_{\text{MLE}} = \dfrac{2n_1+n_2}{2n}$. Note that $l''(\hat{\theta}_{\text{MLE}};\mathbf{x}) < 0.$

3. In the model

$$\mathcal{P} = \{N(\mu_1,\sigma^2)^{\otimes n_1} \otimes N(\mu_2,\sigma^2)^{\otimes n_2} : \theta = (\mu_1,\mu_2,\sigma^2) \in \mathbb{R} \times \mathbb{R} \times \mathbb{R}_+\}$$

the likelihood function and the log-likelihood function are given by

$$L(\theta;\mathbf{x},\mathbf{y}) = \prod_{i=1}^{n_1} f(x_i;\mu_1,\sigma^2) \prod_{j=1}^{n_2} f(y_j;\mu_2,\sigma^2)$$

$$\propto (\sigma^2)^{-\frac{n_1+n_2}{2}} \exp\left(-\frac{1}{2\sigma^2}\left(\sum_{i=1}^{n_1}(x_i-\mu_1)^2 + \sum_{j=1}^{n_2}(y_j-\mu_2)^2\right)\right),$$

$$l(\theta; \mathbf{x}, \mathbf{y}) = -\frac{n_1 + n_2}{2} \ln \sigma^2 - \frac{1}{2\sigma^2} \left(\sum_{i=1}^{n_1} (x_i - \mu_1)^2 + \sum_{j=1}^{n_2} (y_j - \mu_2)^2 \right) + \text{const.}$$

To derive the MLE's the system of the following three equations has to be solved:

$$\frac{\partial l(\theta; \mathbf{x}, \mathbf{y})}{\partial \mu_1} = \frac{1}{\sigma^2} \sum_{i=1}^{n_1} (x_i - \mu_1) = 0 \qquad \frac{\partial l(\theta; \mathbf{x}, \mathbf{y})}{\partial \mu_2} = \frac{1}{\sigma^2} \sum_{j=1}^{n_2} (y_j - \mu_2) = 0$$

$$\frac{\partial l(\theta; \mathbf{x}, \mathbf{y})}{\partial \sigma^2} = -\frac{n_1 + n_2}{2\sigma^2} + \frac{1}{2\sigma^4} \left(\sum_{i=1}^{n_1} (x_i - \mu_1)^2 + \sum_{j=1}^{n_2} (y_j - \mu_2)^2 \right) = 0.$$

The solutions are:

$$\hat{\mu}_{1\text{MLE}} = \bar{x} = \frac{1}{n_1} \sum_{i=1}^{n_1} x_i, \qquad \hat{\mu}_{2\text{MLE}} = \bar{y} = \frac{1}{n_2} \sum_{j=1}^{n_2} y_j,$$

$$\hat{\sigma}^2_{\text{MLE}} = \frac{1}{n_1 + n_2} \left(\sum_{i=1}^{n_1} (x_i - \bar{x})^2 + \sum_{j=1}^{n_2} (y_j - \bar{y})^2 \right).$$

4. We use the following property of a r.v. $Z \sim N(a, \tau^2)$:
$E(Z - a)^3 = EZ^3 - 3EZ^2 a + 2a^3 = 0$ implies $\text{Cov}(Z, Z^2) = EZ^3 - EZ^2 a = 2a\tau^2$. Thus,

$$\begin{aligned}
\text{Cov}(\bar{X}, S^2) &= \frac{1}{n(n-1)} \sum_{i=1}^{n} \sum_{j=1}^{n} \text{Cov}(X_i, X_j^2) - \frac{n}{n-1} \text{Cov}(\bar{X}, \bar{X}^2) \\
&= \frac{1}{n-1} \text{Cov}(X_1, X_1^2) - \frac{n}{n-1} \text{Cov}(\bar{X}, \bar{X}^2) \\
&= \frac{1}{n-1} 2\mu \sigma^2 - \frac{n}{n-1} 2E(\bar{X}) \text{Var}(\bar{X}) \\
&= \frac{1}{n-1} 2\mu \sigma^2 - \frac{n}{n-1} 2\mu \frac{\sigma^2}{n} = 0.
\end{aligned}$$

5. a)

$$ET_n = E(\alpha T_{1n} + (1 - \alpha)T_{2n}) = \alpha ET_{1n} + (1 - \alpha)ET_{2n} = \alpha\theta + (1 - \alpha)\theta = \theta.$$

b)

$$\begin{aligned}
\text{Var} T_n &= \text{Var}(\alpha T_{1n} + (1 - \alpha)T_{2n}) \\
&= \alpha^2 \text{Var} T_{1n} + (1 - \alpha)^2 \text{Var} T_{2n} = \alpha^2 \sigma_{1n}^2 + (1 - \alpha)^2 \sigma_{2n}^2.
\end{aligned}$$

c)

$$V(\alpha) = \alpha^2 \sigma_{1n}^2 + (1 - \alpha)^2 \sigma_{2n}^2, \qquad V'(\alpha) = 2\alpha(\sigma_{1n}^2 + \sigma_{2n}^2) - 2\sigma_{2n}^2,$$

$$V''(\alpha) = 2(\sigma_{1n}^2 + \sigma_{2n}^2) > 0, \qquad \alpha_{\text{opt}} = \frac{\sigma_{2n}^2}{\sigma_{1n}^2 + \sigma_{2n}^2}, \qquad \alpha_{\text{opt}} \in (0, 1).$$

d) The estimator with α_{opt} has variance $\sigma_{1n}^2 \sigma_{2n}^2 / (\sigma_{1n}^2 + \sigma_{2n}^2)$. Since $\mathsf{E}T_n = \theta$ we have by the Chebyshev inequality that for every $\varepsilon > 0$:

$$P(|T_n - \theta| > \varepsilon) \le \frac{1}{\varepsilon^2} \frac{\sigma_{1n}^2 \sigma_{2n}^2}{\sigma_{1n}^2 + \sigma_{2n}^2}.$$

Since $\sigma_{1n}^2 \sigma_{2n}^2 / (\sigma_{1n}^2 + \sigma_{2n}^2) \to 0$, it follows that $T_n \xrightarrow{\mathsf{P}} \theta$.

6. The likelihood function of the uniform distribution $\mathsf{U}[-\theta, \theta]$ is given by

$$L(\theta; \mathbf{x}) = \frac{1}{(2\theta)^n} \prod_{i=1}^n \mathbb{1}_{[-\theta, \theta]}(x_i).$$

Since $\mathbb{1}_{[-\theta, \theta]}(x_i) = 1$ for all i if and only if $\max_i |x_i| \le \theta$, we have

$$L(\theta; \mathbf{x}) = \frac{1}{(2\theta)^n} \mathbb{1}_{[\max_i |x_i|, \infty)}(\theta).$$

The function θ^{-n} is monotone decreasing, thus $\hat{\theta}_{\text{MLE}} = \max_i |x_i|$.

7. a) We have for the likelihood function

$$L(\lambda; \mathbf{x}, \mathbf{y}) = \left(\frac{1}{2\pi\sigma^2}\right)^{\frac{n}{2}} \exp\left(-\frac{\sum_{i=1}^n (x_i - \mu)^2}{2\sigma^2}\right)$$
$$\times \left(\frac{1}{2\pi\lambda\sigma^2}\right)^{\frac{n}{2}} \exp\left(-\frac{\sum_{j=1}^n (y_j - \mu)^2}{2\lambda\sigma^2}\right)$$
$$\propto \lambda^{-\frac{n}{2}} \exp\left(-\frac{\sum_{j=1}^n (y_j - \mu)^2}{2\lambda\sigma^2}\right)$$

and the log-likelihood function

$$l(\lambda; \mathbf{x}, \mathbf{y}) = -\frac{n}{2} \ln\lambda - \frac{\sum_{j=1}^n (y_j - \mu)^2}{2\lambda\sigma^2} + \text{const.}$$

The solution of

$$l'(\lambda; \mathbf{x}, \mathbf{y}) = -\frac{n}{2\lambda} + \frac{\sum_{j=1}^n (y_j - \mu)^2}{2\lambda^2\sigma^2} = 0$$

is given by

$$\hat{\lambda}_{\text{MLE}} = \frac{1}{n\sigma^2} \sum_{j=1}^n (y_j - \mu)^2. \qquad \text{Note: } l''(\hat{\lambda}_{\text{MLE}}) < 0.$$

b) For unknown σ^2 we have

$$L(\lambda, \sigma^2; \mathbf{x}, \mathbf{y}) \propto \frac{1}{\sigma^{2n}\lambda^{\frac{n}{2}}} \exp\left(-\frac{\sum_{i=1}^n (x_i - \mu)^2}{2\sigma^2}\right) \exp\left(-\frac{\sum_{j=1}^n (y_j - \mu)^2}{2\lambda\sigma^2}\right),$$

$$l(\lambda, \sigma^2; \mathbf{x}, \mathbf{y}) = -n \ln \sigma^2 - \frac{n}{2} \ln \lambda - \frac{\sum_{i=1}^{n} (x_i - \mu)^2}{2\sigma^2} - \frac{\sum_{j=1}^{n} (y_j - \mu)^2}{2\lambda\sigma^2} + \text{const.}$$

The MLE is the solution of

$$\frac{\partial l(\lambda, \sigma^2; \mathbf{x}, \mathbf{y})}{\partial \lambda} = -\frac{n}{2\lambda} + \frac{\sum_{j=1}^{n} (y_j - \mu)^2}{2\lambda^2 \sigma^2} = 0$$

$$\frac{\partial l(\lambda, \sigma^2; \mathbf{x}, \mathbf{y})}{\partial \sigma^2} = -\frac{n}{\sigma^2} + \frac{\sum_{i=1}^{n} (x_i - \mu)^2}{2\sigma^4} + \frac{\sum_{j=1}^{n} (y_j - \mu)^2}{2\lambda\sigma^4} = 0,$$

and is given by

$$\hat{\lambda}_{\text{MLE}} = \frac{\sum_{j=1}^{n} (y_j - \mu)^2}{\sum_{i=1}^{n} (x_i - \mu)^2} \quad \text{and} \quad \hat{\sigma}^2_{\text{MLE}} = \frac{1}{n} \sum_{i=1}^{n} (x_i - \mu)^2.$$

8. a) The likelihood function is given by

$$L(\mu; \mathbf{x}, \mathbf{y}) = \prod_{i=1}^{n} f(x_i; \mu, \sigma_1^2) \prod_{j=1}^{n} f(y_j; \mu, \sigma_2^2)$$

$$\propto \exp\left(-\frac{1}{2\sigma_1^2} \sum_{i=1}^{n} (x_i - \mu)^2\right) \exp\left(-\frac{1}{2\sigma_2^2} \sum_{j=1}^{n} (y_j - \mu)^2\right),$$

the log-likelihood function is

$$l(\mu; \mathbf{x}, \mathbf{y}) = -\frac{1}{2\sigma_1^2} \sum_{i=1}^{n} (x_i - \mu)^2 - \frac{1}{2\sigma_2^2} \sum_{j=1}^{n} (y_j - \mu)^2 + \text{const.}$$

and $l'(\mu; \mathbf{x}, \mathbf{y}) = 0$ yields the equality

$$\frac{1}{\sigma_1^2} \sum_{i=1}^{n} (x_i - \mu) + \frac{1}{\sigma_2^2} \sum_{j=1}^{n} (y_j - \mu) = 0.$$

The MLE is

$$\hat{\mu}_{\text{MLE}} = n^{-1} \frac{\sigma_1^2 \sigma_2^2}{\sigma_1^2 + \sigma_2^2} \left(\frac{1}{\sigma_1^2} \sum_{i=1}^{n} x_i + \frac{1}{\sigma_2^2} \sum_{j=1}^{n} y_j \right).$$

b)

$$\text{Var}_\mu \, \hat{\mu}_{\text{MLE}} = n^{-2} \left(\frac{\sigma_1^2 \sigma_2^2}{\sigma_1^2 + \sigma_2^2} \right)^2 \left(\frac{1}{\sigma_1^4} n \text{Var}_\mu X_1 + \frac{1}{\sigma_2^4} n \text{Var}_\mu Y_1 \right)$$

$$= n^{-1} \left(\frac{\sigma_1^2 \sigma_2^2}{\sigma_1^2 + \sigma_2^2} \right)^2 (\sigma_1^{-2} + \sigma_2^{-2}) = n^{-1} \frac{\sigma_1^2 \sigma_2^2}{\sigma_1^2 + \sigma_2^2}.$$

9. For moments of this distribution we have

$$m_k = \mathsf{E}_\theta X^k = \theta \int_\theta^\infty x^{k-2} dx = \begin{cases} \frac{\theta^k}{(1-k)} & \text{for} \quad k < 1 \\ \infty & \text{for} \quad k \geq 1 \end{cases} .$$

Thus, taking $k = \frac{1}{2}$ we obtain by the method of moments with $m_{\frac{1}{2}} = 2\sqrt{\theta}$

$$\hat{\theta}_{\text{MME}} = \frac{1}{4} \left(\frac{1}{n} \sum_{i=1}^n \sqrt{x_i} \right)^2 .$$

10. a) In (5.2) it was stated that the vector $\mathbf{N} = (N_1, N_2)$ has a multinomial distribution

$$p(\mathbf{n}; \theta) = \frac{n!}{n_1! \, n_2! \, n_3!} p_1^{n_1} p_2^{n_2} (1 - p_1 - p_2)^{n_3},$$

where each component is $\text{Bin}(n, p_j)$-distributed and $\theta = (p_1, p_2)$ with $p_1, p_2 \in (0, 1)$, $0 < p_1 + p_2 < 1$, $n_3 = n - n_1 - n_2$. Hence $\mathsf{E}_\theta N_j = np_j$ and $\text{Var}_\theta N_j = np_j(1 - p_j)$. Since N_j is a sum of independent Bernoulli variables $I_j(X_i)$ with $I_1(X_i)I_2(X_i) = 0$ for all i, we have

$$\text{Cov}_\theta(N_1, N_2) = \mathsf{E}_\theta (N_1 - np_1)(N_2 - np_2) = \mathsf{E}_\theta N_1 N_2 - n^2 p_1 p_2$$

$$= \sum_{i=1}^n \sum_{j=1}^n \mathsf{E}_\theta I_1(X_i) I_2(X_j) - n^2 p_1 p_2$$

$$= \sum_{i=1}^n \mathsf{E}_\theta I_1(X_i) I_2(X_i) + \sum_{i=1}^n \sum_{\substack{i=1 \\ i \neq j}}^n \mathsf{E}_\theta I_1(X_i) \mathsf{E}_\theta I_2(X_j) - n^2 p_1 p_2$$

$$= 0 + n(n-1)p_1 p_2 - n^2 p_1 p_2 = -np_1 p_2.$$

Thus, N_1 and N_2 are negative correlated.
b) Compare Example 3.16 on page 48.
c) The likelihood function and the log-likelihood function satisfy

$$L(\theta; \mathbf{x}) \propto p_1^{n_1} p_2^{n_2} (1 - p_1 - p_2)^{n_3} = p_1^{n_1} p_2^{n_2} (1 - p_1 - p_2)^{n - n_1 - n_2},$$

$$l(\theta; \mathbf{x}) = n_1 \ln p_1 + n_2 \ln p_2 + (n - n_1 - n_2) \ln(1 - p_1 - p_2) + \text{const}.$$

We have to solve the system of the two equations $\dfrac{\partial l(p_1, p_2; \mathbf{x})}{\partial p_j} = 0$:

$$\frac{n_1}{p_1} - \frac{n - n_1 - n_2}{1 - p_1 - p_2} = 0$$

$$\frac{n_2}{p_2} - \frac{n - n_1 - n_2}{1 - p_1 - p_2} = 0.$$

The solution is $\hat{\theta}_{\mathrm{MLE}} = (\frac{n_1}{n}, \frac{n_2}{n})$, i.e., the relative frequencies. This estimator is unbiased. From the results under a) we obtain for the variances and covariances of these estimators:

$$\mathrm{Cov}_\theta\, \hat{\theta}_{\mathrm{MLE}} = n^{-2} \begin{pmatrix} \mathrm{Var}_\theta N_1 & \mathrm{Cov}_\theta(N_1, N_2) \\ \mathrm{Cov}_\theta(N_1, N_2) & \mathrm{Var}_\theta N_2 \end{pmatrix}$$

$$= n^{-1} \begin{pmatrix} p_1(1-p_1) & -p_1 p_2 \\ -p_1 p_2 & p_2(1-p_2) \end{pmatrix}.$$

But this is just the inverse of the Fisher information—in other words, the MLE is efficient.

11. The likelihood function of a sample from $\mathsf{U}[0, \theta]$ is:
$L(\theta; \mathbf{x}) = \theta^{-n}\, \mathbb{1}_{[0,\theta]}(\max_i x_i)$. The statistic $T(\mathbf{x}) = \max_i x_i$ is sufficient. It is also minimal sufficient: Since for all $\mathbf{x}$ and $\mathbf{y}$ with

$$L(\theta; \mathbf{x}) = k(\mathbf{x}, \mathbf{y})\, L(\theta; \mathbf{y}) \quad \text{for}\ \ k(\mathbf{x}, \mathbf{y}) > 0$$

we have $\max_i x_i = \max_i y_i$.

The maximizer of $L(\theta; \mathbf{x})$ is $\hat{\theta}_{\mathrm{MLE}} = X_{\max}$. Because $\mathsf{E}_\theta X = \frac{\theta}{2}$, the MME is $\hat{\theta}_{\mathrm{MME}} = 2\overline{X}$.

Distribution and density function of the maximum of i.i.d. r.v.'s Z_i, $i = 1, \ldots, n$ with cdf F and density f are given by

$$F_{\max}(t) = \mathsf{P}(Z_i \leq t,\ \text{for all}\ i) = F^n(t) \quad \text{and} \quad f_{\max}(t) = n f(t) F^{n-1}(t).$$

Applying this to the uniform distribution we get:

$$\mathsf{E}_\theta X_{\max} = \frac{n}{\theta^n} \int_0^\theta x^n dx = \frac{n\theta}{n+1}, \qquad \mathsf{E}_\theta X_{\max}^2 = \frac{n}{\theta^n} \int_0^\theta x^{n+1} dx = \frac{n\theta^2}{n+2}.$$

Thus, the MLE has the bias $-\theta/(n+1)$ and the variance

$$\mathrm{Var}_\theta \hat{\theta}_{\mathrm{MLE}} = \frac{n}{n+2}\theta^2 - \frac{n^2}{(n+1)^2}\theta^2 = \frac{n\theta^2}{(n+2)(n+1)^2}.$$

The MSE of the MLE is

$$\mathsf{MSE}(\hat{\theta}_{\mathrm{MLE}}, \theta) = \frac{n\,\theta^2}{(n+2)(n+1)^2} + \frac{\theta^2}{(n+1)^2} = \frac{2\theta^2}{(n+1)(n+2)}.$$

The bias-corrected estimator is $\tilde{\theta} = \frac{n+1}{n} X_{\max}$. Its MSE is equal to its variance

$$\mathsf{MSE}(\tilde{\theta}, \theta) = \frac{(n+1)^2}{n^2} \mathrm{Var}_\theta \hat{\theta}_{\mathrm{MLE}} = \frac{\theta^2}{n(n+2)}.$$

Since $\mathsf{E}_\theta \overline{X} = \frac{\theta}{2}$, the MME is unbiased. Its variance and MSE is given by

$$\mathsf{MSE}(\hat{\theta}_{\mathrm{MME}}) = \mathrm{Var}_\theta \hat{\theta}_{\mathrm{MME}} = \frac{4}{n} \mathrm{Var}_\theta X_1 = \frac{4}{n}\frac{\theta^2}{12} = \frac{1}{3n}\theta^2.$$

Let us compare $\hat{\theta}_{\text{MME}}$ and $\hat{\theta}_{\text{MLE}}$:

$$\text{MSE}(\hat{\theta}_{\text{MME}}) - \text{MSE}(\hat{\theta}_{\text{MLE}}) = \frac{\theta^2}{3n} - \frac{2\theta^2}{(n+1)(n+2)}$$

$$= \frac{\theta^2}{3n(n+1)(n+2)}(n^2 - 3n + 2) > 0.$$

Thus, for all n the maximum likelihood estimator is better than the method of moment estimator! The comparison with the bias-corrected estimator yields

$$\text{MSE}(\hat{\theta}_{\text{MLE}}) - \text{MSE}(\tilde{\theta}) = \frac{2\theta^2}{(n+1)(n+2)} - \frac{\theta^2}{n(n+2)} = \frac{\theta^2(n-1)}{n(n+1)(n+2)} > 0.$$

Thus, the bias-corrected estimator is not only the best estimator in the class of all unbiased estimators (since $X_{\max}$ is complete), it is also better than the biased MLE.

12. The likelihood function and the log-likelihood function for one observation are

$$L(a;x) = \frac{2}{a}x\exp(-\frac{x^2}{a}) \quad \text{and} \quad l(a;x) = -\ln a + \ln x - \frac{x^2}{a} + \text{const.}$$

Since $l'(a;x) = -\frac{1}{a} + \frac{x^2}{a^2}$ and $l''(a;x) = \frac{1}{a^2} - \frac{2x^2}{a^3}$, we obtain

$$I_X(a) = -E_a l''(a;X) = -\frac{1}{a^2} + E_a\frac{2X^2}{a^3} = \frac{1}{a^2}.$$

Therefore, the Fisher information of the sample is $\frac{n}{a^2}$. The MLE is the solution of $l'(a;\mathbf{x}) = 0$:

$$-\frac{n}{a} + \sum_{i=1}^{n}\frac{x_i^2}{a^2} = 0, \quad \text{i.e.,} \quad \hat{a}_{\text{MLE}} = \frac{1}{n}\sum_{i=1}^{n}x_i^2.$$

From $m_1 = \frac{\sqrt{\pi a}}{2}$ and $m_2 = a$ we derive the following method of moment estimates:

$$\hat{a}_{\text{MME}}^{(1)} = \frac{(2\bar{x})^2}{\pi}, \quad \hat{a}_{\text{MME}}^{(2)} = \hat{m}_2 = \hat{a}_{\text{MLE}}.$$

The MLE (and $\hat{a}_{\text{MME}}^{(2)}$) are unbiased. Their variance (and MSE) is given by the inverse of the Fisher information, i.e., a^2/n. This follows from Theorem 4.3 since the given Raleigh distribution belongs to a one-parameter exponential family with sufficient statistic $\frac{1}{n}\sum_{i=1}^{n}X_i^2$. And this is an efficient estimator for its expectation, i.e., for the parameter a.

7.4 Solutions for Chapter 5: Testing Hypotheses

1. Set $\alpha = P_0(C_1)$. It holds $P_0(|\bar{x} - 10| > 0.5) = 1 - P_0(|\bar{x} - 10| \le 0.5) = 1 - P(9.5 \le \bar{x} \le 10.5)$ and $P_0(|\bar{x} - 10| > 0.8) = 1 - P_0(|\bar{x} - 10| \le 0.8) = 1 - P(9.2 \le \bar{x} \le 10.8)$. The first critical region gives the larger α.

2. Define $C_1 = \{x : p_0(x) < kp_1(x)\}$ and $C_= = \{x : p_0(x) = kp_1(x)\}$. Then
$$\alpha_1 = P_0(C_1), \text{ and } \alpha_2 = P_0(C_1 \cup C_=) = \alpha_1 + P_0(C_=).$$

Further
$$1 - \beta_1 = P_1(C_1), \text{ and } 1 - \beta_2 = P_1(C_1 \cup C_=) = 1 - \beta_1 + P_1(C_=).$$

The Neyman–Pearson test for $\alpha = \lambda\alpha_1 + (1 - \lambda)\alpha_2$ is given by
$$\varphi(x) = \begin{cases} 1 & \text{if } p_0(x) < kp_1(x) \\ \gamma(\alpha) & \text{if } p_0(x) = kp_1(x) \\ 0 & \text{if } p_0(x) > kp_1(x) \end{cases},$$

where
$$\gamma(\alpha) = \frac{\alpha - P_0(C_1)}{P_0(C_=)}.$$

Because
$$\alpha = \lambda\alpha_1 + (1-\lambda)\alpha_2 = \lambda\alpha_1 + (1-\lambda)(\alpha_1 + P_0(C_=)) = P_0(C_1) + (1-\lambda)P_0(C_=)$$

we have $\gamma(\alpha) = (1 - \lambda)$ and
$$\begin{aligned} 1 - \beta &= P_1(C_1) + \gamma(\alpha)P_1(C_=) = P_1(C_1) + (1 - \lambda)P_1(C_=) \\ &= \lambda P_1(C_1) + (1 - \lambda)\left(P_1(C_1) + P_1(C_=)\right) \\ &= 1 - (\lambda\beta_1 + (1 - \lambda)\beta_2). \end{aligned}$$

3. a), b) Each subset $C_1 \subseteq \mathcal{X}$ of the sample space gives one test:

	C_1	$\alpha = P_0(C_1)$	$\beta = 1 - P_1(C_1)$
1	$\mathcal{X}$	1	0
2	$\mathcal{X} \setminus (1,1)$	0.75	0.64
3	$\mathcal{X} \setminus (1,0)$	0.75	0.16
4	$\mathcal{X} \setminus (0,1)$	0.75	0.16
5	$\mathcal{X} \setminus (0,0)$	0.75	0.04
6	$\{(0,0),(0,1)\}$	0.5	0.8
7	$\{(0,0),(1,0)\}$	0.5	0.8
8	$\{(0,0),(1,1)\}$	0.5	0.32
9	$\{(0,1),(1,0)\}$	0.5	0.68
10	$\{(0,1),(1,1)\}$	0.5	0.2
11	$\{(1,0),(1,1)\}$	0.5	0.2
12	$\{(0,0)\}$	0.25	0.96
13	$\{(0,1)\}$	0.25	0.84
14	$\{(1,0)\}$	0.25	0.84
15	$\{(1,1)\}$	0.25	0.36
16	$\emptyset$	0	1

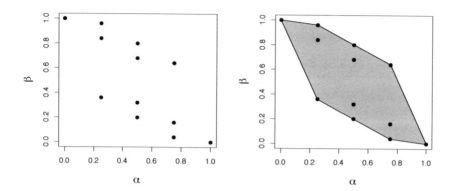

Figure 7.1: The right picture shows the $\alpha\beta$-representations for all nonrandomized tests. The left picture shows the $\alpha\beta$-representations for all tests.

b), c) see Figure 7.1. c) The set of $\alpha\beta$-representations is convex, Theorem 5.1. We take the convex closure of b). The lower boundary line is formed by the $\alpha\beta$-representations of Neyman–Pearson tests φ, because Neyman–Pearson tests are most powerful; compare Figure 5.11. Compare also Problem 2 above. The upper boundary line is related to the tests $1 - \varphi$. d) The Neyman–Pearson test in this case is randomized:

$$\varphi(x_1, x_2) = \begin{cases} 0.2 & \text{for} \quad \{(1,1)\} \\ 0 & \text{for} \quad \{(0,0), (0,1), (1,0)\} \end{cases},$$

with $k = 0.4, \gamma = 0.2$ and $\alpha = P_0(\emptyset) + 0.2P_0((1,1)) = 0.05$ and $\beta = P_1(\{(0,0), (0,1), (1,0)\}) + 0.8P_1((1,1)) = 0.872$.

4. Set $p_0(x) = 1$, $p_1(x) = 2x$ for $0 < x < 1$. The region C_1 of the Neyman–Pearson test $\frac{p_0(x)}{p_1(x)} = \frac{1}{2x} < c$ is equivalent to $\{x : x > k\}$. a) $\alpha = P_0(X > k) = 1 - k$, $\beta = 1 - P_1(X > k) = P_1(X \le k) = \int_0^k 2x \, dx = k^2$; see Figure 7.2. c) Wanted k such that: $\alpha = 4\beta$: $(1 - k) = 4k^2$. Solutions are : $k = -\frac{1}{8} - \frac{1}{8}\sqrt{17}$ and $k = -\frac{1}{8} + \frac{1}{8}\sqrt{17}$. Because $k \ge 0$ we get $k = -\frac{1}{8} + \frac{1}{8}\sqrt{17} = 0.39039$.

5. We have

x	2	3	4	5	6	7	8
$p_0(x)/p_1(x)$	5	6.6667×10^{-2}	33	0.55556	1	0.5	2

a) The Neyman–Pearson test for $\alpha = 0.02$ is

$$\varphi(\mathbf{x}) = \begin{cases} 1 & \text{if} \quad \{3\} \\ 0 & \text{if} \quad \{2, 4, 5, 6, 7, 8\} \end{cases}.$$

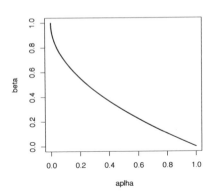

Figure 7.2: The $\alpha\beta$-representations of all Neyman–Pearson tests in Problem 4 in Chapter 5.

b) The Neyman–Pearson test for $\alpha = 0.05$ is

$$\varphi(\mathbf{x}) = \begin{cases} 1 & \text{if} \quad \{3\} \\ \gamma & \text{if} \quad \{7\} \\ 0 & \text{if} \quad \{2,4,5,6,8\} \end{cases},$$

with

$$\gamma = \frac{0.05 - P_0\,(3)}{P_0(7)} = \frac{0.05 - 0.02}{0.1} = 0.3.$$

c) The size of errors of second type for the test in a) $\beta = 1 - P_1(3) = 1 - 0.3 = 0.7$ and for the test in b) $\beta = 1 - P_1(3) - 0.3 P_1\,(7) = 1 - 0.3 - 0.3 * 0.2 = 0.64$.

d) An alternative alpha test for $\alpha = 0.05$ can be

$$\varphi(\mathbf{x}) = \begin{cases} 1 & \text{if} \qquad \{2\} \\ 0 & \text{if} \quad \{3,4,5,6,7,8\} \end{cases}.$$

e) For the test in d) it holds $\beta = 1 - P_1(2) = 1 - 0.01 = 0.99 > 0.64$. The Neyman–Pearson test in b) is better!

6. Set $p_0(\mathbf{x}) = e^{-n} \prod_{i=1}^{n} \frac{1}{x_i!}$ and $p_1(\mathbf{x}) = e^{-2n} \prod_{i=1}^{n} \frac{2^{x_i}}{x_i!}$. Thus

$$\frac{p_0(\mathbf{x})}{p_1(\mathbf{x})} < c \iff e^n \prod_{i=1}^{n} \frac{1}{2^{x_i}} < c \iff \sum_{i=1}^{n} x_i > k.$$

Apply Theorem 5.2 (Neyman–Pearson lemma). The most powerful test is

$$\varphi(\mathbf{x}) = \begin{cases} 1 & \text{for} \quad \sum_{i=1}^{n} x_i > k \\ \gamma & \text{for} \quad \sum_{i=1}^{n} x_i = k \\ 0 & \text{for} \quad \sum_{i=1}^{n} x_i < k \end{cases}, \quad \text{where} \quad \sum_{i=1}^{n} X_i \sim \text{Poi}(n) \text{ under } H_0.$$

7. a) The likelihood ratio $\Lambda^* = \frac{L(\theta_1)}{\max(L(\theta_2), L(\theta_3))}$ is

	z_1	z_2	z_3	z_4
Λ^*	0.4	3.0	0.25	1.33

b), c) The critical regions defined by $\Lambda^* < K$ are

C_1	$\emptyset$	$\{z_3\}$	$\{z_3, z_1\}$	$\{z_3, z_1, z_4\}$	$\{z_3, z_1, z_4, z_2\}$
α	0	0.1	0.3	0.7	1

d) First test: $\alpha = P_0(\{z_1\}) = 0.2$, $\beta = 1 - P_{\theta_2}(\{z_1\}) = 1 - 0.5 = 0.5$. Second test: $\alpha = P_0(\{z_2, z_3\}) = 0.4$, $\beta = 1 - P_{\theta_2}(\{z_2, z_3\}) = 1 - 0.1 - 0.2 = 0.7$. The first test is better when the "true" parameter is θ_2.

8. The exponential distribution belongs to an exponential family $X \sim \mathrm{Exp}(\lambda)$, because $f(x) = \lambda \exp(-\lambda x)$, $x > 0$. For an i.i.d. sample the sufficient statistic is $T(\mathbf{x}) = \sum_{i=1}^{n} x_i$ and $T \sim \mathrm{Exp}(n\lambda)$. We have $P(T < t) = \int_0^t n\lambda \exp(-n\lambda x)dx = 1 - e^{-n\lambda t}$. From Theorem 5.3 follows that the model has an MLR in $-T$. Consider the one-sided test problem $H_0 : \lambda \leq \lambda_0$ versus $H_1 : \lambda > \lambda_0$ and apply Theorem 5.4 (Blackwell theorem). We obtain the UMP test of size α:

$$\varphi(\mathbf{x}) = \begin{cases} 1 & \text{if } T < u \\ 0 & \text{if } T \geq u \end{cases}, \text{ with } u = \frac{-\ln(1-\alpha)}{n\lambda_0}.$$

9. $N(\mu, 1)$ belongs to a one-parameter exponential family with natural parameter μ and sufficient statistics $T(\mathbf{x}) = \sum_{i=1}^{n} x_i$ with $T \sim N(n\mu, n)$. a) We apply Theorem 5.4 (Blackwell theorem) and get the UMP test of size α

$$\varphi(\mathbf{x}) = \begin{cases} 1 & \text{if } \sqrt{n}(\overline{x} - \mu_0) > z_{1-\alpha} \\ 0 & \text{if } \sqrt{n}(\overline{x} - \mu_0) \leq z_{1-\alpha} \end{cases}, \text{ with } \Phi(z_{1-\alpha}) = 1 - \alpha.$$

It is the one sided Z-test. b) The power function is $\pi(\mu) = E_\mu(\varphi(\mathbf{X})) = P_\mu(\sqrt{n}(\overline{X} - \mu_0) > z_{1-\alpha}) = P_\mu(\sqrt{n}(\overline{X} - \mu) > z_{1-\alpha} + \sqrt{n}(\mu_0 - \mu)) = 1 - \Phi(z_{1-\alpha} + \sqrt{n}(\mu_0 - \mu))$; see Figure 7.3 for $n = 20$, $\mu_0 = 1$, $\alpha = 0.05$.
c) We apply Theorem 5.4 (Blackwell theorem) and get the UMP test of size α

$$\varphi(\mathbf{x}) = \begin{cases} 1 & \text{if } \sqrt{n}(\overline{x} - \mu_0) < z_\alpha \\ 0 & \text{if } \sqrt{n}(\overline{x} - \mu_0) \geq z_\alpha \end{cases}.$$

d) Applying Corollary 5.3 with $a = n\mu_0$ we get

$$\varphi(\mathbf{x}) = \begin{cases} 1 & \text{if } \qquad\qquad \text{else} \\ 0 & \text{if } |\sqrt{n}(\overline{x} - \mu_0)| \leq z_{1-\frac{\alpha}{2}} \end{cases}.$$

10. The exponential distribution belongs to an exponential family because the density is $f(x) = \lambda \exp(-\lambda x)$, $x > 0$. The distribution of an i.i.d. sample has the sufficient statistic $T(\mathbf{x}) = \sum_{i=1}^{n} x_i$ and $T \sim \mathrm{Exp}(n\lambda)$ with $P(T <$

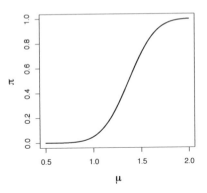

Figure 7.3: The power function of a one-sided Z-test in Problem 9 in Chapter 5.

$t) = \int_0^t n\lambda \exp(-n\lambda x)dx = 1 - e^{-n\lambda t}$. The model has an MLR in $-T$.
a) Apply Theorem 5.5. The test statistic is $T = \sum_{i=1}^n x_i$. b) Apply the Theorem 5.5 with equal tail approximation:

$$\varphi(\mathbf{x}) = \begin{cases} 1 & \text{if} \quad T > k_{\text{upper}}, \ T < k_{\text{lower}} \\ 0 & \text{if} \quad k_{\text{lower}} \leq T \leq k_{\text{upper}} \end{cases} ,$$

where

$$k_{\text{lower}} = \frac{-\ln(1 - \frac{\alpha}{2})}{n\lambda_0}, \quad k_{\text{upper}} = \frac{-\ln(\frac{\alpha}{2})}{n\lambda_0}.$$

11. a) Yes, because:

$$\begin{aligned} p(\mathbf{x}; \lambda) &= \prod_{i=1}^n \frac{(n_i\lambda)^{x_i}}{x_i!} \exp(-n_i\lambda) \\ &= \prod_{i=1}^n \frac{n_i^{x_i}}{x_i!} \exp\left(-\sum_{i=1}^n n_i\lambda\right) \exp\left(\sum_{i=1}^n x_i \ln(\lambda)\right). \end{aligned}$$

b) The sufficient statistic is $\sum_{i=1}^n x_i$. c) We derive the MLE for λ

$$l(\lambda) = -\sum_{i=1}^n n_i\lambda + \sum_{i=1}^n x_i \ln(\lambda) + const$$

$$l'(\lambda) = -N + \sum_{i=1}^n x_i \frac{1}{\lambda}, \text{ with } N = \sum_{i=1}^n n_i, \ \widehat{\lambda}_{\text{MLE}} = \frac{\sum_{i=1}^n x_i}{N}.$$

d) Yes, in $T = \sum_{i=1}^n x_i$; compare Theorem 5.3. e)The test $\varphi^*(\mathbf{x})$ is UMP test of size α iff $\sup_{\lambda \geq 1} \mathsf{E}_\lambda \varphi^*(\mathbf{x}) = \alpha$ and for all tests φ with

$\sup_{\lambda \geq 1} E_\lambda \varphi(\mathbf{X}) \leq \alpha$ and for all $\lambda < 1$ it holds that $E_\lambda \varphi^*(\mathbf{X}) \geq E_\lambda \varphi(\mathbf{X})$. f) Apply Theorem 5.4 (Blackwell theorem). The following test is UMP of size α

$$\varphi(\mathbf{x}) = \begin{cases} 1 & \text{if} \quad T < c \\ \gamma & \text{if} \quad T = c \ , \quad c, \gamma \text{ such that } P_0\,(T < c) + \gamma P_0(T = c) = \alpha \\ 0 & \text{if} \quad T > c \end{cases}$$

where $P_0 = \mathsf{Poi}(N)$.

12. The distribution of the sample belongs to a two-parameter exponential family with the natural parameters $\zeta_1(\theta) = \frac{n}{\sigma^2}\mu$, $\zeta_2(\theta) = -\frac{1}{2\sigma^2}$ and the sufficient statistics: $U(\mathbf{x}) = \bar{x}$, $T(\mathbf{x}) = \sum_{i=1}^{n} x_i^2$; compare Special case 2.1. We can transform the hypothesis $H_0 : \mu = \mu_0$ into $H_0 : \frac{n}{\sigma^2}\mu = \frac{n}{\sigma^2}\mu_0$. Thus the parameter of interest is $\lambda = \frac{n}{\sigma^2}\mu$ and the nuisance parameter is $\vartheta = -\frac{1}{2\sigma^2}$. We apply Theorem 5.8 and obtain the optimal conditional test:

$$\varphi(u, t) = \begin{cases} 1 & \text{if} & \text{else} \\ 0 & \text{if} & c_1(t) \leq u \leq c_2(t) \end{cases}$$

with $E_0\,(\varphi(\mathbf{X}) \mid T = t) = \alpha$ for all t and $E_{\lambda_0}\,(U\varphi(U,T) \mid T = t) = \alpha E_{\lambda_0}\,(U \mid T = t)$ for all t. Let us rewrite the critical region: Remember

$$s^2 = \frac{1}{n-1}\sum_{i=1}^{n}(x_i - \bar{x})^2 = \frac{1}{n-1}\sum_{i=1}^{n}x_i^2 - \frac{n}{n-1}\bar{x}^2 = \frac{1}{n-1}t - \frac{n}{n-1}u^2.$$

Introduce the t-statistic

$$t = F(u) = \frac{\sqrt{n}(u - \mu_0)}{s} = \frac{\sqrt{n}(u - \mu_0)}{\sqrt{\frac{1}{n-1}t - \frac{n}{n-1}u^2}}.$$

It holds for all t that $F(u) \uparrow$. Thus $c_1(t) < u < c_2(t) \iff F(c_1(t)) < t < F(c_1(t))$. The statistic $F(U)$ is t-distributed with $n - 1$ degrees of freedom and independent of T. Thus $P_0\,(U < c_2(t)) = P_0\,(F(U) < k(t)) = \alpha$ implies $k(t) = t_{n-1;\alpha}$. The UMPU α-test is the well-known two-sided t-test:

$$\varphi(\mathbf{x}) = \begin{cases} 1 & \text{if} & \text{else} \\ 0 & \text{if} & -t_{n-1;1-\frac{\alpha}{2}} \leq t \leq t_{n-1;1-\frac{\alpha}{2}} \end{cases}.$$

13. a) The density of the joint sample is

$$h(z)\left(\frac{1}{\sigma^2}\right)^n \exp\left(-\frac{1}{2\sigma^2}\left(\sum_{i=1}^{n}x_i^2 + \frac{1}{2}\sum_{i=1}^{n}y_i^2\right) + \frac{n}{\sigma^2}\mu_1\bar{x} + \frac{n}{2\sigma^2}\mu_2\bar{y}\right).$$

Thus the distribution of $\mathbf{Z}$ belongs to a 3-parameter exponential family.
b) The sufficient statistics are $T_1 = \sum_{i=1}^{n} x_i^2 + \frac{1}{2}\sum_{i=1}^{n} y_i^2$, $T_2 = \bar{x}$, $T_3 = \bar{y}$.
c) Parameter of interest is $-\frac{1}{2\sigma^2}$. The nuisance parameters are $\frac{n}{\sigma^2}\mu_1$, $\frac{n}{2\sigma^2}\mu_2$.
d) The hypothesis can be transformed. $H_0 : \sigma^2 \leq \Delta_0 \iff H_0 : -\frac{1}{2\sigma^2} \leq$

$-\frac{1}{2\Delta_0}$. e) We apply Theorem 5.7. UMP α-similar test for $H_0 : \sigma^2 \leq \Delta_0$ versus $H_1 : \sigma^2 > \Delta_0$ is

$$\varphi(\mathbf{x}) = \begin{cases} 1 & \text{if } T_1 > u(t_2, t_3) \\ 0 & \text{if } T_1 \leq u(t_2, t_3) \end{cases}$$

with $E_0\left(\varphi(\mathbf{x}) \mid T_2 = t_2, T_3 = t_3\right) = \alpha$ for all t_2, t_3. We transform the critical region by using $(n-1)s^2 = \sum_{i=1}^{n} (x_i - \bar{x})^2 = \sum_{i=1}^{n} x_i^2 - n\bar{x}^2$ and we obtain

$$(n-1)s_x^2 + (n-1)\frac{1}{2}s_y^2 = t_1 - nt_2^2 - \frac{1}{2}nt_3^2.$$

Thus

$$t_1 < u(t_2, t_3) \iff (n-1)s_x^2 + (n-1)\frac{1}{2}s_y^2 < c.$$

Using $(n-1)S_x^2 + \frac{1}{2}(n-1)S_y^2 \sim \chi^2_{2n-2}\sigma^2$ we get

$$\varphi(\mathbf{x}) = \begin{cases} 1 & \text{if } (n-1)s_x^2 + \frac{1}{2}(n-1)s_y^2 > \Delta_0\chi^2_{2n-2;1-\alpha} \\ 0 & \text{if } (n-1)s_x^2 + \frac{1}{2}(n-1)s_y^2 \leq \Delta_0\chi^2_{2n-2;1-\alpha} \end{cases}.$$

7.5   Solutions for Chapter 6: Linear Model

1. See Figure 7.4. $\hat{\alpha} = 1.56$, $\hat{\beta} = 0.176$.

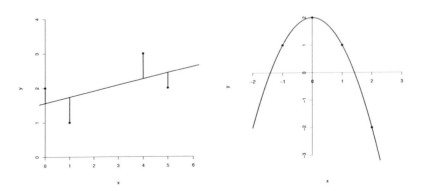

Figure 7.4: Solution to Problems 1 and 2.

2. We write the model in the form $Y = X\beta + \varepsilon$, where X is the $n \times 1$ vector consisting of ones, say $X = \mathbb{1}_n$, and $\beta = m$. The covariance matrix Σ is the diagonal matrix with the elements σ^2. By the Gauss–Markov theorem the best linear unbiased estimator is given by $\hat{m} = (X^{\mathsf{T}}\Sigma^{-1}X)^{-1}X^{\mathsf{T}}\Sigma^{-1}Y$. Since $\Sigma^{-1} = \mathrm{Diag}(\sigma_1^{-2}, \ldots, \sigma_n^{-2})$, we obtain

$$\hat{m} = (\mathbb{1}_n^{\mathsf{T}}\Sigma^{-1}\mathbb{1}_n)^{-1}\mathbb{1}_n^{\mathsf{T}}\Sigma^{-1}Y = \left(\sum_{i=1}^{n}\sigma_i^{-2}\right)^{-1}\sum_{i=1}^{n}\sigma_i^{-2}Y_i.$$

3. The normal equations have the form:

$$\sum_i y_i - n\beta_1 + \beta_2 \sum_i x_i + \beta_3 \sum_i x_i^2 = 0$$

$$\sum_i y_i x_i - \beta_1 \sum_i x_i + \beta_2 \sum_i x_i^2 + \beta_3 \sum_i x_i^3 = 0$$

$$\sum_i y_i x_i^2 - \beta_1 \sum_i x_i^2 + \beta_2 \sum_i x_i^3 + \beta_3 \sum_i x_i^4 = 0$$

or, using the matrix

$$X = \begin{pmatrix} 1 & 1 & 1 \\ 1 & 0 & 0 \\ 1 & 1 & 1 \\ 1 & 2 & 4 \end{pmatrix},$$

we get

$$(X^TX)\beta = X^T\mathbf{y} = \begin{pmatrix} 4 & 4 & 4 \\ 4 & 6 & 10 \\ 6 & 10 & 18 \end{pmatrix} \quad \beta = \begin{pmatrix} 2 \\ -2 \\ -6 \end{pmatrix}.$$

The solution is $\hat{\beta}_1 = 2$, $\hat{\beta}_2 = 0$ and $\hat{\beta}_3 = -1$, that is, the estimated regression function is given by $2 - x^2$ (see Figure 7.4).

4. The distributions $N(\alpha + \beta x_i, \sigma^2)$ constitute a three-parameter exponential family:

$$f_i(y_i) = \frac{1}{\sqrt{2\pi\sigma^2}} \exp\left(-\frac{1}{2\sigma^2}(y_i - \alpha - \beta x_i)^2 \right)$$

$$= \frac{1}{\sqrt{2\pi\sigma^2}} \exp\left(-\frac{1}{2\sigma^2}y_i^2 + \frac{\alpha}{\sigma^2}y_i + \frac{\beta}{\sigma^2}y_i x_i - \frac{\beta^2}{2\sigma^2}x_i^2 \right.$$

$$\left. -\frac{\alpha\beta}{\sigma^2}x_i - \frac{\alpha^2}{2\sigma^2} \right),$$

thus we obtain for the sample $\mathbf{y}$

$$\prod_{i=1}^n f_i(y_i) \propto A(\theta)\exp\left(-\frac{1}{2\sigma^2}\sum_{i=1}^n y_i^2 + \frac{\alpha}{\sigma^2}\sum_{i=1}^n y_i + \frac{\beta}{\sigma^2}\sum_{i=1}^n y_i x_i \right)$$

where $A(\theta) = \frac{1}{\sqrt{2\pi\sigma^2}} \exp\left(\frac{\beta^2}{2\sigma^2}\sum_{i=1}^n x_i^2 - \frac{\alpha\beta}{\sigma^2}x_i - \frac{\alpha^2}{2\sigma^2} \right)$ is independent on $\mathbf{y}$. A minimal sufficient statistic is given by

$$T(\mathbf{y}) = \left(\sum_{i=1}^n y_i^2, \sum_{i=1}^n y_i, \sum_{i=1}^n x_i y_i \right).$$

5. a) Let us start with one observation at each point: The design matrix is

$$X = \begin{pmatrix} 1 & 0 & 0 \\ 1 & 1 & 1 \\ 1 & 1.5 & 2.25 \end{pmatrix}.$$

This matrix is of full rank, β is estimable and therefore also γ. Let us check, whether it is possible to estimate γ by observing the response at less data points. To do this we choose rows of X and have to find a matrix L such that $L^T \tilde{X} = F$. Here $\tilde{X}$ is the matrix consisting of the chosen rows and

$$F = \begin{pmatrix} 1 & -1 & 0 \\ 0 & 1 & -1 \\ -1 & 0 & 1 \end{pmatrix}.$$

Consider for example $\tilde{X} = \begin{pmatrix} 1 & 0 & 0 \\ 1 & 1.5 & 2.25 \end{pmatrix}$. One can show that the system of equation $L^T \tilde{X} = F$ has no solution (for L). Thus, one needs at least one observation at each point.

b) The R-procedure lm yields the following output

```
Coefficients:
(Intercept)              x           I(x^2)
   3.0667            0.7556          0.6444
```

Hence the estimate for γ is $\hat{\gamma} = (2.311, 0.111, -2.422)$ (rounded).

6. a) We have to minimize

$$\sum_{i=1}^{n} (y_i - \alpha - \beta(x_i - \bar{x}))^2$$

$$\sum_{i=1}^{n} (y_i \pm \bar{y} - \alpha - \beta(x_i - \bar{x}))^2 = s_{yy} + n(\bar{y} - \alpha)^2 + \beta^2 s_{xx} - 2\beta s_{xy}$$

with
$s_{xy} = \sum_{i=1}^{n}(y_i - \bar{y})(x_i - \bar{x})$, $s_{xx} = \sum_{i=1}^{n}(x_i - \bar{x})^2$ and $s_{yy} = \sum_{i=1}^{n}(y_i - \bar{y})^2$.
As already shown in Special case 6.3, the minimum is attained at

$$\hat{\alpha} = \bar{y} \quad \text{and} \quad \hat{\beta} = \frac{s_{xy}}{s_{xx}}.$$

Note that $\hat{\beta}$ satisfies

$$\hat{\beta} = \sum_{i=1}^{n} c_i y_i \text{ with } c_i = \frac{x_i - \bar{x}}{s_{xx}}, \quad \sum_{i=1}^{n} c_i = 0, \quad \sum_{i=1}^{n} c_i (x_i - \bar{x}) = 1.$$

b) $E_\theta \hat{\alpha} = E_\theta \bar{Y} = \alpha - \beta \frac{1}{n} \sum_{i=1}^{n}(x_i - \bar{x}) = \alpha,$

$E_\theta \hat{\beta} = \sum_{i=1}^{n} c_i E_\theta Y_i = \alpha \sum_{i=1}^{n} c_i + \beta \sum_{i=1}^{n} c_i(x_i - \bar{x}) = \beta.$

Both estimators are linear functions of normally distributed r.v.'s. Hence both estimators are normally distributed. Furthermore,

$$
\begin{aligned}
\mathsf{Cov}_\theta\left(\hat{\alpha}, \hat{\beta}\right) &= \mathsf{Cov}_\theta\left(\frac{1}{n}\sum_{i=1}^{n} Y_i, \sum_{j=1}^{n} c_j Y_j\right) \\
&= \frac{1}{n}\sum_{i=1}^{n}\sum_{j=1}^{n} c_j \mathsf{Cov}_\theta(Y_i, Y_j) = \frac{\sigma^2}{n}\sum_{j=1}^{n} c_j = 0
\end{aligned}
$$

and

$$
\mathsf{Var}_\theta\, \hat{\alpha} = \mathsf{Var}_\theta \overline{Y} = \frac{\sigma^2}{n},
$$

$$
\mathsf{Var}_\theta \widehat{\beta} = \mathsf{Var}_\theta\left(\sum_{i=1}^{n} c_i y_i\right) = \sigma^2 \sum_{i=1}^{n} c_i^2 = \frac{\sigma^2}{s_{xx}}\sum_{i=1}^{n} c_i(x_i - \overline{x}) = \frac{\sigma^2}{s_{xx}}.
$$

Thus,

$$
\begin{pmatrix} \hat{\alpha} \\ \hat{\beta} \end{pmatrix} \sim \mathsf{N}_2\left(\begin{pmatrix} \alpha \\ \beta \end{pmatrix}, \sigma^2 \begin{pmatrix} \frac{1}{n} & 0 \\ 0 & s_{xx}^{-1} \end{pmatrix}\right).
$$

c) Let $\delta > 0$ be arbitrarily fixed. Then we get by the Chebyshev inequality:

$$
\mathsf{P}_\theta\left(|\hat{\alpha} - \alpha| > \delta\right) \le \frac{\mathsf{Var}_\theta \hat{\alpha}}{\delta^2} = \frac{\sigma^2}{n\delta^2} \to 0 \qquad \text{as } n \to \infty
$$

and

$$
\mathsf{P}_\theta\left(|\hat{\beta} - \beta| > \delta\right) \le \frac{\mathsf{Var}_\theta \hat{\beta}}{\delta^2} = \frac{\sigma^2}{s_{xx}\delta^2}.
$$

It follows that a sufficient condition for the consistency of $\hat{\beta}$ is that s_{xx} tends to infinity, as $n \to \infty$.

d) The residuals are given by

$$
y_i - \hat{y}_i = y_i - \hat{\alpha} - \hat{\beta}(x_i - \overline{x}),
$$

considered as r.v.'s, a linear combination of normal variables. Thus they are normal, too. It is enough to calculate the covariance for $i \ne j$

$$
\begin{aligned}
&\mathsf{Cov}_\theta(Y_i - \hat{\alpha} - \hat{\beta}(x_i - \overline{x}), Y_j - \hat{\alpha} - \hat{\beta}(x_j - \overline{x})) \\
=\ &\mathsf{Cov}_\theta(Y_i, Y_j) + \mathsf{Var}_\theta\, \hat{\alpha} + \mathsf{Var}_\theta\, \hat{\beta} x_i x_j - \mathsf{Cov}_\theta\left(\hat{\alpha}, \hat{\beta}\right)(x_i + x_j - 2\overline{x}) \\
&- \mathsf{Cov}_\theta(\hat{\alpha}, Y_j) - \mathsf{Cov}_\theta(\hat{\alpha}, Y_i) \\
&- (x_j - \overline{x})\mathsf{Cov}_\theta(\hat{\beta}, Y_i) - (x_i - \overline{x})\mathsf{Cov}_\theta(\hat{\beta}, Y_j) \\
=\ &-\frac{\sigma^2}{n} - \frac{\sigma^2}{s_{xx}}(x_i - \overline{x})(x_j - \overline{x}).
\end{aligned}
$$

Here we use $\mathsf{Cov}_\theta(\hat{\alpha}, Y_j) = \frac{1}{n}\sum_{i=1}^{n}\mathsf{Cov}_\theta(Y_i, Y_j) = \frac{\sigma^2}{n}$ and $\mathsf{Cov}_\theta(\hat{\beta}, Y_j) = \sum_{i=1}^{n} c_i\mathsf{Cov}_\theta(Y_i, Y_j) = c_i\sigma^2$. The residuals are not independent.

e) The fitted line has the form $\hat{m}(x) = \hat{\alpha} + \hat{\beta}(x - \overline{x})$. As a linear combination of normal r.v.'s, it is normally distributed for fixed x. Since the parameter estimators are unbiased, we have $\mathsf{E}_\theta \, \hat{m}(x) = \alpha + \beta(x - \bar{x})$. Furthermore, because of the independence of $\hat{\alpha}$ and $\hat{\beta}$ we obtain immediately

$$\mathsf{Var}_\theta \, \hat{m}(x) = \frac{\sigma^2}{n} + \frac{\sigma^2}{s_{xx}}(x - \bar{x})^2.$$

Thus,

$$\hat{\alpha} + \hat{\beta}(x - \bar{x}) \sim \mathsf{N}\left(\alpha + \beta(x - \bar{x}), \sigma^2\left(\frac{1}{n} + \frac{(x - \bar{x})^2}{s_{xx}}\right)\right).$$

f) Since $\hat{m}(x)$ is normally distributed, it follows that the ratio

$$\frac{\hat{m}(x) - m(x)}{\sigma d_x} \sim \mathsf{N}(0, 1), \qquad d_x^2 = \frac{1}{n} + \frac{x - \bar{x}}{s_{xx}}.$$

If σ^2 is replaced by the estimate (6.25), which has in this model the form

$$\hat{\sigma}^2 = \frac{1}{n - 2} \sum_{i=1}^n (y_i - \hat{y}_i)^2,$$

then

$$\frac{\hat{m}(x) - m(x)}{\hat{\sigma} d_x} \sim t_{n-2}.$$

Thus, a confidence interval with coverage probability $1 - \alpha$ for the value of the regression line at the point x is given by

$$[\hat{m}(x) - t_{n-2;1-\frac{\alpha}{2}} \, \hat{\sigma} \, d_x, \ \hat{m}(x) + t_{n-2;1-\frac{\alpha}{2}} \, \hat{\sigma} \, d_x].$$

g) A new observation at x is predicted by $\hat{Y} = \hat{m}(x) + \varepsilon$. With the same method as in f) we can construct an interval. The only difference is that $\hat{Y}$ is normally distributed with variance $\sigma^2(1 + d_x^2)$. Thus, the prediction interval has the form

$$[\hat{m}(x) - t_{n-2;1-\frac{\alpha}{2}} \, \hat{\sigma} \, \sqrt{1 + d_x^2}, \ \hat{m}(x) + t_{n-2;1-\frac{\alpha}{2}} \, \hat{\sigma} \, \sqrt{1 + d_x^2}].$$

h) The prediction interval is longer. It takes the randomness of the new observation into account.

7. Compute the statistic of the F-test in (6.36) for testing a single parameter: With $V = (X^{\mathsf{T}}X)^{-1}$ and $G_j = (0, \ldots, 1, \ldots, 0)$ we obtain immediately that the test statistic has the form

$$F = \frac{\hat{\beta}_j^2}{\hat{\sigma}^2 V_{jj}}.$$

That is, for testing $H_0 : \ \beta = 0$ vs. $H_1 : \ \beta \neq 0$ in the simple linear regression model the F-test has the form: Reject H_0 iff

$$\frac{s_{xx}\hat{\beta}^2}{\hat{\sigma}^2} > F_{1,n-2;1-\alpha}.$$

This test is equivalent to the corresponding t-test: The null hypothesis is rejected, iff

$$\frac{\sqrt{s_{xx}}|\hat{\beta}|}{\hat{\sigma}} > t_{n-2;1-\frac{\alpha}{2}}.$$

8. Under the conditions $\sum_{i=1}^{I} \alpha_i = 0$ and $\sum_{j=1}^{J} \beta_j = 0$ we have an unknown parameter $\boldsymbol{\beta} = (\mu, \alpha_1, \ldots, \alpha_{I-1}, \beta_1, \ldots, \beta_{J-1})^{\mathsf{T}}$ of dimension $p = I + J - 1$. The solution of the minimization of

$$\sum_{i=1}^{I-1} \sum_{j=1}^{J-1} (y_{ij} - \mu - \alpha_i - \beta_j)^2 + (y_{Ij} - \mu + \sum_{r=1}^{I-1} \alpha_r - \beta_j)^2)$$

$$+ (y_{IJ} - \mu + \sum_{r=1}^{I-1} \alpha_r + \sum_{l=1}^{J-1} \beta_l)^2$$

is given by $\quad \hat{\mu} = \frac{1}{IJ} \sum_{i=1}^{I} \sum_{j=1}^{J} y_{ij} = \bar{y}_{..}$,

$$\hat{\alpha}_i = \bar{y}_{i.} - \bar{y}_{..}, \qquad \bar{y}_{i.} = \frac{1}{J} \sum_{j=1}^{J} y_{ij}, \qquad \hat{\beta}_j = \bar{y}_{.j} - \bar{y}_{..}, \qquad \bar{y}_{.j} = \frac{1}{I} \sum_{i=1}^{I} y_{ij}.$$

As MLE we obtain as usual

$$\hat{\sigma}^2_{\text{MLE}} = \frac{1}{IJ} \sum_{i=1}^{I} \sum_{j=1}^{J} (y_{ij} - \hat{y}_{ij})^2$$

where $\hat{y}_{ij} = \hat{\mu} + \hat{\alpha}_i + \hat{\beta}_j$.

An unbiased estimator is given by $\hat{\sigma}^2 = \dfrac{IJ}{(I-1)(J-1)} \hat{\sigma}^2_{\text{MLE}}$. (Note that $IJ - (I + J - 1) = (I - 1)(J - 1)$.)

9. Let $\hat{\boldsymbol{\beta}}_A$ be the estimate for the parameter $\boldsymbol{\beta}$ under the hypothesis $H_{A0} \colon \alpha_1 = \alpha_2 \cdots = \alpha_I = 0$. Then

$$\|\mathbf{y} - X\hat{\boldsymbol{\beta}}_A\|^2 = \sum_{i=1}^{I} \sum_{j=1}^{J} (y_{ij} - \bar{y}_{.j})^2.$$

As in the previous problem

$$\text{SSRes} = \|\mathbf{y} - X\hat{\boldsymbol{\beta}}\|^2 = \sum_{i=1}^{I} \sum_{j=1}^{J} (y_{ij} - \bar{y}_{i.} - \bar{y}_{.j} + \bar{y}_{..})^2.$$

The difference of these quadratic distances, the numerator of the test statistic, is

$$\text{SSA} = \sum_{i=1}^{I} \sum_{j=1}^{J} (\bar{y}_{i.} - \bar{y}_{..})^2.$$

Thus, the F-statistic is given by

$$F = \frac{\text{SSA}}{\text{SSRes}} \frac{(I-1)(J-1)}{(I-1)}.$$

Bibliography

J. Arbuthnott. An argument for divine providence, taken from the constant regularity observed in the birth of both sexes. *Philos. Trans. R. Soc. London*, pages 186–190, 1712.

A. C. Atkinson. *Robust Diagnostic Regression Analysis*. Springer, 2000.

H. Augier, L. Benkoel, J. Brisse, A. Chamlian, and W. K. Park. Necroscopic localization of mercury-selenium interaction products in liver, kidney, lung and brain of Mediterranean striped dolphins (*Stenella coeruleoalba*) by silver enhancement kit. *Cell. and Molec. Biology*, 39:765–772, 1993.

O. Barndorff-Nielsen. *Information and Exponential Families*. Wiley, 1978.

D. M. Bates and D. G. Watts. *Nonlinear Regression Analysis and Its Applications*. Wiley, 1988.

M. J. Bayarri and J. O. Berger. P values for composite null models (with discussion). *Journal of the American Statistical Association*, 95:1127–1142, 2000.

J. O. Berger. *Statistical Decision Theory and Bayesian Analysis*. Springer, 1985.

J. O. Berger and R. L. Wolpert. *The Likelihood Principle*. IMS Lecture Notes-Monograph Series 6, 1988.

P. J. Bickel and K. A. Doksum. *Mathematical Statistics*. Pearson, 2007.

L. D. Brown. *Fundamentals of Statistical Exponential Families*. IMS Lecture Notes-Monograph Series 9, 1986.

H. Bunke and O. Bunke. *Statistical Methods of Model Building*. Wiley, 1986.

D. R. Cox. Regression models and life-tables (with discussion). *J. R. Statist. Soc. B*, 34:187–220, 1972.

D. R. Cox. *Principles of Statistical Inference*. Cambridge University Press, 2006.

A. C. Davison. *Statistical Models*. Cambridge University Press, 2003.

A. C. Davison and D. V. Hinkley. *Bootstrap Methods and their Application*. Cambridge University Press, 2003.

J. Devore and R. Peck. *Statistics: The Exploration and Analysis of Data*. Duxbury Press, 1993.

N. R. Draper and H. Smith. *Applied Regression Analysis*. Wiley, 1966.

E. J. Dudewicz and S. N. Mishra. *Modern Mathematical Statistics*. Wiley, 1988.

B. Efron and R. J. Tibshirani. *An Introduction to the Bootstrap*. Chapman & Hall/CRC, 1993.

P. P. B. Eggermont and V. N. LaRiccia. *Maximum Penalized Likelihood Estimation, Part I*. Springer, 2001.

P. P. B. Eggermont and V. N. LaRiccia. *Maximum Penalized Likelihood Estimation, Part II*. Springer, 2009.

T. S. Ferguson. *Mathematical Statistics: A Decision Theoretic Approach*. Academic Press, New York, 1967.

D. A. Freedman. *Statistical Models*. Cambridge University Press, 2009.

P. H. Garthwaite, I. T. Jolliffe, and B. Jones. *Statistical Inference*. Oxford University Press, 2002.

N. C. Giri. *Multivariate Statistical Analysis*. Marcel Dekker Inc., 2004.

P. E. Greenwood and M. S. Nikulin. *A Guide to Chi-Squared Testing*. Wiley, 1996.

A. Gut. *Probability: A Graduate Course*. Springer Texts in Statistics, 2005.

A. Hald. *A History of Mathematical Statistics from 1750 to 1930*. Wiley, 1998.

F. R. Hampel, E. M. Ronchetti, P. J. Rousseeuw, and W. A. Stahl. *Robust Statistics*. Wiley, 1986.

W. Härdle. *Applied Nonparametric Regression*. Cambridge University Press, 1990.

T. Hastie, R. Tibshirani, and J. Friedman. *Elements of Statistical Learning*. Springer, 2001.

C. C. Heyde and E. Seneta. *Statisticians of the Centuries*. Springer, 2001.

P. J. Huber. Robust estimation of a location parameter. *Annals of Mathematical Statistics*, 35:73–101, 1964.

I. A. Ibragimov and R. Z. Has'minski. *Statistical Estimation*. Springer, 1981.

A.V. Ivanov. *Asymptotic Theory of Nonlinear Regression*. Kluwer, 1997.

J. Jurečková and P. K. Sen. *Robust Statistical Procedures*. Wiley, 1996.

R. E. Kass and W. V. Paul. *Geometrical Foundations of Asymptotic Inference*. Wiley, 1997.

T. Kollo and D. von Rosen. *Advanced Multivariate Statistics with Matrices*. Springer, 2005.

S. Konishi and G. Kitagawa. *Information Criteria and Statistical Modeling*. Springer, 2008.

H. Läuter and R. Pincus. *Mathematisch-statistische Datenanalyse*. Akademie-Verlag Berlin, 1989.

E. L. Lehmann. The Fisher, Neyman–Pearson theories of testing hypothesis: One theory or two? *Journal of the American Statistical Association*, 88: 1242–1249, 1993.

E. L. Lehmann. *Nonparametrics: Statistical Methods Based on Ranks.* Prentice Hall, 1998.

E. L. Lehmann. *Elements of Large Sample Theory.* Springer, 1999.

E. L. Lehmann and G. Casella. *Theory of Point Estimation.* Springer, 1998.

E. L. Lehmann and J. P. Romano. *Testing Statistical Hypotheses.* Springer, 2005.

F. Liese and K.-J. Miescke. *Statistical Decision Theory.* Springer, 2008.

B. W. Lindgren. *Statistical Theory.* Chapman & Hall, 1962.

K. V. Mardia, J. T. Kent, and J. M. Bibby. *Multivariate Analysis.* Academic Press, 1979.

P. McCullagh and J. A. Nelder. *Generalized Linear Models.* Chapman & Hall/CRC, 1990.

A. B. Owen. *Empirical Likelihood.* Chapman & Hall/CRC, 2001.

L. Pardo. *Statistical Inference Based on Divergence Measures.* Chapman & Hall/CRC, 2006.

Y. Pawitan. *In All Likelihood Statistical Modelling and Inference Using Likelihood.* Oxford University Press, Oxford, 2001.

W. S. Peters. *Counting for Something.* Springer, 1987.

C. R. Rao. *Linear Statistical Inference and its Applications.* Wiley, 1973.

C.R. Rao and H. Toutenburg. *Linear Models.* Springer, 1995.

C. P. Robert. *The Bayesian Choice.* Springer, 2001.

M. Rudolf and W. Kuhlisch. *Biostatistik: Eine Einführung für Biowissenschaftler.* Pearson, 2008.

M. J. Schervish. *Theory of Statistics.* Springer, 1995.

G. R. Shorack and J. A Wellner. *Empirical Processes with Applications to Statistics.* Wiley, 1986.

S. M. Stigler. *The History of Statistics: The Measurement of Uncertainty before 1900.* The Belknap Press of Havard University Press, 2000.

A. W. van der Vaart. *Asymptotic Statistics.* Cambridge University Press, 1998.

J. Verzani. *Using R for Introductory Statistics.* Chapman & Hall/CRC, 2005.

M.P. Wand and M. C. Jones. *Kernel Smoothing.* Chapman & Hall/CRC, 1995.

E. Weber. *Grundriss der biologischen Statistik.* VEB Gustav Fischer Verlag Jena, 1972.

H. Witting. *Mathematische Statistik.* B. G. Teubner, Stuttgart Germany,

1978.

H. Witting. *Mathematische Statistik I, Parametrische Verfahren bei festem Stichprobenumfang.* Teubner, Stuttgart, 1987.

L. Zhu. *Nonparametric Monte Carlo Tests and Their Applications.* Springer, 2005.

Index

algorithm
    independent Monte Carlo, 132
    parametric bootstrap, 134
alpha
    level, 151
    p-value, 144
    significance level, 142
    similar, 173
    size, 140, 151, 153, 157
    test, 151
    UMPU α-test, 165
analysis of variances (ANOVA), 188, 210, 217, 218
    one-way model, 188
asymptotic efficiency, 102

Bernoulli distribution
    likelihood function, 28
    MLE, 80
beta
    size, 140
bias, 92
binomial distribution
    alpha-level, 152
    as exponential family, 16, 21
    Cramér–Rao bound, 99
    Fisher information, 37
    logit function, 160
    MLE, 79
    MSE, 95
    Neyman–Pearson test, 149
    Rao–Blackwell theorem, 105
    test
      error of first and second type, 141
    UMP test, 160
    UMPU test, 171

Blackwell theorem, 156

Cauchy distribution
    asymptotic behavior, 114
    minimal sufficiency, 62
    MLE, 82
censored data, 41
Chebyshev inequality, 113
completeness, 107, 165, 174
    exponential family, 108
    normal linear model, 208
    uniform distribution, 108
conditional expectation, 105
consistency, 112
correlation, 76
covariance matrix
    order between, 102
Cramér–Rao bound, 97, 98
    binomial distribution, 99
    exponential distribution, 102
    multivariate, 102

delta method, 116
distribution
    binomial, 10, 51
    chi-squared distribution, 212
    F-distribution, 212
    hypergeometric, 10
    multinomial, 14, 53
    normal distribution, 11
    two-parameter exponential, 17
    uniform distribution, 23
double exponential distribution
    MLE, 84
dummy variables, 186

efficiency, 100, 102
    asymptotic efficiency, 116